Jane's fighting ships

Jane's Fighting Ships 1900

(facsimile edition)

THIRD YEAR OF ISSUE

ALL THE WORLD'S FIGHTING SH

ALL THE WORLD'S
FIGHTING SHIPS

TOUS
LES NAVIRES DE
GUERRE DU MONDE

ALLE KRIEGSSCHIFFE
DER WELT

NAVI DI BATTAGLIA
APPARTENENTI A TUTTI
LE NAZIONI

全世界
戰艦

ВСѢ ВОЕННЫЯ
СУДА НА
СВѢТѢ

TODOS LOS BUQUES
GUERREROS DEL
MUNDO

VERLDENO
STRIDO FARTYG

COSMOPOLITAN NAVAL ANNUAL

BY

FRED. T. JANE

Author of "The Imperial Russian Navy," etc., etc. Inventor of "The Jane Naval War Game" (Naval Kriegspiel)

PUBLISHED FOR

THE NAVAL SYNDICATE

BY

ENGLAND
SAMPSON LOW, MARSTON & CO., LTD.
ST. DUNSTAN'S HOUSE, FETTER LANE
LONDON, E.C

FRANCE
LE YACHT
55, RUE DE CHÂTEAUDUN
PARIS

ITALY
RIVISTA MARITTIMA
ROME

AND IN
UNITED STATE
RUSSIA, S
SWEDEN

1900

THIRD YEAR OF ISSUE

(CORRECTED TO THE END OF JUNE 1900, WITH ADDENDA TO THE DATE OF PUBLICATION)

Fred T. Jane

REPUBLISHED BY NIMBLE BOOKS LLC

Nimble Books LLC
1521 Martha Avenue
Ann Arbor, MI, USA 48103
http://www.NimbleBooks.com
wfz@nimblebooks.com
+1.734-330-2593

Version 1.0; last saved 2011-03-02.

Printed in the United States of America
ISBN-13: 978-1-60888-119-2

The paper used in this publication meets the minimum requirements of the American National Standard for Information Sciences—Permanence of Paper for Printed Library Materials, ANSI Z39.48-1992. The paper is acid-free and lignin-free.

THIRD YEAR OF ISSUE

ALL THE WORLD'S FIGHTING SH

ALL THE WORLD'S
FIGHTING SHIPS

全世界
戰艦

TOUS
LES NAVIRES DE
GUERRE DU MONDE

ВСѢ ВОЕННЫЯ
СУДА НА
СВѢТѢ

ALLE KRIEGSSCHIFFE
DER WELT

TODOS LOS BUQUES
GUERREROS DEL
MUNDO

NAVI DI BATTAGLIA
APPARTENENTI A TUTTI
LE NAZIONI

VERLDENO
STRIDO FARTYG

COSMOPOLITAN NAVAL ANNUAL

BY

FRED. T. JANE

Author of "The Imperial Russian Navy," etc., etc. Inventor of "The Jane Naval War Game" (Naval Kriegspiel)

Published for

THE NAVAL SYNDICATE

BY

ENGLAND
SAMPSON LOW, MARSTON & CO., Ltd.
St. Dunstan's House, Fetter Lane
LONDON, E.C

FRANCE
LE YACHT
55, Rue de Châteaudun
PARIS

ITALY
RIVISTA MARITTIMA
ROME

AND IN
UNITED STATE
RUSSIA, S
SWEDEN

1900

THIRD YEAR OF ISSUE

(CORRECTED TO THE END OF JUNE 1900, WITH ADDENDA TO THE DATE OF PUBLICATION)

LONDON: PRINTED BY WILLIAM CLOWES AND SONS, LIMITED,
STAMFORD STREET AND CHARING CROSS.

ALPHABETICAL INDEX—"INDEX II."

ALPHABETICAL INDEX—"INDEX II."

PREFACE.

SEVERAL distinct changes are embodied in this year's "Fighting Ships." Directly reproduced photographs have been substituted for most of the old pen-and-ink drawings of ships of any importance. Should this new departure meet with favour, it will be extended to the remainder of the illustrations.

For ships of little importance, silhouettes have been adopted instead of small pen-and-ink sketches. These silhouettes are usually produced from small photographs taken at a distance sufficient to prevent any distortion. They will probably be found more useful than the sketches for recognition. All the silhouettes (other than those in Index 1) are on a uniform scale of 1 inch = 160 feet (25 m/m = *circa* 50 metres). Index I. is, for convenience, upon a much smaller scale, *i.e.* 1 inch = *circa* 500 feet (25 m/m = *circa* 150 metres). This Index has been thoroughly revised and brought up to date. In order to obviate any difficulty in finding it immediately, it is now placed at the very end of the book. The old name for it, "Index I.," is retained to avoid confusion.

All or most of the **?**'s have disappeared. This does not mean that the system is given up, but that the questionable illustrations have been mostly verified. As before, anything doubtful will be queried. When plans or illustrations as given in this book differ from those usually current, and are not prefaced by a **?**, their authenticity may be relied upon. The ideal of this book is absolute fidelity, and should any errors creep in they are small accidental ones—if facts by any chance are guessed at, the guess or doubtful information is plainly acknowledged.

The number of plans has again been greatly increased, and as before, the uniform war-game plan scale has been followed, 1 inch = 100 feet (5 m/m = 6 metres). As before, each elevation carries its own scale, being vertically divided into 25 feet sections (metres 7·60), so that immediate comparison of armoured areas is possible in a way which ordinary plans do not admit of. While every effort has been made to secure accurate plans of new ships, the older ironclads have also received considerable attention—merited not so much by their relative importance as by the now very general interest taken in the re-armament problem. Mention may be made of the fact that the sections into which these plans are divided correspond to the naval war game unit of damage—one minute's fire of a 6 inch (15 c/m) quick-firer using common shell (*obus ordinaire*) against unarmoured structures.

A silhouette index of the principal liners and merchant ships has been added. It has unfortunately been found impossible to make this liner index fully complete, but with the kind assistance of readers and users of the book I hope to get it so next year. The task proved more difficult than had been anticipated.

A certain number of flat-bottomed coast-defence gunboats have been deleted altogether. Otherwise the term "fighting ships" has been as liberally translated as heretofore, and taken to mean any warship that is in commission or in a condition to be commissioned. Many warships of little or no fighting value are constantly met with on the seas, and if this book failed to afford a means of naming all of them it would fall short of its purpose.

No fresh matter in four languages has been added. The preface in French, German and Italian as well as English served no very useful purpose, so it is done away with. Nearly every-one now-a-days is able to read English sufficiently well to gather the meaning: any difficulties are caused by technical terms. Of these a glossary is given in French, German, Italian, Japanese, Russian, Spanish and Swedish. Readers of any other nationalities can have glossaries in their particular languages inserted if they will be kind enough to co-operate; and the present glossary can be considerably extended in the future if any utilitarian object is to be secured thereby.

The general arrangement of ships by appearance remains the same. An arrangement by classes, as suggested by one or two correspondents, was carefully considered, but eventually discarded in favour of the original style. The book is intended solely for practical purposes: and for these, arrangement by appearance, as in Index I., seems best.

The classification of ships has undergone a change (fully explained a page or two further on). It is no longer easy to differentiate between battleships and armoured cruisers, and the inclusion of such widely different ships as the *Fürst Bismark* and *Chiyoda* in one general "armoured cruiser" class is obviously unsatisfactory. Nor can speed be held to justify an "armoured cruiser" class, since many of the newest battleships are as swift or swifter than fifty per cent. of the armoured cruisers usually so called.

Consequently, all ships are now divided into two general groups:—(*a*) "armoured ships"—*i.e.*, ships with armour on both belt and guns—and (*b*) "cruisers"—ships which have *not* got armour on *both*. The ***Italia, Khrabry, Rossia, Chiyoda***, and ***Koning landen*** are a few ex these are now placed category.

Even upon a system sometimes difficult to and, again, ships like thin belts and a hu closely approach the sequence of this they a battleship lists.

The classing of *Friedrich III.* and *Ro* rank of armoured shi two of explanation. because the ship ca piercing the vitals of ***Royal Sovereign's*** bec ment and the seconda of action by shell att necessarily thin in plac the class of the *Fuji* a the *Bouvet* and *Charl* rating of the *Drake* a or, rather, as on a par ship, may again at What applies to the *K* however, in another w classes. At ordinary to deliver a vital belt owing to the great armour; and between ***Friedrich III.*** there of only two main guns In the matter of prot off, while she has a gr is impossible. therefo fighting value as lower than that of the *Kaiser Friedrich III*.

To take another case, the *Fürst Bismark* (usually called an armoured cruiser) is an obviously better all-round ship than the *Camperdown* (usually called a first-class battleship), with her hull practically unprotected against 6-inch shell fire, without submerged tubes, and with an inferior speed. Despite her powerful big guns, the *Camperdown* can no more be classed with the *Royal Sovereign* and *Trafalgar* than those ships can be classed with the *London* or the *Shikishima*. Often, of course, these varieties of armoured ships are built for different purposes; but purposes are obvious without need of index figures. The rating in this book is purely a matter of a general fighting value index.

This matter has been alluded to at length in order to make clear the position taken. The policy of *All the World's Fighting Ships* always has been, and always will be, "Facts without theories." The inclusion of anything that may seem to touch on theory requires, therefore, an explanation of its *raison d'être*.

A few other additions to this year's issue do not require reference here; they are common to all other naval annuals, and will, no doubt, be useful for occasional reference.

A great many readers have sent in suggestions, some of which have been adopted, some not. All, however, have been very carefully considered. Quite an unexpected number of correspondents have written suggesting the substitution of "Great Britain" or "British Empire" for "England." The latter is, however, retained, because this is not merely a British book but a cosmopolitan one, used in all nations and in every navy, and "England" is the most general term for the British Empire outside the United Kingdom.

The system under which countries are named in this book having caused some enquiry, it may be stated that when it is generally recognisable the national name is usually put as a heading, in accordance with the cosmopolitan rôle of the book. When the national name is not put, the heading is in French, because that is the language of the Postal Union. It may be noted here that Sweden and Norway are now grouped together, though, as the two fleets are separately administered, each is kept distinct.

In conclusion, I wish to thank those numerous correspondents who have sent corrections and snap-shot photographs—in particular Mr. W. Nye*, several of whose excellent snap-shots will be found in the following pages. For the information of others who may be disposed to place their cameras at the disposal of *Fighting Ships*, I may mention that quarter-plate photos are a quite convenient size. The views should be broadside ones (*par le travers*), not taken from too near. They should be as sharp as possible, and the less pictorial or artistic effect the better. This book is purely utilitarian, and "pictures" are not required as illustrations.

All communications should be addressed to me c/o The Naval Syndicate, 8 Marmion Buildings, Southsea, Portsmouth, England.

FRED. T. JANE

PORTSMOUTH, *June*, 1900.

* Mr. Nye is requested to communicate again, his address having been unfortunately lost.—F. T. J.

For assistance in connection with this issue, special thanks are tendered to:—

Kapitan His Imperial Highness Grand Duke Alexander of Russia; Sir W. G. Armstrong, Whitworth & Co., of Elswick; Capitaine de frégate M. le Marquis de Balincourt; "W. A. B."; Herr Ingenior C. G. Björkman; Major C Field; Löjtnant Erik Hägg; "A. L. K."; Lieut. Haynes; Mr. Kondo; "L. M."; Mr. Nye; Mr. C. de Grave Sells; Lieut.-Commander Takerabe; Mr. C. V. Travers, and many others.

For kind help in connection with the Glossary, I must especially thank:—

Contra-almirante Pascual Cervera y Topete (Spanish), Commander Kawashima and Lieut.-Commander Takeshita (Japanese), Mr. A. W. Bolton, R.N., and Gospodin Vassily Jantchevetchsky (Russian), Herr Ingenior C. G. Björkman (Swedish).

NOTICE.

In deference to the wishes and suggestions of several correspondents, sheets of corrections will in future be issued to subscribers at 12*s*. 6*d*. per annum, post free. The first will be issued about the end of August, and subsequent ones every ten weeks, or oftener, as may be necessary. The sheets will include naval intelligence of every shape and form, photographs of new ships as procured, and various other naval matters of interest, which do not always admit of being treated at length in the annual issue. Special attention will be paid to corrections for the silhouette indices. These will be issued so that they can be stuck on over the present. The first issue will contain a photograph of the *Kaiser Wilhelm II.* and one or two other vessels that were not received in time for inclusion in this issue of the annual. Should a sufficient number of subscriptions be sent in, the sheets will be issued monthly as a journal.

Orders may be sent to any of the publishers of FIGHTING SHIPS, to any bookseller, or direct to the office of the *Naval Syndicate*, 8 Marmion Buildings, Southsea, Portsmouth.

CONTENTS.

GLOSSARY OF TECHNICAL TERMS.

ENGLISH.	FRENCH.	GERMAN.	ITALIAN.	RUSSIAN.	RUSSIAN.	SPANISH.	SWEDISH.
Above water	1 Au dessus de l'eau	1 Ober wasser	1 Sopracquei	1 Надводные	(*Nadvodnie*)	1	1 Öfvervattens
Armour	2 Cuirasse	2 Panzer	2 Corazzato	2 Броня	(*Bronia*)	2 Coraza Blindage	2 Pansar
Armoured cruiser	3 Croiseur cuirassé	3 Gepanzerter Kreuzer	3 Incociatori corazzati	3 Броненосный-крейсеръ	(*Bronenosnii-kreiser*)	3 Crucero acorazado	3 Pansarkryssare
Ahead	4 Avant	4 Vorwärts	4 A prora (davanti)	4 Впередъ	(*Vpered*)	4 A proa	4 Framåt
Astern	5 Arrière	5 Zückwärts	5 A poppa	5 Назадъ	(*Nazad*)	5 A popa	5 Back
Amidships	6 Au milieu	6 Mittschiffs	6 Media	6 Въсередину	(*Vseredinu*)	6 Crujia	6 Midskepps
Barbettes	7 Barbettes	7 Brustwerk	7 Barbetta	7 Барбеты	(*Barbeti*)	7 Torres-barbetas	7 Barbettorn
Bases (of barbettes)	8 Audessous la tourelle	8 Trunk (Schacht)	8	8 Внизу	(*Vnizu*)	8 Platformas	8 Fasta torn
Battery	9 Batterie	9 Batterie	9 Batteria	9 Баттарея	(*Battereya*)	9 Bateria	9 Batteri
Battleship	10 Cuirassé d'escadre	10 Schlachtschiff	10 Carazzati	10 Броненосецъ	(*Bronenosets*)	10 Acorazado	10 Slagskepp
B. (Beam)	11 Largeur (B.)	11 Breite (B.)	11 Larghezza (B.)	11 Ширина (B.)	(*Shirina* (B.))	11 Manga	11 Bredd (B.)
Belt	12 Ceinture	12 Gürtel	12 Cintura	12 Поясъ	(*Poyass*)	12 Faja	12 Gördelpansar (citadel eller [gördel)
Bow	13 Avant	13 Bug	13 Prora	13 Носъ	(*Noss*)	13 Proa	13 Bog
Boilers	14 Generateurs	14 Kessel	14 Caldaie	14 Котлы	(*Kotli*)	14 Caldeŕas	14 Ångpannor
Bulkheads	15 Traverses	15 Querwände	15 Traversi	15 Переборки	(*Pereborki*)	15 Mampanos	15 Tvärpansar
Bridge	16 Pont	16 Brücke	16 Ponte	16 Мостъ	(*Most*)	16 Puente	16 Brygga
Bunkers	17 Soutes	17 Bunker	17 Carbonili	17 Угольные ящики	(*Ugolnie yastchiki*)	17 Carboneras	17 Kolboxar
Capacity	18 Approvisement	18 Capacität	18 Capacita	18 Поинѣчение	(*Poinyestchenie*)	18 Capacidad	18 Kapacitet
Casemates	19 Casemates	19 Casematten	19 Casemetti	19 Казематъ	(*Kazemat*)	19 Casamatas	19 Kasematter
Coal	21 Charbon	21 Kohlenvorrat	21 Carbon	21 Уголь	(*Ugol*)	21 Carbon	21 Kol
Complement	22 Équipage	22 Besatsung	22 Equipaggio	22 Комплектъ	(*Komplekt*)	22 Dotacion	22 Besättningsstyrka
C.T. (conning tower)	23 Block-haus (C. T.)	23 Commando Thurm	23 Torri di commando	23 Рулевая башня	(*Roolevaia bashnia*)	23 Torse o' redueto de mando	23 Styrtorn
Cruiser	24 Croiseur	24 Kreuzer	24 Incrociatori	24 Крейсеръ	(*Kretser*)	24 Crucero	24 Kryssare
Deck	25 Pont	25 Deck	25 Ponte	25 Палуба	(*Palooba*)	25 Cubierta	25 Däck
Destroyer	26 Contre-torpilleur	26 Zerstörer	26 Caccia-torpedinieri	26 Контръ-миноносецъ	(*Kontr-minonosets*)	26 Caratorpedenos	26 Torpedobåtsförstörare
Displacement	27 Déplacement	27 Wasserdrängung	27 Dislocomento	27 Водоизмѣщение	(*Vodoizmiestchenie*)	27 Desplazamiento	27 Deplacement
Draught	28 Tirage	28 Tiefgang	28 Immersione (D.)	28 Углубление (D.)	(*Oogloobienie* (D.))	28 Puntal (D.)	28 Djupgående (D.)
Ends	29 Extrémités	29 Ende	29 Extremite	29 Край	(*Krai*)	29	29 Stäfvar
Fighting tops	30 Hunes	30 Marsen	30 Coffe militari	30 Пушечный марсъ	(*Poostchechnii mars*)	30 Cofas militares	80 Militärmaster

No.	English	Japanese	
1	Above water		水上
2	Armour	*Kotetsu*	甲鉄
3	Armoured Cruiser	*Kotetsu-Junyo-Kan*	甲鉄巡洋艦
4	Ahead	*Omote*	艦前
5	Astern	*Tomo*	艦後
6	Amidships	*Chuō*	中央
7	Barbettes	*Rohoto*	露砲塔
8	Bases (of Barbettes, etc.)	*Hoto-no-Kibu*	砲塔ノ基部
9	Battery	*Hodai*	砲臺
10	Battleship	*Shento-Kan*	戰闘艦
11	B. (Beam or breadth)	*Haba*	巾
12	Belt	*Kotai*	甲帯
13	Bow	*Kanshu*	艦首
14	Boilers	*Kikan*	汽罐
15	Bulkheads	*Kakuheki*	隔壁
16	Bridge	*Kankio*	艦橋
17	Bunkers	*Shekitanko*	石炭庫
18	Capacity	*Yorio*	容量
19	Casemates	*Kuku-Hodai*	穹窖砲臺
20	c/m (*Centimetre*)		センチメーター
21	Coal	*Shekitan*	石炭
22	Complement	*Teiin*	定員
23	C. T. (Conning tower)	*Shireito*	司令塔
24	Cruiser	*Junyo-Kan*	巡洋艦
25	Deck	*Kanpan*	甲板
26	Destroyer	*Kuchiku-tei*	驅逐艇
27	Displacement	*Hai-sui-Tonsu*	排水噸數
28	Draught	*Kissui*	吃水
29	Ends	*Hashi*	両端
30	Fighting tops	*Shento-Shorow*	戰闘檣樓

GLOSSARY OF TECHNICAL TERMS.

ENGLISH.	FRENCH.	GERMAN.	ITALIAN.	RUSSIAN.	RUSSIAN.	SPANISH.
31 Funnels	31 Cheminées	31 Schornsteine	31 Fumaioli	31 Трубы	(*Trubi*)	31 Chimeneas
32 Guns	32 Artillerie	32 Geschütze	32 Cannoni	32 Орудія	(*Orudiia*)	32 Cañones
33 Hoods	33 Carapaces	33 Cupoles	33 Spalti	33 Покрышки	(*Pokrishki*)	33 Carapachos
34 Kts. (Knots)	34 Noends (Kts.)	34 See-milen (kts.)	34 Nodi	34 Узелъ	(*Uzell*)	34
35 L. (Length)	35 Longeur (L.)	35 Länge (L.)	35 Lungo (L.)	35 Длина (L.)	(*Dlina* (*L.*))	35 Eslora (L.)
36 Liquid fuel	36 Combustile liquide	36 Petroleum	36 Petrolio	36 Жидкое топливо	(*Jhidkoe toplivo*)	36 Combustibile liqu
37 Lower deck	37 Faux pont	37 Zwischen-deck	37 Coperta inferiore	37 Нижняя палуба	(*Nijhniaia paluba*)	37 Cubierta baja
38 Machinery	38 Machines	38 Maschinen	38 L'apparato motore	38 Механизмъ	(*Mekhanizm*)	38
39 Main-deck	39 Pont de batterie	39 Batterie deck	39 Coperta centrale	39 Главная палуба	(*Glavnaia paluba*)	39 Cubierta principa
40 Masts	40 Mâts	40 Masten	40 Alberi	40 Мачты	(*Machti*)	40 Palos
44 Port (side)	44 Babord	44 Backbord	44 Sinistra	44 Лѣвый	(*Lievii*)	44 Babor
45 Protection to vitals	45 Protection des machines	45 Protektion an Schutz der Maschinen	45 Prot	45 Защита	(*Zastchita*)	45 Protection de las v
46 Ram	46 Eperon	46 Widder(schiff) Ramsparn	46 Navi speroni	46 Таранъ	(*Tarann*)	46 Espolon
47 Screens	47 Écrans	47	47 Scudi	47 Ширмы	(*Shirmi*)	47 Mantelete
48 Shields	48 Boucliers	48 Schützschildern	48 „	48 Щитъ	(*Stchit*)	48 Mantelete
49 Speed	49 Vitesse	49 Geschwindigkeit	49 Velocita	49 Скорость хода	(*Skorost khoda*)	49 Velocidad
50 Starboard	50 Tribord	50 Steuerbord	50 Diritta	50 Правая сторона	(*Pravaia storona*)	50 Estribor
51 Stern	51 Arrière	51 Heck	51 Poppa	51 Корма	(*Korma*)	51 Popa
52 Submarine	52 Sousmarin	52 Unterwasser	52	52 Подводная-лодка	(*Podvodnaia lodka*)	52 Submarino
53 Submerged	53 Submergé	53 Unterwasser	53 Subacquei	53 Подводный	(*Podvodnii*)	53 Sumerfido
54 Torpedo	54 Torpille	54 Torpedo	54 Siluri	54 Мина	(*Mina*)	54 Torpedero
55 Torpedo boat	55 Torpilleur	55 Torpedo boot	55 Torpediniere	55 Миноносецъ	(*Minonosets*)	55 Torpedero
56 Torpedo tubes	56 Tubes lance torpille	56 Torpedo lancier-röhre	56 Tubi lancia silure	56 Минные апараты	(*Minnie aparati*)	56 Tubos lanza torped
57 Tons	57 Tons	57 Tonnen	57 Tonnelate	57 Тонны	(*Tonni*)	57 Toneladas
58 Trials	58 Essais	58 Probefahrten	58 Prove	58 Испытанія	(*Ispitaniia*)	58 Pruebas
59 Turrets	59 Tourelles	59 Thürme	59 Torri	59 Башня	(*Bashnia*)	59 Torres
60 Upper deck	60 Pont de gaillards	60 Oberdeck	60 Coperta superiore	60 Верхняя палуба	(*Verkhniaia paluba*)	60 Cubierta alta
61 Water line	61 Flotation	61 Wasserlinie	61 Linea d'acqua	61 Ватерлинія	(*Vaterliniia*)	61 La linea de flotacio

No.	English	Japanese (romanized)	Japanese
61	Waterline	*Suishen*	水線
60	Upper deck	*Jokanpan*	上甲板
59	Turrets	*Yihoto*	圓砲塔
58	Trials	*Shi Unten* (*Kokoromi Unten*)	試運轉
57	Tons		噸
56	Torpedo tubes	*Suirai Hatsushakuan*	水雷發射管
55	Torpedo-boat	*Suirai-Tei*	水雷艇
54	Torpedo	*Suirai*	水雷
53	Submerged	*Suichu*	水中
52	Submarine	*Shen-sui*	潜水
51	Stern	*Kanbi*	艦尾
50	Starboard (side)	*Ugen*	右舷
49	Speed	*Sokurioku*	速力
48	Shields	*Tate*	楯
47	Screens	*Kakusho*	スクル
46	Ram	*Shoto*	撞頭
45	Protection to vitals	*Yobu-Bogio*	要部防禦
44	Port (side)	*Sagen*	左舷 左
43	Normal	*Kitei*	規定
42	m/m (Millimetre)		ミリメータ
41	M. (*Metre*)		ミーター
40	Masts	*Hobashira*	檣
39	Main (-deck)	*Chukan-pan*	中甲板
38	Machinery		機械
37	Lower (-deck)	*Gekan-pan*	下甲板
36	Liquid fuel	*Yakitai-Nenrio*	液体燃料
35	L. (Length)		長

ALPHABETICAL INDEX—"INDEX II."

BRÉSIL.

The names "Almirante" and "Marechal" are to be done away with in all Brazilian ships. The Quinze de Novembro (p. 29) will again be known as Republica, and the Vinte Quatro de Maio as the Aquidaban (p. 31). The military masts of Riachuelo (p. 33) are to be removed.

CHINE.

The four destroyers (p. 41) are now in the possession of England, France, Germany, and Russia respectively. For Fei Ting, on the silhouette (p. 42), *read* Fei Ying, as printed below it.

ENGLAND.

p. 69. The Leopard, Otter, Avon, and Bittern are **3e**, *not* **3c.** The aftermost funnel of all the Laird destroyers is nearer the stern than the silhouettes show.

Four large torpedo boats, to replace others, are building at Thornycroft's.

p. 82. Prince George is to be distinguished by black anchor davits.

The Blonde is to be fitted with Laird boilers.

p. 93. Gossamer is to have water-tube boilers.

p. 98. Glory on trial, 30 *hours*, 10,587 I.H.P. = 16·78 kts.; *full power*. ? *hours*, 13,747 = 18·12 kts.

p. 109. The recently reported inability of the Powerful and Terrible to steam at their proper speed is fictitious. Somewhat similar reports about the Europa (p. 106) and Highflyer (p. 105) have also little foundation in fact.

ESPAGNE.

All, except a few of the modern ships, are to be sold out of the service. Prosperina (p. 110) has only 2 funnels. To Rio de la Plata (p. 113) add *Estramadura* (1900), completing for sea.

FRANCE.

p. 129. For class index (II.) *read* (III.), Bouvines, etc.

p. 144. Troude, etc., for (3) *read* (4).

ITALIA.

p. 163. The amidships funnels of *Fulmine* have been united, 3 funnels. In appearance is just like *Brazen*, *Electra*, etc. (see building, will have the same funnels. *Dardo*, etc., are just li *Hai Lung*, etc. (p. 41).

p. 168. A ship of Elba type, the Bascilicata, is building. For (*read* (3).

p. 173. The Pisa and Venezia will not be commenced yet, Benedetto Brin type. The Regina Margherita will be launched 20,

JAPON.

Three destroyers fitted with turbines are building at Newcastle.

p. 189. The Hatsuse has *shorter masts* than the Shikishima.

RUSSIE.

p. 218. A battleship, *Imperator Nikolai II.*, of the Alexander building.

p. 219. Variag has 30 boilers. On trial with 29 boilers she expected with forced draught. Pamiat Azova (p. 219) will have fi present rig. The Bayan was launched .

SVERIGE.

p. 228. Three 4000-ton protected cruisers, with guns in superpo are projected.

TURQUIE.

pp. 232, 233. It is proposed to rebuild Hamidieh and severa nothing is yet definitely settled.

ADDENDA to 14th July, 1900.

Photo, Steinitz, Kiel.

(**RUSSIE**, p. 203.)

YAKUMO. *Photo, Steinitz, Kiel.*

(**JAPON**, p. 187.)

(Note that the positions of some 3 inch (75 m/m) guns are changed from original design.)

DEUTSCHLAND.

"*C*" (p. 59) launched and named WITTELSBACH.

ENGLAND.

Amphion (p. 102) is having rig altered to that of Arethusa.
Ariadne, Diadem, Camperdown fitted for wireless telegraphy.

CLASSIFICATION OF SHIPS.

Uniform throughout the book.

Abbreviation. ARMOURED.

(I.) *1st class.* Ships carrying A guns, well protected, well protected secondary armaments, protection to vitals = *aaa* or more, and with submerged torpedo tubes.

(II.) *2nd class.* Powerful ships lacking one or other of the above requirements.

(III.) *3rd class.* Ships lacking heavy armour-piercing guns, or else two of the requirements for a class I. ship.

(IV.) *4th class.* Ships lacking three or all of the requirements for a 1st class battleship.

(V.) *5th class.* Coast defence ships unable to keep the sea; very small ironclads, or ships of no fighting value

PARTIALLY ARMOURED.

(1.) *1st class.* Carrying armour-piercing guns, or a very large quickfiri some vertical armour either on guns *or* water-line. Equival class IV. armoured.

(2.) *2nd class.* Of about 4000 tons, modern, fast, and well-protected vitals.

(3.) *3rd class.* Smaller or else older cruisers.

(4.) *4th class.* Lightly armoured and protected.

(5.) *5th class.* Of very slight fighting value.

(6.) *6th class.* No fighting value.

(X.) = Rams. (T.) = Torpedo ships. (T.X.) = Torpedo rams. Torpedo if over 100 tons and modern.

NOTE TO ILLUSTRATIONS.

Before the name of every ship or series of ships is her class according to the classification on page 17. Immediately following her name is the date of launch, and if reconstructed that date also. The place of building is given in the case of a vessel not built in the country to which she belongs.

Every possible care has been taken to render the illustrations of the ships accurate portraits. Occasionally the salient and characteristic features of vessels have been slightly accentuated. The drawings do not pretend to anything of an artistic nature: utilitarian distinctness is the one thing aimed at.

All vessels of the same nationality that resemble each other are placed together. The method followed is made sufficiently clear by Index 1. Index 2 allows of any particular ship being looked up.

EXPLICATION DES ILLUSTRATIONS.

Devant le nom de chaque navire, ou série de navires, on indique la classe dont il fait partie selon le système uniforme expliqué à page 17. Immédiatement après le nom du navire, on dit quand il a été lancé, et s'il a eu transformation on en donne la date. On nomme le chantier où il a été construit s'il ne l'a pas été dans son propre pays.

Les plus grands soins ont été employés pour que les illustrations des navires soient des représentations exactes; à remarquer que quand cela a été possible, les points saillants en ont été légèrement accentués.

Les dessins n'ont aucune prétention artistique; nous avons taché de leur donner seul la clarté qui a été notre but principal.

Tous les navires semblables de la même nation sont placés ensemble. Le système que nous avons adopté est clairement démontré à l'Index

ERLÄUTERUNG DER ZEICHNUNGEN.

Vor dem Namen jedes Schiffes oder jeder Schiffsserie steht die Classe, wozu das Schiff gehört nach der auf Seite 17 erklärten Klassification.

Gleich nachher steht das Datum an welchem das Schiff vom Stapel gelassen wurde, und auch das Datum seines Umbaues (sollte derselbe stattgefunden haben). Der Bauort eines Schiffes wird in dem Falle gegeben, wo es im Ausland ist.

Ganz besondere Sorgfalt wurde darauf gelegt, dass jede Illustration eines Kriegsschiffes, soweit dies zulässig, authentisch sei. Hierbei muss jedoch erwähnt werden, dass die hervorragenden und characteristischen Kennzeichen eines Kriegsschiffes accentuirt wurden.

Alle Schiffe, die Aehnlichkeit miteinander haben, sind zusammengestellt.

Maste und Schornsteine sind auf der See viel deutlicher als die Geschütz-Anordnung zu

SPIEGA DELL ZIC

Innanzi al nome di navi, viene indicata la secondo il sistema indica diatamente dopo il nom fu varata, e se trasforma caso che una nave non proprio paese, si fa m stata costruita.

Abbiamo presa la più che le illustrazioni delle autentiche per quanto è che, ove ciò ci è stato p caratteristici delle navi accentuati. I disegni pretenzione artistica; scopo al quale abbiamo

L' ordinamento rich spiega. Tutt' i bastim sono messi insieme. Il n

NOTE TO ILLUSTRATIONS.

Where any distinguishable difference between ter ships exists it is noted: where the difference practically not to be detected, one ship of the be does duty for all.

Masts and funnels being more easily distin- ishable at sea than the arrangement of guns, ere two ships otherwise identical differ in the mber of their funnels or masts, this fact is paid re attention to than the similarity of type. t otherwise, ships of the same type are kept gether as much as possible; and where sister ps are placed on different pages the fact is fully ted.

EXPLICATION DES ILLUSTRATIONS

1er. L'Index 2nd offre le moyen de se renseigner sur chacun des bâtiments.

Quand il existe une différence notable entre deux navires du même type nous l'avons indiqué.

Les mâts et les cheminées se distinguent plus facilement en mer que la disposition d'artillerie; si deux navires semblables offrent quelque différence dans le nombre des cheminées ou des mâts, on y a fait plus d'attention qu'à la ressemblance de type. Du reste, les navires du même type ont été placés ensemble autant que possible, et si deux navires de la même classe se trouvent sur deux pages différentes, nous en avons tenu compte.

ERLÄUTERUNG DER ZEICHNUNGEN

erkennen; wo also zwei Schiffe, die sonst vollkommen ähnlich sind, sich allein durch die Zahl ihrer Schornsteine oder Masten unterscheiden lassen, wird auf diesen Punkt mehr Gewicht gelegt, als auf die Gleichartigkeit des Typus. Im Uebrigen wurden Schiffe, welche demselben Typus angehören, soweit es möglich, zusammengestellt: wo solches nicht auf der nämlichen Seite geschehen, wurde darauf besonders hingewiesen.

SPIEGA DELLE ILLUSTRAZIONI.

è chiaramente dimostrato all' Indice I. Indice 2[o] offre facilità di confronto riguardo ad alcuna nave in particolare.

Ove esista qualche distintivo particolare fra navi del medesimo tipo, se ne fa debita menzione.

Gli alberi ed i fumaiuoli si distinguono più facilmente in mare che la disposizione dei cannoni; ove due navi, altrimenti identiche, differiscono nel numero dei fumaiuoli o degli alberi, si fa di questo maggior caso, che della similarità di tipo. Per altro, i bastimenti dello stesso tipo sono messi insieme per quanto è possibile; e se due di queste navi si trovano rappresentate su due pagini separate, se ne fa menzione speciale.

Approximate maximum thickness of armour netrated by armour-piercing shot (projectiles rupture):—

Approximate maximum thickness of armour to be penetrated by common shell (obus ordinaires):—

Range Circa								*Range Circa*				
00 ls.	4000 yds.	3000 yds.	2000 yds.	1000 yds.	500 yds			5000 yds.	4000 yds.	3000 yds.	2000 yds.	1000 yds.
b	*a*	*a*	*aa*	*aa–aaa*	*aaa–aaaa*	**A** =	guns of >20,000 foot tons (*i.e.*, 6,200 metre tnx) initial energy	*f*	*e*	*e*	*d*	*d*
c	*c*	*b*	*b*	*a*	*aa*	**B** =	,, >14,000 foot tons (= *circa* 4,340 metre tnx dinamodi) initial energy...	*f*	*f*	*e*	*e*	*d*
d	*d*	*c*	*c*	*c*	*b*	**C** =	,, >6,000 foot tons (= *circa* 1,860 metre tnx) initial energy	*f*	*f*	*e*	*e*	*e*
e	*e*	*d*	*d*	*d*	*dc*	**D** =	,, >2,000 foot tons (= *circa* 620 metre tnx) initial energy	*f*	*f*	*e*	*e*	*e*
		f	*e*	*e*	*de*	**E** =	,, >800 foot tons (= *circa* 250 metre tnx) initial energy				*f*	*f*
				f	*ef*	**F** =	,, >400 foot tons (= *circa* 125 metre tnx) initial energy					

An * (thus: D*) = Quick fire. * = À tir rapide. * = Schnelllader-Kanonen. * = A tiro rapido.

NOTATION OF ARMOUR.

For rapid comparison and reference, all armour is notated upon the *a, b, c* system of the Naval War Game. The principle is, roughly, that at 2000 yards armour c: successfully attacked with shot from the gun correspondingly lettered, while at closer range a grade higher may be attempted. The *approximate* penetrations are on pa opposite. Below are the values of the principal kinds of armour worked out in comparison with iron armour, and the notation letters used in this book.

IN INCHES.

		Krupp. Inches.		Harvey nickel: *Acier special durci*; Ternil. Inches.		Harvey or *acier durci*. Inches.		Compound. Inches.		Early Compound. Inches.		Steel. Inches.		Iron. Inches.		
f	=	1	=	1¼	=	1½	=	2	=	2¼	=	2½	=	3	=	*f*
e	=	2	=	2½	=	3	=	4	=	4½	=	5	=	6	=	*e*
d	=	3	=	3¾	=	4½	=	6	=	6¾	=	7½	=	9	=	*d*
c	=	4	=	5	=	6	=	8	=	9	=	10	=	12	=	*c*
b	=	5	=	6¼	=	7½	=	10	=	11¼	=	12½	=	15	=	*b*
{ *a*	=	6	=	7½	=	9	=	12	=	13½	=	15	=	18	=	*a* }
{ *a*	=	7	=	8¾	=	10½	=	14	=	15¾	=	17½	=	21	=	*a* }
{ *aa*	=	8	=	10	=	12	=	16	=	18	=	20	=	24	=	*aa* }
{ *aa*	=	9	=	11¼	=	13½	=	18	=	20¼	=	22½	=	27	=	*aa* }
{ *aaa*	=	10	=	12½	=	15	=	20	=	22½	=	25	=	30	=	*aaa* }
{ *aaa*	=	11	=	13¾	=	16½	=	22	=	24¾	=	27½	=	33	=	*aaa* }
aaaa	=	12	=	15	=	18	=	24	=	27	=	30	=	36	=	*aaaa*
aaaaa	=	14	=	17½	=	21	=	28	=	31½	=	35	=	42	=	*aaaaa*

IN MILLIMETRES.

		Krupp. m/m.		Harvey nickel: *Acier special durci*: Terni. m/m.		Harvey. m/m.		Compound. m/m.		Old Compound. m/m.		Steel. m/m.		
f	=	25	=	31	=	38	=	50	=	57	=	63	=	
e	=	50	=	63	=	76	=	101	=	114	=	127	=	
d	=	76	=	95	=	114	=	152	=	171	=	190	=	
c	=	101	=	127	=	152	=	203	=	228	=	254	=	
b	=	127	=	159	=	190	=	254	=	286	=	318	=	
{ *a*	=	152	=	190	=	228	=	304	=	342	=	381	=	
{ *a*	=	177	=	222	=	266	=	355	=	400	=	444	=	
{ *aa*	=	203	=	254	=	304	=	406	=	456	=	508	=	
{ *aa*	=	228	=	286	=	342	=	457	=	514	=	571	=	
{ *aaa*	=	254	=	318	=	380	=	508	=	571	=	635	=	
{ *aaa*	=	279	=	340	=	418	=	558	=	628	=	698	=	
aaaa	=	304	=	381	=	456	=	609	=	685	=	761	=	
aaaaa	=	355	=	444	=	533	=	710	=	798	=	889	=	1

The slopes of protective decks are reckoned at double the resisting value of vertical armour (1 / = 2 ‖). For example, a 2½ inch steel deck is valued at = *e*, a 76 m/m Krupp deck would be = *a*. In the data of ships the water-line belt, armour deck, and coal are all reckoned up together under the head of *Protection to Vitals*. 1 therefore, visible at a glance whether belt attack is feasible.

1 inch = *circa* 25 millimetres.

Displacements are given indifferently in English or metric tons. The difference, where displacements are concerned, is negligable (1000 tons Eng. = 1016 metre tx).

Horse powers are also given indifferently (10,000 Eng. H.P. = 9862 *chevaux*).

Note.—Ships whose names are printed in italics are not likely to be ready for sea before June 1901 at the earliest.

Knots.	Kilom.	Knots.	Kilom.	Knots.	Kilom.
·5 =	·9	11 =	19·8	21 =	37·8
1 =	1·8	12 =	21·6	22 =	39·6
2 =	3·6	13 =	23·4	23 =	41·4
4 =	7·2	14 =	25·2	24 =	43·2
5 =	9·0	15 =	27·0	25 =	45·0
6 =	10·8	16 =	28·8	26 =	46·8
7 =	12·6	17 =	30·6	27 =	48·6
8 =	14·4	18 =	32·4	28 =	50·4
9 =	16·2	19 =	34·2	29 =	52·2
10 =	18·0	20 =	36·0	30 =	54·0

m.m.	Inch.		Inch.	
50 =	1·96	*i.e.*	2	appro
75 =	2·94	„	3	„
100 =	3·93	„	4	„
150 =	5·90	„	6	„
152 =	6·00	„	6	„
200 =	7·87	„	8	„
250 =	9·84	„	10	„
300 =	11·91	„	12	„
350 =	13·78	„	14	„
400 =	15·74	„	15¾	„

ARGENTINA.

(*Revised by "X.X."*)

ARTILLERIE.

	Nominal Calibre.			Length in Cal.	Muzzle Velocity.		Muzzle Energy Approximate.		Projectile.	
	in.	c/m			ft. secs.	metre sec.	ft. tons.	metre tnx.	lbs.	kgs.
B	10	25·4	Elswick (1) . .	—	2207	672	16250	4963	500	226·8
B	9·4	24	Krupp (2) 32 tons	47	2500	762	14800	4456	350	158·7
C*	8	20·3	Elswick, Q.F. .	45	2660	810	10300	3190	210	95·2
C	8	21	Elswick (3) . .	40	2286	696	7330	2270	210	95·2
C	8	20	? (4) . .	—	2120	646	6000	1858	200	90·6
C	10	25·4	? (5) . .	30	2030	618	12330	3819	450	204
D*	5·9	15	Canet, Q.F. (6) .	50	2750	838	4645	1438	88	40
D*	6	15	Elswick, Q.F. .	40	2500	762	4334	1342	100	45·3
D	6	15	?	30	1960	579	2550	789	100	45·3
D*	4·7	12	Canet, Q.F. (6) .	50	2756	840	2440	755	46·3	21
E*	4·7	12	Elswick, Q.F. (7.)	45	2570	783	2060	638	45	20·4
E*	4·7	12	Elswick, Q.F. .	40	2150	655	1440	446	45	20·4
F*	3	7·6	Vickers. Maxim .	46	2100	640	—	—	14	6·35

(1) Garibaldi; (2) Libertad and Independencia; (3) 25 de Mayo; (4) Andes and Plata only; (5) Patagonia only; (6) A. Brown only; (7) Sarmiento.

CORRIENTES, ENTRE RIOS, MISSIONES, SANTA FÉ.

(Yarrow, Poplar, 1896.)

Guns: 1 F* (14 pdr.*) + 3 6 pdr.* + 2 Maxims. Torpedo tubes, 18 inch, 1 in bow, 2 on deck abaft funnels, one abaft the other. Armour: side and deck ½-inch (1·26 c/m) steel over machinery. Machinery: I.H.P. 4200. *Trial:* Missiones, 27·1 kts. (mean); Corrientes, 27·4 kts.: Entre Rios, 26·7 kts.; Santa Fé, 26·5 kts. Boilers: 6 Yarrow water tube. Draught (*max*); 5 feet (1·5 metres), 2 screws. Coal capacity: 80 tons. L.: 190 ft. = 58 m.

Santa Fé wrecked 1897. Raised and repaired (new hull) 1898–9.

Le Santa Fé a fait naufrage 1897. Refondu 1898.

The torpedo boats are:—

2 Thornycroft, 110 tons, speed (trial) 24·5 kts., 150 feet long, 3 tubes (1890).
6 Yarrow 85 " " " 24 " 130 " 2 " (1890).
4 Yarrow 52 " " " 20 " 100 " 3 " (1880).
—
12
—

Water-tube boilers to the six 130 feet boats. Also 8 3rd class boats Yarrow, 17 kts (1890), and 2 Thornycroft (1881), speed 17 kts.

ARGENTINA.

INDEPENDENCIA. *Photo by Rimattié, Buenos Aires.*

(IV.) INDEPENDENCIA (1891 Birkenhead) (1890 Birkenhead).

L.: 230 ft. = 70 m. Sea speed: 12·5 kts.

2336 Tons. Complement 225.

Guns: 2B (9·4 in. Krupp) + 4E* (4·7 in. Elswick Q.F 2 above water torpedo tubes.

Armour: Deck (50 m/m) = *e*. Belt (5 ft. wide) *c* (2(forward, *c*; aft, *d*. Barbettes, *c*; bases, *c*. Shields to B guns, *d*. Compound armour.

Machinery: 4 double-ended boilers. I.H.P. 2100 = 13· draught, 2780 = 14·2 kts. (trial) Coal: normal 230. Max. 3 at 10 kts.

Libertad now has a high top-mast, yard carried high up on

Independencia, short top-mast—s.l. on bridge.

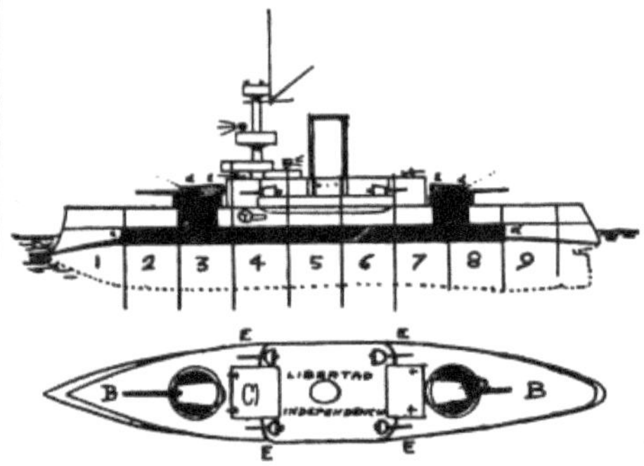

(4.) PATAGONIA (1885 Trieste).

L.: 192 ft. = 58·5 m.

Guns: 1C + 3D + 6*. Armour: 0. Sea speed: 8 kts.

1442 tons. Comp. 210. Partial steel deck over vitals = *f*. Gun shields, *f*. Coal capacity, 350 tons. The C gun is a 10-in.

(IV.) ALMIRANTE BROWN (1880 London) (1897 La Seyne).

L.: 240 ft. = 73·5 m. Sea speed: 11 kts.

4267 tons. Complement 350.

Guns: 10D* (5·9 in. Canet Q.F.) + 4D* (4·7 in. Canet Q.F.) + 2 3-pdr.* Two above water torpedo tubes.
Armour: Deck (35 m/m) = *f*. Belt, *e*. Battery, *e*. Bulkheads, *e* (225 m/m) compound armour.
Machinery: I.H.P. 4500 = 14 kts. (trial). Coal capacity 650 tons = *circa* 3000 kts. at 10 kts.
(Re-armed 1897-8.)

LOS ANDES 1898.

(V.) EL PLATA (1874 Birkenhead) & LOS ANDES (1875 Birkenhead).

L.: 187 ft. = 57 m. Sea speed: 6 kts.

1677 tons. Complement 178.

Guns (1898): 2C (8 in. 20 c/m) + 2 6-pdr.* (47 m/m) + 4* (25 m/m).

Armour on turret *d*, belt *e*—*f*. Draught, 10 ft. 2 screws.

Dif. = Funnel.

ARGENTINA.

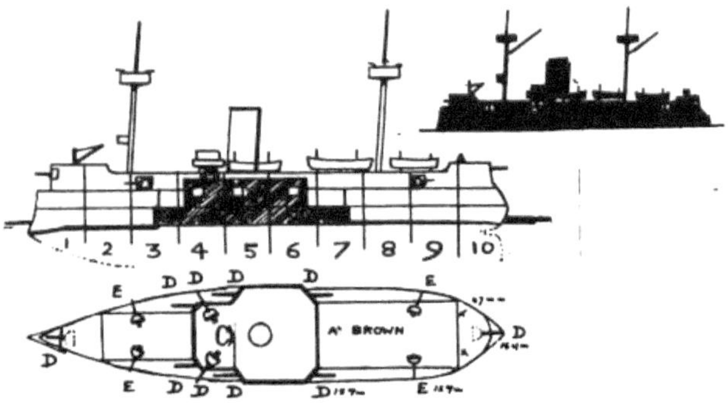

ALMIRANTE BROWN plan.

Trials—

S. Martin (1899), 8 *hrs. n.d.*, 8286 I.H.P. = 18·07 kts. 2 *hrs. forced d.*, 12,436 I.H.P. = 19·
Belgrano (1898), 6 „ „ 8600 „ = 18·09 „ 2 „ „ 13,000 „ = 20·

(Belgrano on trial displaced 7182 tons (metric) and mean draught was 23¾ feet (7 m. 25).

BELGRANO. *Photo by*

(For particulars see next page.)

(III.) GARIBALDI (1895) & GENERAL PUEYRRDON (ex Garibaldi II.) (1897): SAN MARTIN (1896): GENERAL BELGRANO (ex Varese).

Guns: Garibaldi & Pueyrrdon, 2B (10-in. 25·4 c/m) + 10D* + 6E* + 10* (6-pdrs. 57 m/m) + 10* (1-pdr. 37 m/m). Belgrano, 2B + 14D* + 2F* + 10* (57 m/m) + 8* (37 m/m). S. Martin, 4C* + 10D* + 6E* + 12* (57 m/m) + 10* (37 m/m).
All these ships purchased from Italy, where they were built.

Circa 6900 tons. Complement 450. L. B. D. 328 × 59½ × 25 ft. = 100 × 18·1 × 7·6 m.

Four torpedo tubes lower deck behind *c* armour.

All armour *c*, except ends of belt, which are *d*. Shields to all guns *e*. Bulkheads *c*. Protection to vitals, belt + coal = *b*. Lower deck + coal = *b*. There is a screen *f* down centre of battery and across abaft mast. Armour: acier nickelé, special process, Société de Terni.

Machinery. Engines (*Garibaldi*, Maudslay): 8 cylindrical boilers (but *Pueyrrdon*, Belleville): 2 screws; I.H.P. nat. 8600: forced d. 13384 = 19·4 kts. trial: heating surface = 6500 ft.: coal cap. 1000 tons + petroleum fuel.

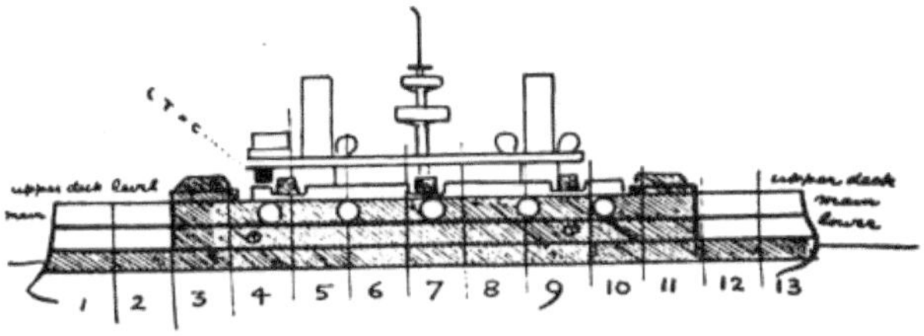

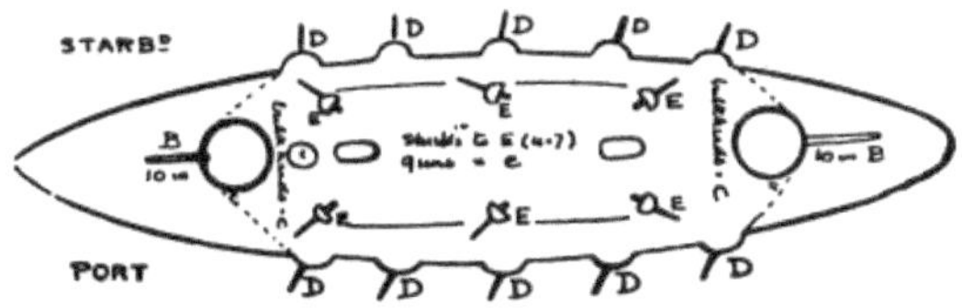

GARIBALDI & PUEYRREDON.

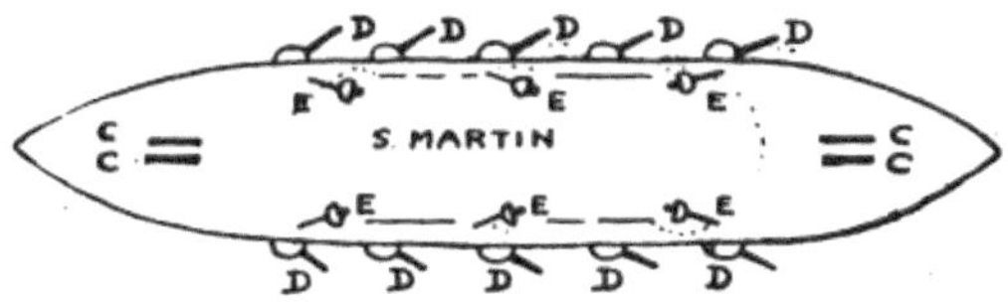

ARGENTINA.

(6) ARGENTINA (1884 Trieste).

L.: 192 ft. = 58·5 m.

820 tons. Complement 120.

Guns: 1D (6 in.) + 4F (2·7 in.) + 6*.
I.H.P. 850 = 12 kts. (trial).
Coal capacity 220 tons. 1 screw. Sea speed: 10 kts.

(6) PARANA & URUGUAY (1874 Birkenhead).

L.: 142 ft. = 43·5 m

550 tons. Complement 90,

Guns: Parana, 4F (10 c/m) + 2*. Uruguay, 2E (4·7 m., 12·5 c/m) + 2*. Sea speed: 9 kts. Uruguay copper sheathed

(6) REPUBLICA & CONSTITUCION (1873 Birkenhead).

L.: 103 ft. = 31·5 m. Guns: 1C + 2*. Sea speed: 5 kts.

(T.) ESPORA (1890 Birkenhead).

L.: 210 ft. = 64 m

Guns: 5* Armour: *prot.* Sea speed: 17·5 kts.

PATRIA. 1070 tons. Complement 100. 5 torpedo tubes. Machinery: I.H.P. nat. 3336 = 17·6 kts. (trial), forced draught 5048 = 20·5 kts. (trial). Boilers, 4 locomotive. Coal capacity 288 tons = *circa* 3000 miles at 10 kts. Can cruise about 7 to 8 days. Mean draught 10 ft. 2 screws.

ESPORA. 615 tons. Complement 60. Machinery: I.H.P. forced draught 3448 = 20 kts. (trial). Coal: 100 tons = 1200 kts. at 10 kts. 30 tons more can be carried. 4 locomotive boilers. Armour: belt over engines, steel ½ to 1½ in. (1·29 c/m to 2·1 c/m). 5 18-in. torpedo tubes. Guns: 2F* (14-pdr.*), 1 8-pdr.*, 2 3-pdr. 2 screws.

(T.) PATRIA (18

L.: 250 ft. =

Guns: 2E* + 6*. Armour:

(2.) NUEVE DE JULIO (1892 Elswick).

L.: 350 ft. = 106·5 m.

Guns: 4D* + 8E* + 12*. Armour: *prot.* Sea speed: 19 kts.

3500 tons. Trial speed = 22·7 kts. Prot. deck = *c.* (110 m/m). 5 torpedo tubes. 2 screws.

(2.) VINTE-CINCO DE MAIO (1890 Elswick).

L.: 330 ft. = 100·5 m. 3200 tons. Complement 310. 3 torpedo tubes. Guns: 2C (8 in. 21 c/m Elswick) + 8E* (4·7 in. Q.F. 12 c/m + 12 3-pdr.* (47 m/m) + 12* 37 m/m). Prot. deck = *c.* Engine hatches *c.* Trial speed = 22 kts. Coal capacity normal 300 *max.* 600 = *circa.* 7000 miles at 10 kts. 2 screws.

Sea speed: 19 kts.

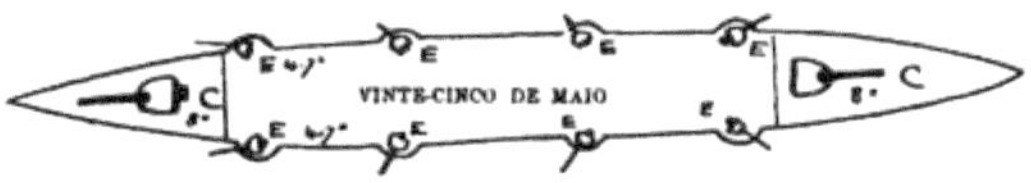

Photos, Elswick.

Photo, Elswick.

ARGENTINA.

(2.) BUENOS AIRES (1895 Elswick).

L. : 400 ft. = 122 m.

Guns: 2C* + 4D* + 6E* + 20*. Armour: *prot* Sea speed

4500 tons. Complement 400. Armour deck = c. Protection to vi =b. C. T. *d*. Gun shields *d*. Torpedo tubes, above water: 1 in broadside. Machinery: 2 pairs 3-stage compound direct acting, by Tennant. Boilers: return tube, 8 double ended, 8 single ended. I.H. pressure 13.000 = 23·2 kts. (trial), forced draught 18,000 I.H.P. = 24 kts. 1000 tons = 10,000 miles. 2 screws.

6 *hours natural draught trial* = 23 *kts. mean*.

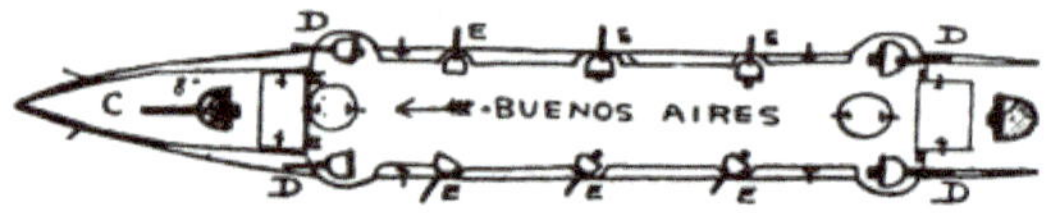

(6) PRESIDENTE SARMIENTO (1897 Birkenhead).

2750 tons. I.H.P. 2000 = 13 kts. Boilers: 2 cylindrical return tube, 2 Niclausse. 1 screw. G 5E* (4·7 in.) + 2F* (14-pdr.) + 4 6-pdr.*, 4 3-pdr.* + 4*. 2 torpedo tubes, above water, bow and s (Completing for sea.)

The Argentine navy also contains: One transport of 2700 tons, 8 kts. speed; one of 1000 tons, two of about 400 tons, and six small river gunboats.

BRESIL.

ARTILLERIE.

Notation.	Calibre.			Length calibres.	Muzzle Velocity *circa.*		Approximate Muzzle Energy.		Weight of Projectile.	
	Inches.	c/m			ft. secs.	met. sec.	ft. tons.	mets. tnx.	lbs.	kgs.
B	9·2	23	Elswick . . .	45	2700	823	19000	5884	380	174·5
B	9·4	24	Canet [1]	36	2500	762	14250	4413	374	169 5
B *or* **O**	9·4	23	Whitworth (altered by Elswick) [2]	?	?	?	?	?	?	?
D*	6	15	Elswick Q.-F. . .	40	2500	762	4330	1341	100	45·3
D*	6	15	" . .	50	2640	804	4540	1498	100	45·3
E*	4·7	12	" . .	40	2150	655	1440	449	45	20·4
DE*	4·7	12	" . .	50	2630	801	2160	652	45	20·4

[1] Deodoro. [2] 24 de Maio.

N.B.—The illustration of Tammandare has not been verified this year.

"Destroyers."

Eight boats of 26 kts., *pro.*

TORPEDO BOATS.

5 Schichau boats (1893), 152 ft. long, 130 tons, speed (trial) 28 kts ; 3 tubes.
3 Thornycroft boats (1891), 150 " 150 " " 25–26 kts., 4 tubes.
5 Yarrow boats (1882), 100 " 52 " " 20 kts., 2 tubes.

13

Also 1 wooden 3rd class of 25 kts. trial speed (1893), and 6 others : speed 17 kts. in 1886.

The Brazilian navy also includes wooden ships : Tonelero (1400 tons) and Paysandu (1900 tons), 6 small steel or iron gunboats, 5 wood or composite gunboats (all of from 700–250 tons) ; all of no fighting value.

Armed merchantmen : 1 (Andrada) 2600 tons, speed 14 kts., 2E* + 2F* guns ; 4 (Mercurio, Jupiter, Meteoro, and Marte) of 1000 tons, armed with 6-pdr. or 3-pdr. Hotchkiss, and 3 very small steamers.

Two transports of 1400 tons (paddle wheel).

One old ironclad, used as a floating battery ; 3 monitors of 400 tons (river), and two river monitors of 500 tons (building).

(T.) PIRATINY (ex Ericsson's "Destroyer").

One *submarine* gun. Armour over gun = *b*. Speed 8 kts.

(T.) GUSTAVO SAMPAIO
(1898 Elswick).

L.: 197 ft. = 60 m.

Guns: 6*. Armour: 0. Sea speed: 16 kts.

(T.) TIRADE
(1892 Elswic
L.: 171 ft. + 5
Guns: 4E* + 3*. Armo
speed: 12 kts.

(4.) QUINZE DE NOVEMBRO
(1892 Elswick).

L.: 225 ft. = 68·5 m.

Guns; 6E* + 4*. Armour: *prot.* Sea speed: 15 kts.

(At present [1899] this ship is carrying only one yard on each mast: no sails.)

(3.) BENJAMIN C
(1892 La Seyne)

L.: 236 ft. = 7

Guns: 4D* + 8E* +
prot. Sea speed: 18·

GUSTAVO SAMPAIO. 500 tons. Complement 95. Guns: 2F* (3·5 in. = 9 c/m) + 4 3-pdr.* Torpedo tubes: 1 in bow, one on each broadside. Machinery: I.H.P. 2500 = 18 kts. (trial). Coal capacity 150 tons = 1000 miles. 2 screws. Draught (*max.*) 9 ft. (= 2·74 m.).

QUINZE DE NOVEMBRO (*ex* Republica). 1300 tons. Armour deck = *d*. Torpedo tubes: 1 in bow, 1 in stern, 1 each broadside. Coal capacity 170 tons. *Max.* trial speed 17·5 kts. 2 screws. Draught (*max.*) 15 ft. (= 4·57 m.).

TIRADENTES. 800 tons. Complement 105. Draught (*max.*) 12 ft. (=
Max. trial speed 14·7 kts. Coal capacity 100 tons. Torpedo tubes 2.

BENJAMIN CONSTANT. 2750 tons. Complement 380. The E* g
deck in battery, protected by *e* shields. Armour deck = *e*. *Max.* trial speed 15 kt
260 tons = *circa* 3000 kts. at 10 kts. 4 torpedo tubes: 1 in bow, 1 in stern, 1 on ea

(IV.) MARECHAL DEODORO & MARECHAL FLORIANO (1898 La Seyne).

3162 tons. Complement 200.

L. B. D. $267\frac{1}{2} \times 48 \times 13\frac{1}{8}$ (mean) feet = $81 \cdot 5 \times 14 \cdot 6 \times 13 \cdot 1$ (moyen) m.

Guns: 2B (9·2 in. Elswick) + 4E* (4·7-in. Elswick) + 2 Howitzers (mortiers) (6-in. = 15 c/m calibre) + 2F* (12-pdr.) + 4 6-pdrs.* (57 m/m) + 4* + 2 torpedo tubes (Elswick), submerged.

Armour: Belt (6 ft. deep) *d*——*aa*——*d*; all the under water part of belt is *c* or less. Deck = *f*. Prot. to vitals, belt + deck + coal = *aaa*. Turrets, *b*. Casemates to E guns, *e*. C. T. *d*.

Machinery: I.H.P. nat., 3400 = 14kts. (*max.* estimated). Boilers, 8 Lagraufel D'Allest. Coal: 236 tons.

To be completed 1899. Design modified during construction.

Armour (*acier special durci*): Belt $\frac{350 - 100 \text{ m/m}}{13 \cdot 7 - 4''}$ Deck curved behind belt $\frac{35 \text{ m/m}}{1\frac{1}{2}''}$ Turrets $\frac{203 \text{ m/m}}{8''}$

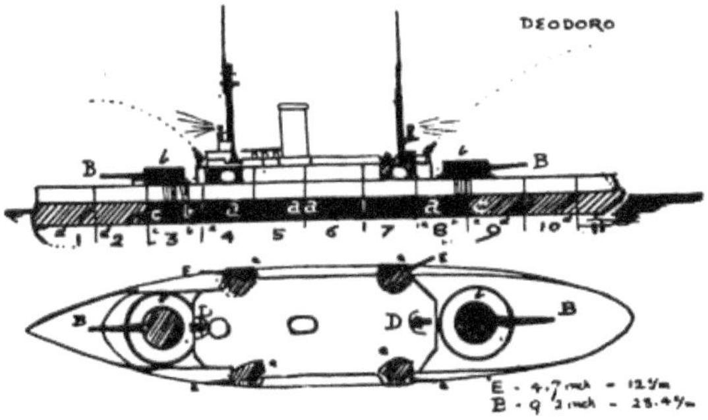

DEODORO. Photo by Bar, Toulon.

(III.) VINTE-QUATRO DE MAIO (*ex* AQUID Poplar) (1897 Stettin).

Guns: 4B (9·2-in. Whitworth) + 4E* + ?*

5000 tons. Complement 388.

Torpedo tubes: 1 in stern, 1 on each quarter. Also 2 submerg forward.

L.B.D. 280 × 52 × 20·5 ft. = 85·5 × 16 × 6·2 m.

Armour deck = *f*. Partial belt (7 ft. deep) *b* to *d*. Redoubt deck) turrets, turret glacis, and bulkheads *b*. C.T. *b*.

Machinery: I.H.P. nat. 5270, forced 6200 = 15·8 kts. (trial *max* 8 boilers. 2 screws. Coal capacity 800 tons = 4000 miles at 10 kts

(2.) BARROZO (1896 Elswick).

Guns: 6D* (50 *cal.*) + 4DE* (50 *cal.*) + 10*. Armour: *prot.* Sea speed: 17·5 kts. (?)

3450 tons. Complement 300.

L. B. D. 330 × 44 × 18½ ft. = 100 × 13·4 × 5·65 m.

Armour deck = *c*. Gun shields *c*: one bow torpedo tube, one on each quarter.

Machinery: I.H.P. 7500 (forced draught) = 20 kts. (trial). Two screws. Coal capacity 700 tons = *circa* 4000 miles at 10 kts.

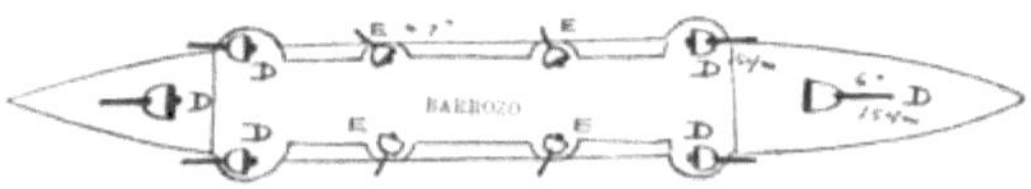

.) TUPY (ex Caramuru): TIMBIRA (ex Tupy) & TAMAYO (ex Timbira) (Kiel, 1896).

1030 tons. Complement .

L. B. D.: 259 × 30 × 10¼ = 79 × 9 × 3·1 m.
Guns: 2E* (4·7 Elswick) + 6 6 pdr* + 4*. Torpedo tubes, 1 in bow, 1 each broadside.
Armour: Deck = *f*.
Machinery: I.H.P. 6000 = 23 kts. (trial), 2 screws.
Coal capacity:

TIMBIRA.

TUPY AND TAMAYO

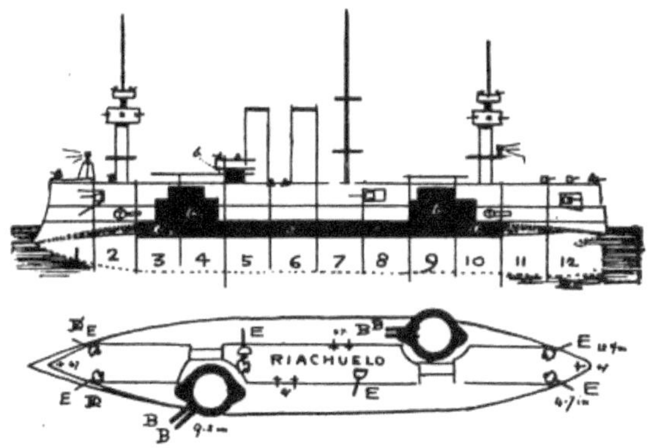

BRÉSIL.

(III.) RIACHUELO (1883 Poplar) (1895 La Seyne)

Guns: 4B + 6E* + 2*. Armour: *b*. Sea speed: 12 kts.

5900 tons. Complement 390.

L.B.D. 305 × 52 × 22 ft. = 93 × 16 × 6·7 m.

Armour = Vinte Quarto de Maio's.

Machinery: I.H.P. nat. 6900 = 16 kts. (*trial*), forced draught 7300 = 16·7 (trial) 2 screws. Coal capacity 840 tons = *circa* 4000 kts. at 10 kts.

(Military masts to be removed.)

?

(2.) ALMIRANTE TAMANDARÈ (1890).

L.: 294 ft. = 89·5 m.

Guns: 10D* + 2E*. Armour: *prot.* + *e* (D guns). Sea speed: 14 kts.

4537 tons. Complement 450.

Armoured deck = *f*. Casemates for D* guns = *e*. Four torpedo tubes. Machin Engines by Maudslay, London. I.H.P. 7500 = 17 kts. trial. Coal capacity 750 = *circa* 4000 miles at 10 kts

BULGARIE.

(T.) NADIEJDA (1898) (France).

715 tons. Complement .

L. B *D.* (***mean***): 228 × 26¾ × *10½* ft. = 69·50 × 8·16 × *3·20* m.

Guns: 2E*. 3 torpedo tubes.

Machinery: Triple expansion. 2 screws. Speed ***trial***: 18·85 kts.

CHILI.

ARTILLERIE.

Nominal Calibre.			Length in Cal.	Muzzle Velocity, *circa*.		Muzzle Energy Approximate.		Projectile.	
in.	c.m.			ft. secs.	metre sec.	ft. tons.	metre tox.	lbs.	kgs.
9·4	24	Canet, 23 ton (1) .	36	2230	679	12420	3846	374	169·5
8	20·3	Elswick, Q.F. (2) .	45	2660	810	10300	3190	210	95·2
8	20·3	Elswick, Q.F. (3) .	40	2570	783	9600	2972	210	95·2
8	20·3	Elswick (4) . . .	32	2120	646	6000	1858	200	90·6
6	15	Elswick, Q.F. . .	45	2600	792	4700	1455	100	45·3
5·9	15	Canet, Q.F.	45	2460	749	3700	1146	88	40·0
8	20·3	Armstrong 14 ton(5)	30	1950	594	5500	1703	200	90·6
4·7	12	Canet, Q.F. . .	45	2460	749	1940	590	46·3	21·0
4·7	12	Elswick, Q.F. . .	40	2150	655	2000	619	45	20·4
3	7·6	Elswick, Q.F. (12 pr.)	40	2200	670	420	130	12	5·4

(1) Capitaõ Prat ; (3) Esmeralda ; (4) B. Encalada, Cochrane and Huascar.

DESTROYERS.

(T.) CAPITA
CAPITA
GAMER
TENIEN
& GUA
RIQUE

(Birkenhe

& CAPITAN
& TENIENTE

L : 210 ft

Guns : 6*. Sea sp

300 tons. Complement 65. Normand boilers. Engines by Laird. 2 scr
90 tons = 3000 miles at 13 kts. 2 torpedo tubes (18 in.). I.H.P. 6000. Trial 1897
30·1 kts., Gamero 30·4 kts., Serano 30·3 kts., Riquelme 30 kts. Guns : 1F* (12-pdr. Q.

TORPEDO BOATS.

6 Yarrow boats (1896),	152 ft. long,	140 tons,	speed (trial) 27·5 kts.,	3 tube	
1 ,, (1886).	125 ,,	70 ,,	,, 20 ,,	4 ,,	
8 ,, (1881),	100–86 ,,	35–25 ,,	,, 20–18 ,,	4 ,,	
2 + 2 Chili (1898–1900),	— ,,	130 ,,	,, 25·5 ,,	3 ,,	
17 + 2					

Also 1 3rd class boat, 60 ft., 15 tons, 19 kts. trial speed (1892) and 3 boats of 5 tons d
All are 1-screw boats. The 140-ton Yarrow boats, Ingeniero Hyatt class, carry 40 to

The Chilian Navy also contains :—Pilcomayo, 600-ton gunboat (1864); and 12 stea
Sud America de Vapores are held at disposal for armed transports (2300—1300 ton
Also 4 or 5 smaller steamers.

CHILI.

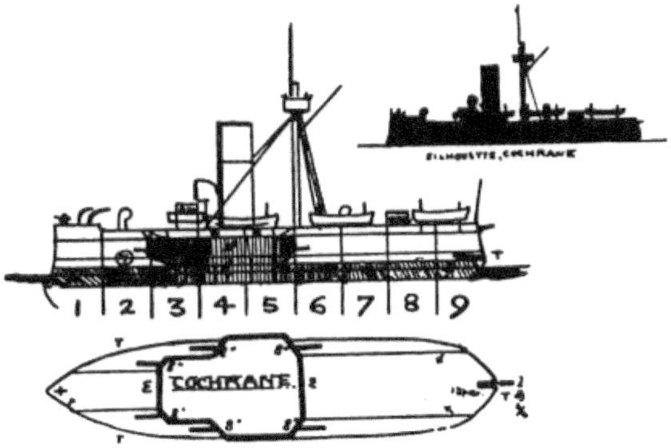

(V.) ALMIRANTE COCHRANE (Hull, 1874).

3500 tons. Complement 300.

L. B. D.: 210 × 46 × 22 feet = 64 × 14 × 6·70 metres.

Guns: 6D (20 c/m, 30 cals.) + 4E* + 1Z (13-pdr.) + 8*. 3 torpedo tubes.

Armour (iron): Belt *d - f*. Battery *d*. Battery bulkheads *e*.

Machinery: 2 screws. I.H.P. 2920 = 11·3 kts. (1891). Coal: 250 tons = 500 miles at 10 kts. (Now used as gunnery ship.) (Re-rigged 1898.)

(V.) HUASCAR (Birkenhead, 1865).

1870 tons. Complement 135.

L. B. D.: 200 × 35½ × 16½ feet = 61 × 10·7 × 5 metres.

Guns: 2D (20 c/m, 30 cals.) + 2E* (12 c/m) + 7*.

Armour: Belt *f*. Turret *e*.

Machinery: 1 screw. I.H.P. 1100. Speed, *circa* 8 kts. Coal: 250 tons = 800 kts. at 10 kts.

(4.) PRESIDENTE PINTO & PRESIDENTE ERRAZURIZ

(La Seyne, 1890).

Sheathed and coppered.

2100 tons. Complement 171.

L. B. D.: 262 × 35¾ × 20 feet = 82 × 10·8 × 6 metres.

Guns: 4D* (15 c/m Canet in sponsons + 2E* (12 c/m) + 4* (57 m/m) + 6*. 3 torpedo tubes.

Armour: Deck 2¼ in. = *e*. Gun shields *f*. Cellulose belt.

Machinery: Horizontal triple expansion. Boilers, 4 cylindrical. 2 screws. I.H.P. *natural* 3550 = 17 kts., *induced* 5400 = 19 kts. (*trial*). Coal: 200 tons.

(6.) JENERAL BAQUEDANO (Elswick 1

Sheathed and Coppered.

2500 tons. Complement .

L (*p.p*). B. *D.* (*mean*): 240 × 45⅞ × *18* feet = 73 × 14 × 4·

Guns: 4E* (12 c/m 45 cal. Elswick) + 2F* (12-pdr.) + machine guns. 1 torpedo tube (18 inch).

Armour: 0.

Machinery: 1 screw. Boilers: Belleville. I.H.P. 1500 =
Coal: 300 tons.

(Sea-going training ship.)

(T.) ALMIRANTE SIMPSON (1896 Birkenhead).

L.: 240 ft. = 73 m.

Guns: 2E* + 4*. Armour: ½ *prot.* Sea speed: 18·5 kts.

800 tons. Comp. . Draught (*max.*) 13·1 ft. = 4 m. Armour belt over machinery, 1 inch steel (2·5 c/m) Gun shields, *d.* 3 torpedo tubes (1 in bow, 1 on each beam). 2 screws. I.H.P. 4500 = 21·5 kts. (trial *max.*) 4 Normand boilers. 2 sets triple expansion engines by Laird. Coal capacity, 100 tons = 4000 miles at 11 kts. (*nominal*). Can remain at sea, under war conditions, *circa* 6 days.

ALMIRANTE SIMPSON (CHILIAN)

(T.) ALMIRANTE LYNCH & ALMIRANTE COND
Birkenhead).

L.: 230 ft. = 70 m.

Guns 7*. Armour: 0. Sea speed: 15·5 kts.

750 tons. Comp. 65. Draught (*max.*) 11 ft. = 3·3 m. Guns, 3F* (14
3 pdr. (37 m/m) 5 torpedo tubes (1 in bow, 1 on each beam, 1 on each q
machinery 1 inch (2·5 c/m) steel. Trial speed 21 kts. Recent trial 18·5 kts.

ALMIRANTE LYNCH & ALMIRANTE CONDELL (CHILIAN)

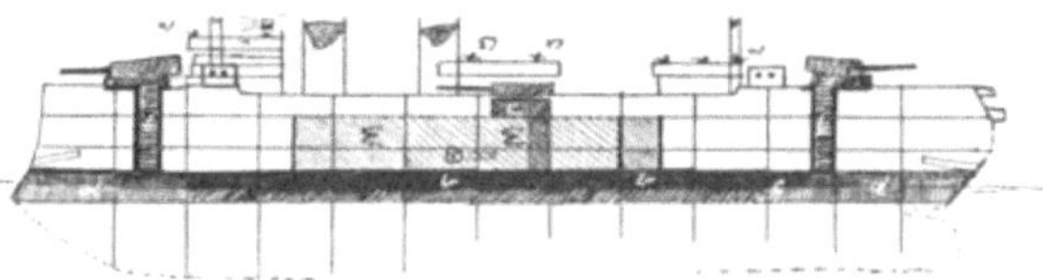

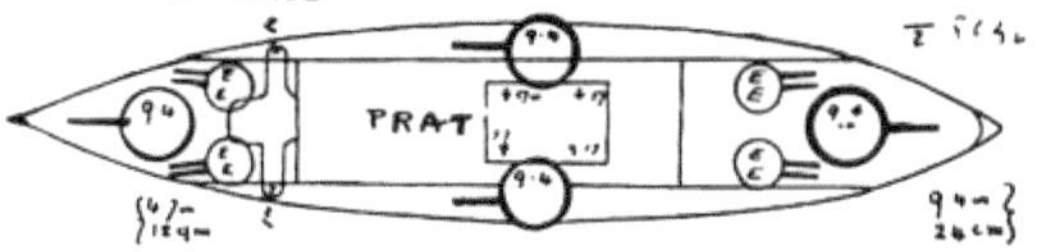

(III.) CAPITANO PRAT (1890 La Seyne).

L.: 328 ft. = 100 m.

Guns: 4C + 8E* + 10*. Sea speed: 15 kts.

7000 tons. Complement 480.

B. 60⅔ ft. = 18·5 m. D. (*max.*) 26·2 ft. = 8 m.

Armour: belt (2 metres wide) *b*—*c*. Redoubt *e*. Barbettes *c* (10½-in. steel). Deck on top of belt = *e* (flat). Shields to big guns *f*. No armour on E guns.

Torpedo tubes 4 (1 in bow, 1 in stern, 1 on each broadside).

Machinery: I.H.P. forced 12150 = 18·3 kts. (trial). 2 screws. Coal capacity *normal* = 400 tons, *max.* = 1100 = *circa* 4000 kts at 8 kts.

(2.) MINISTRO ZENTENO (1896 Elswick).

L: 330 ft. = 100·5 m.

Guns: 8D* + 14*. Armour: *prot.* Sea speed: 19 kts.

3450 tons. Complement 187. Draught: 20 ft. = 6 m.

Armour deck = *c*. Gun shields *f*. C.T. *e*. 3 torpedo tubes.

Machinery: I.H.P. 7500, forced draught = 20 kts. (*max.* trial). Coal capacity *normal* 500 tons = *circa* 4000 kts. at 12 kts.

Photo, Elswick.

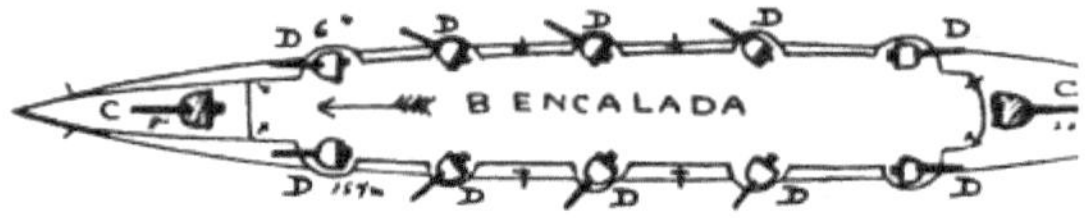

(2.) BLANCO-ENCALADA (1893 Elswick).

L.: 370 ft. = 113 m.

Guns: 2C + 10D* + 12*. Armour: *prot.* Sea speed: 19·5 kts.

4400 tons. Complement 400. Draught: 20 ft. = 6 m.

Armour deck = *c*. C.T. *e*. Gun shields *f*. 5 torpedo tubes (1 in bow). Carries a 2nd class torpedo boat 18 m. (60 ft.) long.

Machinery: I H P. 11,000 and 14,500 = 22·78 kts. (*max.* trial). 2 screws. Coal capacity *normal* 900 tons = *circa* 3500 kts. at 14 kts.

(1.) ESMERALDA (1896 Elswick).

L.: 436 ft. = 133 m.

7000 tons. Complement 500.

B. 53 ft. = 16 m., D. 28 ft. = 8·5 m. (?).

Guns: 2C* (8 in. 20·3 c/m Q.F.) + 16D* (6 in. Q.F.) + 8F*.

2 submerged torpedo tubes + 1 above water tube in stern.

Armour: deck = *c*; belt (7 ft. wide, 350 ft. = 106 m. long) *c*. Bulkheads and C. T. *c*.

Machinery: I.H.P. 18,000 = 23 kts. (trial). At sea, Pacific 1898 = 21·2 kts. Engines: 4 cyl. triple expans. by Humphrys & Tennant. 2 screws. Coal capacity *normal* 550 tons, *max.* 1350 = kts. at kts.

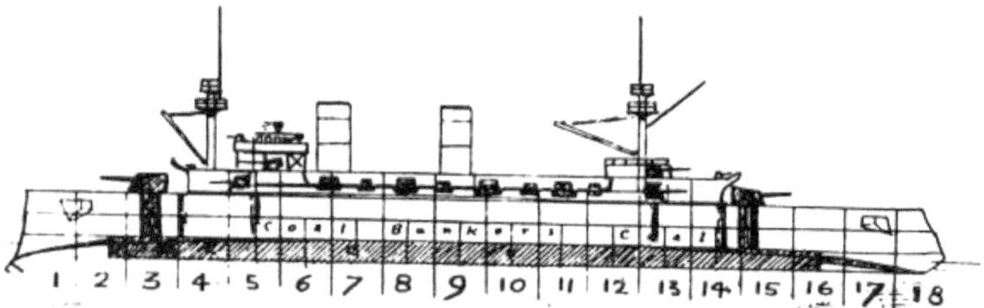

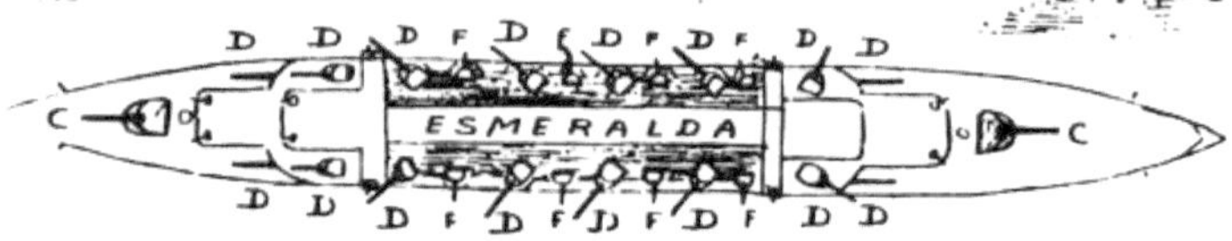

B. ENCALADA

ESMERALDA

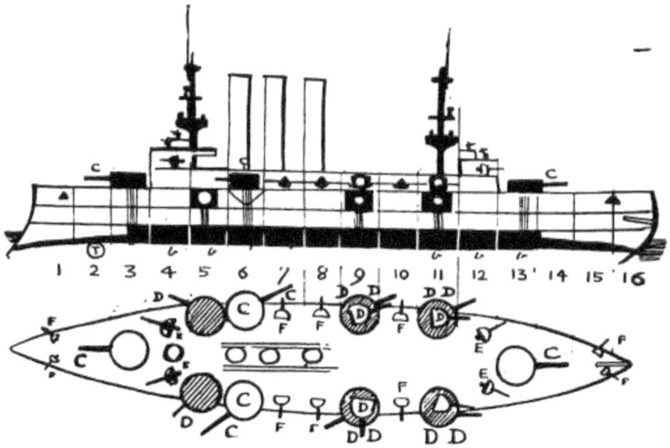

(III.) O'HIGGINS (1897 Elswick), 8500 tons. Complement.

L.B.D. 407 × 62 × 22 (*mean*) ft. = 124 × 18·9 × 6·70 (*moyen*) m.
Guns: 4C* (8 in. = 20·3 c/m Q.F.) + 10D* (6 in. Q.F.) + 10F* (3 in. Q.F.) + 10 6 pdr. (57 m/m).
Torpedo tubes: 2 submerged, 3 above water.
Armour deck: *d*. Belt (7 ft., *i.e.*, 2·1 m. wide): *b—c*. Cofferdam. *Prot. to vitals.* Belt + Coal + urved deck: = *aaa*. C gun turrets: *a* in front, sides and backs *c*. Hoists: *c*. Casemates and D gun urrets: *c*. Hoists: *c*. Shields to F guns: *d*.
Machinery: 2 sets, each of 8000 I.H.P. (=16,000 H.P.), 30 Belleville boilers (in 3 groups of 10), 2 screws. ingines by Humphrys & Tennant. Speed: 24 hours' trial 10,000 H.P. = 19 kts.; 6 hours 16,000 H.P., ean speed = 21·5 kts Coal capacity, *normal*: 700 tons.
La quille a été posée en avril 1896, le lancement a été effectué en mai 1897. Le navire est entrée en rvice en mai 1898.

CHINE.

(Revised by N. W. M.)

L'ARTILLERIE.

	Nominal Calibre.			Length in Cal.	Muzzle Velocity, *circa.*		Muzzle Energy Approximate.		Projectile.	
	in.	c/m			ft. sec.	metre sec.	ft. tons.	metre tnx.	lbs.	kgs.
C✲	8	20·3	Elswick	45	2660	810	10300	3190	374	169·5
C	11	28	M.L. Armstrong (1).	13	1360	414	7015	2172	540	244·8
D✲	5·9	15	Krupp	50	2600	792	5250	1625	110·2	50
D	8·2	21	Krupp	13	1320	402	2600	805	198·4	90
E✲	4·7	12	Elswick	45	2570	783	2060	638	45	20·4
E✲	4·1	10	Krupp	40	2460	749	1296	401	30	13·6
E✲	4·7	12	Krupp	—	—	—	—	—	—	—
F✲	3	7·6	Elswick . . .	40	2200	670	420	130	12	5·4

(And others of ancient type in very bad condition.) (1) = in gunboats.

HAI LUNG

DESTROYERS.

(T) HAI-LUNG & HAI-NJU & HAI-CHING & HAI-HOHA (Sc
200 tons. Complement 90. Guns: 6* (47 m/m). Machinery: I.H.P.
Coal capacity: 67 tons. Torpedo tubes 2.
L.B.D. 194 × 21 × ? feet = 59 × 6·40 × ? metres.

TORPEDO BOATS.

2 Schichau (1895), 128 ft. long, 120 tons, 2 tubes, speed
? 12 Stettin (1883–87) ? „ ? „ preser

14

CHINE.

(T.) { KWANG-CHING
KWANG-TING

L. : 236 ft. = 72 m.

Guns : 3E*. Armour : *prot.* Sea speed : *circa* 10 kts.

(T.) FEI-YING (1895 Elbing).

L. : 257 ft. = 78·5 m.

Guns : 2E* + 4*. Armour : 0.
Sea speed : 19·5 kts. (1896.)

(5.) { FEI-TING (γ)
HAI-CHANG-CHING (λ)
HO-WEI (β)
LUNG-SHAN (α)
TCHE-TIEN (δ)

(1877–81 Elswick).

L. : 115 ft. = 35 m.

Guns : 1C + 2Z. Armour : 0.
Sea speed : (?) 4 kts.

(3.) YANG-PO : KAI TCHI (1884).
2500 tons.

Guns : 3D (21-c/m Krupp) + 7E (12-c/m). 1 screw
Armour : Deck = *f* (50 m.m.).

PAO-MIN & FEE-CHEN (1887). 1600 tons.

Guns : 2 15-c/m + 4 12-c/m. 1 screw.

KWANG CHING class. 1030 tons. 2 screws. 4 torpedo tubes. Armour deck 1 in. (2·5 c/m); steel shields *f.* Machinery ? Speed : one of this class at battle of Asan made 12 kts.

α, β, γ, δ, λ. 420—340 tons.

FEI YING. 850 tons. I.H.P. 4500 = 22 kts. (trial). Yarrow tubulous boilers.

Hai-Tien. (Both identical.) *Photo, Elswick.*

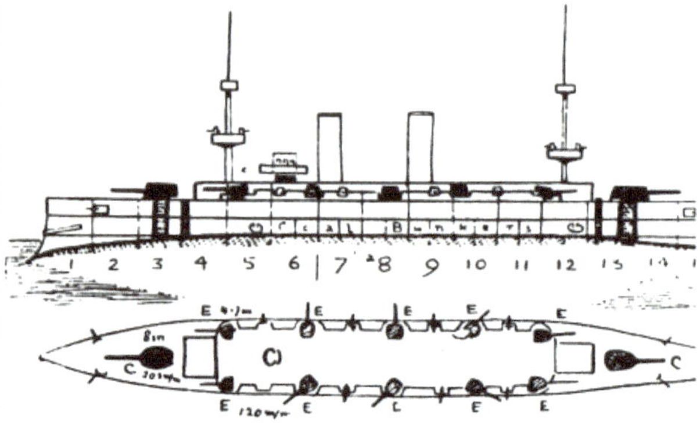

(2) HAI-CHI (1898, Elswick) & HAI-TIEN (1897, Elswick). 4300 tons. L.B.D. 360 × 46½ × 17 (*mean*) ft. = 109·7 × 14·2 × 5·18 (*moyen*) m/m.
Guns: 2C* (8-in. Elswick Q.F.) + 10E* (4·7-in. Q.F.) + 12*. Torpedo tubes: 5 above water (1 in bow).
Armour: deck = *a*. C.T. *c*. Shields to C guns *e*, to E guns *d*, to F *e*.
Machinery: I.H.P. 17,000 = 24 kts. 2 screws. Coal capacity (*maximum*) 1000 tons.

(2) HAI YUNG & HAI SHEW (1897, Stettin) & HAI-SHEN (1898, Stettin).

3000 tons. Complement 244. L.B.D. 328 × 41 × ft. = 100 × 12·5 × m/m.

Guns: 3D* (15 c/m Krupp Q.F.) + 8E* (10 c/m Krupp Q.F.) + 12*. Torpedo tubes, 1 *submerged* in bow + 2 *above water*.

Armour deck = *c*.

Machinery: I.H.P. 7500 = 19·5 kts. Coal capacity 200 tons (*normal*). 500 (*maximum*).

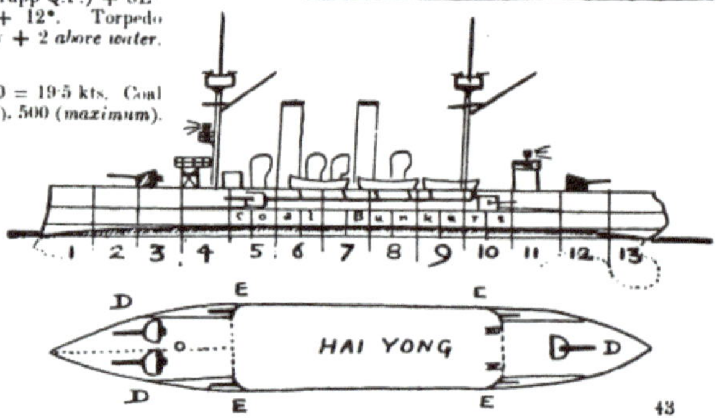

NAN YIN & NAM SHUI (1884 Kiel) & FU-CHING

2200 tons.

Guns: 2D (21-c/m Krupp) + 8 (12-c/m). 1 screw. Armour: D. 2 ins. on slopes = *f*.

DANEMARCK.

Illustrations of Gorm, Lindormen, Drogden not verified since 1896.

ARTILLERIE.

	Nominal Calibre.			Length in Calibres.	Muzzle Velocity, *circa*.		Muzzle Energy Approximate.		Projectile.	
	n.	c/m		cals.	ft. sec.	metre sec.	ft. tons.	metre tnx.	lbs.	kgs.
A	14	85·5	Krupp (1)	22	1620	493	21000	6504	1157½	525
B	9·4	24	Krupp (2)	40	2500	762	14800	4583	352	161
B	12	30·5	Krupp (3)	19	1670	509	14100	4367	725	329
C	10·2	26	Krupp	32	2000	609	12800	3964	450	204
C	10·2	26	Krupp	19	1640	499	8500	2632	450	204
C	8·2	21	Krupp	35	2020	615	6750	2080	238	108
D	10	26	Armstrong M.L. (4) .	17	1460	445	5900	1827	400	181·4
D	10	26	Armstrong M.L. (5) .	15	1370	417	5100	1579	400	181·4
D	5·9	15	Krupp	32	1850	563	2780	861	112	50·8
D	9	—	Armstrong M.L. . .	14	1370	417	3250	1006	250	113·5
E*	4·7	12	Krupp Q.F. (2) . .	40	2460	749	1850	573	45	20·4
E	5·9	15	Krupp	23	1650	502	1460	453	86	39
E	5·9	15	Krupp	19	1650	502	1420	440	86	39
E	4·7	12	Krupp	27	1720	524	1300	402	45	20·4
E	4·7	12	Krupp	22	1550	472	1150	356	45	20·4
F☆	3·3	8·4		40	2380	725	700	217	18	8

(1) Tordenskjold. (2) Skjold. (3) Helgoland. (4) Gorm. (5) Gunboats.

DESTROYERS.

4 *pro.*

TORPEDO BOATS.

First Class.

2 Copenhagen (1893), 140 ft. long, 114 tons, Speed *max.* 22 kts., 4 tubes.
3 + 2 „ (1896-8), 151 „ 142 „ „ 22·9 „ 3 „

5 + 2

Second Class.

1 Copenhagen (1891), 119 ft. long, 82 tons, Speed *max.* 18 kts., 2 tubes.
2 Thornycroft (1888), 137 ,, 95 ,, 1888 (*trial*) 22·5 kts., 4 tubes.
2 Thornycroft (1887), 131 ,, 90 ,, 1887 ,, 23·5 ,, 4 ,,
1 Thornycroft (1884), 114 ,, 64 ,, 1884 ,, 18·7 ,, 2 ,,
1 Thornycroft (1883), 111 ,, 60 ,, 1884 ,, 20·7 ,, 2 ,,

7

Third Class.

1 Thornycroft (1881), 110 ft. long, 50 tons.
1 Normand (1880), 100 ,, 37 ,,
10 Thornycroft (1889–1882), of from 25 to 15 tons.

12

1 Spar torpedo boat (1879).
2 *building* at Copenhagen, 44 tons, 1 tube.

The Danish Navy contains in addition to ships illustrated, St. Thomas (wood) of 1600 tons. 8E (12-c/m 22 cal.) (like Fyen).

Ingolf, 1000 tons, barque rig, swan bow, 1 funnel, 2E (15-c/m) guns; and Diana, 560 tons (wood), 6 small guns; and Absolon, 530 tons, 2E guns, 60 m/m belt (iron), (1 funnel, schooner rig)

DANEMARCK.

(V) SKJOLD (1896).

2160 tons. Complement 210.

Guns: 1B (24 c/m Krupp) + 3E* (12 c/m Krupp) + 4 3-pdr.* + 2*.
Torpedo tubes: 4.
Armour: Deck = *d*. Belt *a*. Turret B gun *b*. E gun turrets *od*. Bulkheads *b*.
Machinery: I.H.P. 2200 = 13 kts.
Coal capacity:
L.B.D.: 226¼ × 38 × 17½ ft. (*max.*) = 69 × 11·5 × 5·3 m.

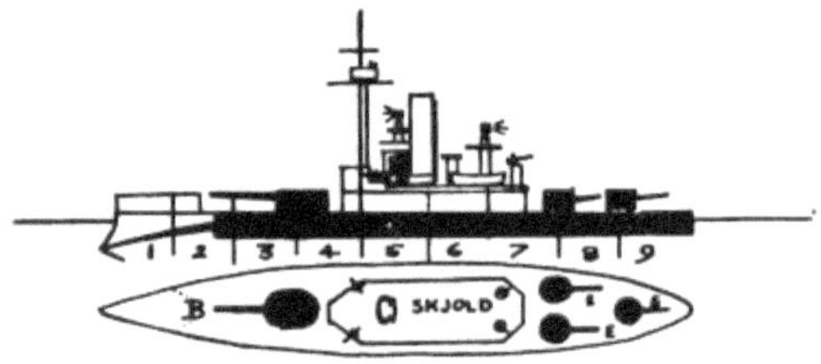

(IV.) HERLUF TROLLE (1899).
(Completing).

3470 tons. Complement .

L.B.D: 272 × 49 × 16 feet = 83 × 15 × 4·87 m.
Guns: 2B (24 c/m) + 4D* (15 c/m) + 10* (57 m/m) + 4*.
Torpedo tubes: 3 *submerged* (one of them in bow).
Armour: Belt *a* – *e*. Deck = ?. Turrets *a*. Casemates *a* (152 m/m). Protection to vitals = *aaaa*.
Machinery: I.H.P. 4200 = 16·5 kts.

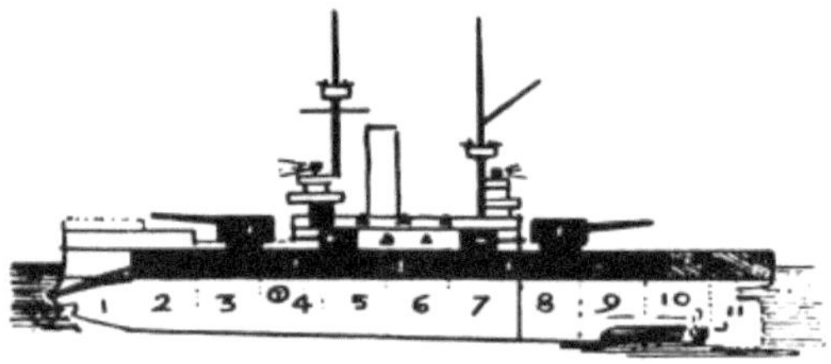

(III.) *New ship* (building).

5317 tons. Complement .

DANEMARCK.

STORE BELT
LILLE BELT
ÖRESUND
MOEN
FALSTER
} 1875

L.: 85 ft. = 26 m.

Guns: 1C + 2Z. Armour: 0. Speed: 4 kts.

(Store Belt and Lille Belt now carry mast *before* the funnel.)

DROGDEN (1872).

Guns: 1D. Armour: 0. Speed: 4 kts.

(5.) HEKLA (1890).

L.: 233 ft. = 71 kts.

Guns: 2D + 4*. Armour: *prot.* Speed: 14 kts.
1290 tons. Complement 155.
Deck 1¼ in. (= 3·5 c/m)

(V). GORM (1870).

2350 tons. Complement 160.
L.B.D. 231 × 40 × 16 ft. = × 70·5 × × 4·9 m.
Guns: 2C (10 in. Armstrong M.L.) + 4F (9 c. Krupp) + 6*. I.H.P. 1600. Speed: 8 kts.
Armour: Belt *e* - *f*. Turret *d*.

(V.) LINDORMEN (1868).

2080 tons. Complement 140.
L.B.D. 216 × 40 × 15 ft. = 66 × 12 × 4·5 m.
Guns: 2D (9 in. M.L.) + 4F (9 c/m Krupp) + 5*. Speed 8 kts.
Armour: belt *f*. Turret *f*.

(5.) HEIMDAL (1894).

L.: 257 ft. = 78·5 m.

Guns: 2E* + 4*. Armour: *prot.* Sea speed: 13 kts.
Deck 1¼ in. (= 3 c/m) steel. 1290 tons.
Complement 155.

Photo. West.

(III.) HELGOLAND (1878) (1896)

L.: 257 ft. = 78·5 m. B.: 59 ft. = 18 m. D.: (*max.*) 21 f

5370 tons. Complement, 350. Sea speed: 10·5 kt

Guns: 1B (30·5 c.m.) + 4C (26 c.m.) + 5E (12 c.m. 27 calibres) + tubes.

Armour: Belt *c* - *e*. Battery and turret *d*. Bulkheads *e*

Machinery: I.H.P. 4000 = 12 kts. (*trial*). Coal capacity 200 ton 8 kts. 2 screws.

(V.) ODIN (1872) (1898–9).

L.: 237 ft. = 72 m.

4000 tons. Complement, 220. Sea speed: 8 kts.

Guns: 4B (24 c/m) + 4F + 8*.

Armour: Belt *d* - *f*; Battery *b*; Bulkheads *f*.

Machinery 1 screw; I.H.P. 2300. Coal capacity, 180 tons = ? at 8 kts.

DANEMARCK.

(IV.) IVER HVITFELD (1886).

L.: 242 ft. = 74 m. B.: 49½ ft. = 15 m. D.: 20 ft. = 6 m.

3300 tons. Complement 298. Sea speed: 13 kts.

Guns: 2C (26 c/m) + 4E (12 c/m = 4·7 ins. of 22 cal.) + 12 3 pdr.* 4 torpedo tubes

Armour: Deck = *e*. Belt (164 × 6½ ft. = 50 × 2 m.) *a*. Barbettes *a*. Bulkheads (flat) *c*. C.T. *d*

Machinery: I.H.P. 5100 = 15·6 kts. (*trial*). 2 screws. Coal capacity 250 tons = ?

On transforme ce navire.

(3.) VALKYRIEN (1888).

L.: 269 ft. = 82 m

Guns: 2C (21 c/m) + 6D + 4*. Armour: *prot.* Sea speed: 15 kts.

2900 Tonnen gross. Besatzung 300.

Panzerdeck = *e*. 5 torpedo tubes (1 in stern)

Maschinen. 5300 Pferdestärken = 17·5 kts. (*trial*). 2 screws.

DANEMARCK.

Photo, Daniellsen, Copenhagen.

(TX.) TORDENSKJOLD (1880).

L. : 221 ft. = 67·5 m.

2430 tons. Complement 220. Sea speed : 12·5 kts.

Guns : 1A (35·5 c/m = 14 inch) + 4E (12 c/m) + 8*. 4 torpedo tu

Armour : Deck = *d*. Barbette : *c*. C.T. *b*. I.H.P. 2600 = 14 kts. (cap. 170 tons. 2 screws.

(4.) FYEN (1882).

L. : 226 ft. = 69 m.

Guns : 4D (15 c/m) + 14E (15 c/m 19 cal.)

Armour : *prot.* deck = *f*. Sea speed : 8 kts.

Déplacement 2580.

Tx. Equipage 405.

(5.) GEISER (1892).

L. : 257 ft. = 78·5 m.

Guns : 2E* + 4*. Armour : *prot.* Sea speed : 15 kts.

(= Heimdal. See p. 46.)

DEUTSCHLAND.

(All revised 1900.)

GESCHUTZE DEUTSCHLANDS KRIEGSFLOTTE.

	Nominal Calibre.		—	Length in Calibres.	Muzzle Velocity, *circa*.		Muzzle Energy Approximate.		Projectile.	
	in.	c/m		cals.	ft. secs.	metre secs.	ft. tons.	metre tons.	lbs.	kgs.
A	11	28	Brandenburg . .	40	2362	720	21750	6740	562 shell. 474	...
B	11	28	Brandenburg . .	35	2133	650	17740	5494	474	...
B	9·4	24	K.F. III. . . .	40	2500	762	17000	5265	474	...
B	9·4	24	Heimdall . . .	35	2500	762	14800	4583	352	161
B	12	30·5	Gunboats . . .	19	1670	509	14100	4367	725	329
C*	8·2	21	New ships . . .	40	2500	762	11300	3500	250	113
C	10·2	26	Baden *class*. . .	17	1580	481	7200	2230	400	181
C	10·2	26	Deutschland . .	19	1640	500	8500	2632	450	204
C	8·2	21	..	35	2020	615	6750	2080	238	108
C	9·4	24	Oldenburg . . .	26	1657	505	9000	2790	...	...
D*	5·9	15	New ships . . .	40	2460	750	3700	1146	88	40
D	5·9	15	Irene	32	1850	564	2780	861	112	50·8
D	5·9	15	Kaiser	27	1624	495	2055	636	112	51
E	5·9	15	Old ships . . .	19	1500	457	1130	350	76	34·5
E*	4·1	10·5	New ships only .	40	2460	750	1296	401	30	13·6
E*	4·1	10·5	...	35	2330	710	1000	310	26½	12
E	4·1	10·5	...	32	1526	465	870	270	40	18
F*	3·4	8·7	Heimdall . . .	30	1575	480	600	186	20	9
F*	3·3	8·4	New ships . . .	40	2380	725	700	217	18	8

The German Navy contains, of no fighting value, the ships following: Friedrich der Grosse and Preussen, ironclads, struck off list. 6770 tons. 4 C guns. 2 D.

9 ships (training service): Stein, Gneisenau, etc., *circa* 3000 tons, about 14 E (15 c/m) guns.

Zeiten, 950 tons. 5 gunboats, *circa* 500 tons, and a number of miscellaneous ships.

Merchant auxiliaries: 4 Hamburg-American liners, 10,000 tons, 19 kts. *sea speed*. (Fürst Bismark, Normannia, Augusta Victoria, Columbia). 8 Nord Deutsches Lloyd (Spree and Havel, 9000 tons, 19 kts. *sea speed*; K. Wilhelm der Grosse and K. Friedrich, 13,000 tons, 22 kts.; Lahn, 7500 tons, 18·5 kts.; Saale, Adler and Trave, 5000 tons, 16 kts. sea speed).

Division Boot. (D 5 has 2 masts.)

TORPEDO BOOTE.

(All built at Elbing unless otherwise stated.)

Divisions Boote.				*Max. trial.*	
1 (D 10) Thornycroft	(1898),	212 ft. long,	300 tons,	Speed 28·5 kts.,	2 tubes.
2 (D 9)	(1894),	200 ,,	380 ,,	,, 26 ,,	2 ,,
2 (D 7 & 8)	(1890),	190 ,,	350 ,,	,, 29·5 ,,	3 ,,
2 (D 6 & 5)	(1889),	190 ,,	320 ,,	,, 22·5 ,,	3 ,,
2 (D 4 & 3)	(1888),	184 ,,	300 ,,	,, 20 ,,	3 ,,
2 (D 2 & 1)	(1887),	180 ,,	250 ,,	,, 19 ,,	3 ,,
1 (D 11)	(pro.)	—	350 ,,		
11 + 1					

First Class.				*Max. trial.*	
6 boats (S 87–82)	(1898),	152 ft. long,	140 tons,	Speed 26 kts.,	3 tubes.
8 ,, (S 81–74)	(1894),	154 ,,	125 ,,	,, 25 ,,	3 ,,
10 ,, (S 74–66)	(1893),	154 ,,	140 ,,	,, 24 ,,	3 ,,
2 ,, (G 88, 89)	(1898),	154 ,,	160 ,,	,, 26 ,,	3 ,,
8 ,, (G 90–97)	(1900),	157 ,,	155 ,,	,, 25 ,,	3 ,,
36					

Second Class.				*Max. trial.*	
62 (S 65–1)	(1892–1884),	*circa* 140 ft. long,	85 tons,	Speed 21 kts.,	2 tubes.
2 (K 1, H 1) Kiel	(1887),	118 ,,	85 ,,	,, 22 ,,	2 ,,
2 Thornycroft,	(1884),	117 ,,	80 ,,	,, 20 ,,	2 ,,
1 Yarrow	(1884),	120 ,,	65 ,,	,, 19 ,,	2 ,,
1 Gaarden	(1885),	124 ,,	88 ,,	,, 18 ,,	2 ,,
6 (V 10·5) Stettin	(1884),	124 ,,	90 ,,	,, 16 ,,	? ,,
4 (V 4–1) Stettin	(1884).	120 ,,	75 ,,	,, ? ,,	? ,,
5 (W) Bremen	(1893),	? ,,	90 ,,	,, 23 ,,	3 ,,
93					

Third Class.				*Max.*	
4 (W 6–3) Bremen	(1884),	100 ft. long,	50 tons,	Speed 18 kts.,	2 tubes.
1 Thornycroft	(1884),	63 ,,	15 ,,	,, 15·5 kts.,	2 tubes.
15 ?	?	? ,,	13 ,,	,, ? ,,	? ,,
20					

(? seront rayés de la liste de la flotte.)

(V.) BIENE (1876) & BASILISK (1878) & CAMÄLEON (1871) & CROCODIL (1879) & HUMMEL (1881) & NATTER (1880) & MUECKE (1877) & SALAMANDER (1880) & SKORPION (1877) & VIPER (1876) & WESPE (1876.)

L.: 143 ft. = 44·5 m.

Guns: 1B (30·5 c/m) + 2F (8·18 m.) + 2*. Armour: *d* – *e*. Sea speed: 6 kts.

1109 tonnen gross. Besatzung 88.

(5.) BREMSE (1884) & BRUMMER (1883).

L.: 203 ft. = 62 m.

Guns: 1C (24 c/m) + 1F (8·7 c/m) + 2*. Armour: deck = *f* (65 m/m). Sea speed: 12 kts. 866 tonnen gross. Besatzung 78. 3 torpedo tubes. 1 screw. Coal: 65 tons.

(T.) WACHT (1887) & JAGD (1888).

L.: 275 ft. = 84 m. B.: 31½ ft. = 9·5 m. Dr.: 14 ft. = 4·2 m.

Sea Speed: 17 kts.

1250 tons. Complement 141.
Guns: 4F* (8·7 c/m) + 6*.
Torpedo tubes: 3.
Armour: deck 25 m/m (1").
Machinery: 2 sets comp. with 6 locomotive boilers. I.H.P. (forced draught) 4000 = 19 kts. 2 screws.
Coal: 130 tons.

DEUTSCHLAND.

SIEGFRIED.

HEIMDALL : BEOWULF : FRITHJOF.

HILDEBRAND.

HAGEN.

(IV.) SIEGFRIED (1889): HEIMDALL (1892): BEOWULF (1890): FRITHJOF (1891): HAGEN (1893): HILDEBRAND (1892).

L. B. D.: 239 × 49 × 18 ft. = 73 × 15 × 5·5 m.

3496 tons. Complement, 276.

Guns: 3B (24 c/m 35 cal.) + 8F* (8·7 c/m) + 6*. 4 torpedo tubes (1 submerged in bow; 3 above water.)

(Siegfried has 6F* only instead of 8F*.)

Armour Belt: Siegfried, Beowulf and Frithjof, *c*. Heimdal, Hagen, Hildebrand *b*. Barbettes *d*. Hoods *e*. Hoists *f*. Deck *e*.

Machinery: 2 sets triple expansion. Cylindrical boilers. I.H.P. 4800 = 14 - 15·5 kts. *trial*. 2 screws. Coal capacity, 225 tons.

(All these are to be reconstructed and lengthened by 5 metres).

(IV.) ODIN (1894).

3600 tons. Équipage, 276.

L. B. D.: 253 × 51 × 18 ft. = 77 × 15·5 × 5·5 m.

Guns: 3B (24 c/m 35 cal.) + 10F* (8·7 c/m) + 8*. 4 torpedo tubes (1 in bow submerged, 3 above water.)

Armour: Deck = *d*. Partial belt (2 metres wide) *b*. Barbettes *b*. Hoods *d*. Hoists *f*.

Machinery: = Hagen, etc. I.H.P. 5500 = 16 kts. (*trial*). Coal capacity, 270 tons =

ODIN. *Photo by Renard.*

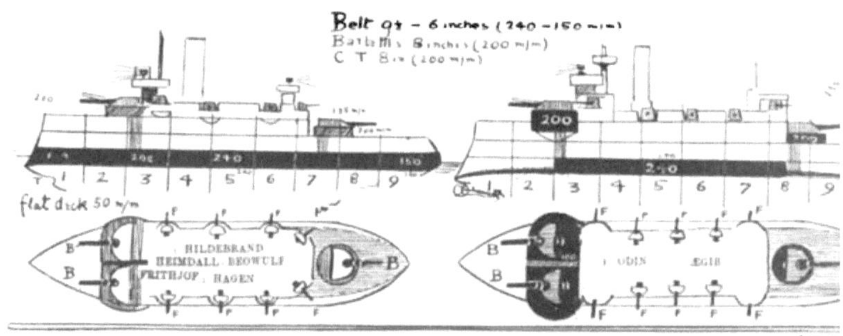

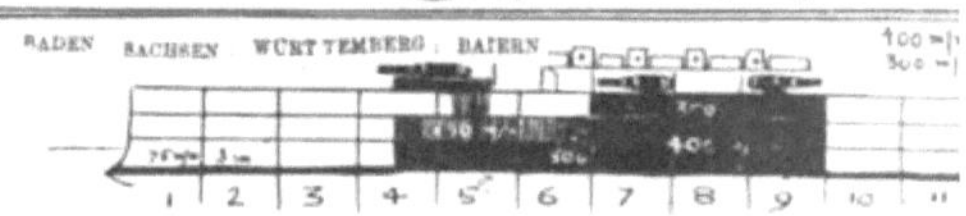

BADEN. *Photo by Renard.*

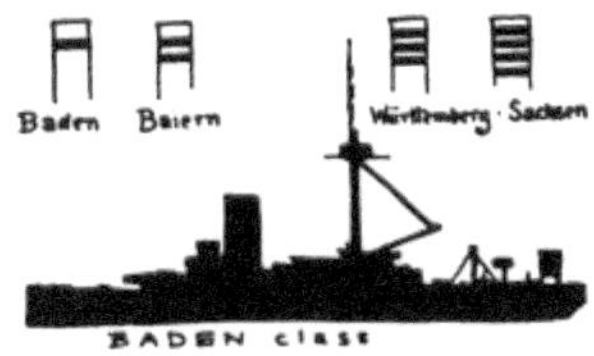

(IV.) BADEN (1880) (1897): SACHSEN (1877) (1898): WÜRTTEMBERG (1878) (1898): BAIERN (1878) (1897).

7400 tons. Complement 389.

L. B. D.: 298 × 59 × 21 ft. = 91 × 18 × 6·5 m.

Guns: 6C (26 c/m, 17 calibres) + 8F* (8·7 c/m) + 14*. 6 torpedo tubes (2 submerged, 4 above water).

Armour (*iron*): deck = *d.* Belt and citadel (135 × 12 ft. = 41 × 3·5 m.) *b - d.* Barbette *b.* No shields to C guns.

Machinery: 2 sets triple expansion. Dürr water-tube boilers. 2 screws. I.H.P. (forced draught) 6200 = 17·3 kts. (*trial* Baiern 1897–8).

Coal capacity, 700 tons.

Baden 1 ring round funnel. Baiern 2. Würtemburg 3. Sachsen 4.

(Plan, see p. 53.)

(T.) HELA (1895).

2000 tons. Complement, 168.

L.: 328 ft. = 100 m. B.: 36 ft. = 11 m. D.: 15½ ft. = 4·7 m.

Guns: 4F* (8·7 c/m) + 12*. Torpedo tubes: 4.

Armour: 75 m/m deck = *c.*

Machinery: 2 sets triple expansion with 6 locomotive boilers. 2 screws. I.H.P. 6000 = 21 kts. *nearly*, on *trial.*

Sea speed: 19 kts. Coal capacity: 350 tons = 3000 miles at 11 kts.

DEUTSCHLAND.

(5.) GEIER (1894) (*sheathed*).

Tonnage, 1776. Complement, 160. L.: 249½ ft. = 76 m. B.: 35 ft. = 10½ m. D.: 17 ft. = 5·2 m. Guns (1898): 8E* (4·1 in. = 10·5 c/m) + 4* (rev.) Torpedo tubes: 2.

Armour: deck = ε. Machinery: 2 sets triple expansion with 4 cylindrical boilers; 2 screws; I.H.P., 2960 = 16·3 kts. Coal capacity: 400 tons = 3000 miles at 10 kts.

&

(5.) BUSSARD (1890) & FALKE (1891).

Tonnage, 1580. Complement, 160. L.: 256 ft. = 78 m. B.: 33 ft. = 10 m. D. (*max.*) 16½ ft. = 5 m. Guns: 8E (10·5 c/m) + 4*. Torpedo tubes: 2. Armour: partial deck = ε. Machinery: 2 sets compound with 4 cylindrical boilers: 2 screws. I.H.P., 2800 = 15·5–16 kts. on trial. Coal capacity: 400 tons = 4000 miles at 8 kts.

&

(5.) { CONDOR, CORMORAN, SEE-ADLER } (1892)

1640 tons. Complement, 160. L.: 246 ft. = 75 m. B.: 33 ft. = 10 m. D. 16½ ft. = 5 m. Guns (1898) 8E* = 10·5 c/m + 4*. Torpedo tubes: 2. Armour: partial deck = ε. Machinery: 2 sets triple expansion with 4 cylindrical boilers; 2 screws. I.H.P., 2900 = 16 kts. (*trial.*) Sea speed, 12 kts. Coal capacity, 400 tons = 4000 miles at 8 kts.

(6.) SPERBER & SCHWALBE (—1888)

1120 tons. Complement, 117. L.: 203 ft. = 62 m. B.: 30¼ ft. = 9·2 m. D. (*max.*): 14 ft. = 4·2 m. Guns: 8E (10·5 c/m) + 5*. Torpedo tube: 1. Armour: 0. Machinery: 2 sets comp. with cylindrical boilers. 2 screws. I.H.P., 1500 = 13·5 kts. on trial. Coal capacity, 360 tons = 3000 miles at 11 knots.

(6.) MARIE & SOPHIE (—1881).

2169 tons. Complement, 296. L.: 226 ft. = 69 m. Guns: 10D (15c/m) + 2F* + 10*. Armour: 0. 1 screw. Sea speed, 10–11 kts. Coal, 250 tons.

Note.—Geier, Bussard and Falke will be re-rigged as See-Adler.

(T.) METEOR (1890) & COMET (1892).

L.: 262 ft. = 80 m. B. 31½ ft. = 9·6 m. D.: 13 ft. = 4 m.

946 tons. Complement 115.

Guns: 4F* (8·7 c/m) + 6*.
Torpedo tubes: 1 in bow submerged.
Armour: deck = *f*.
Machinery: 2 sets comp. with 4 locomotive boilers. 2 screws. I.H.P. 4500 and 5000 = 21 × 23 kts. on trial.
Coal: 90 tons.

(IV.) OLDENBURG (1884).

L.: 246 ft. = 75 m. B.: 59 ft. = 18 m. D.: 20 ft. = 6 m.

5300 tons. Complement 356.

Guns: 8C (24 c/m) + 2F* (8·7 c/m) + 18*.
Torpedo tubes: 4.
Armour: Belt *a*–*b*. Flat deck = *f*. Central battery, bulkheads, and barbettes *c*. C. T. *d*.
Machinery: 2 sets. 4 cylinder comp. with cylindrical boilers. 2 screws. I.H.P. 3900 = 13·5 kts. on trial.
Sea speed, *circa* 10 kts. Coal: 450 tons.
(Reported troublesome boilers.)

JAGUAR.

(6.) ILTIS (1898) & JAGUAR (1898) TIGER (1899) & LUCKS (1900).

(Sheathed and coppered.)

900 tons. Complement 120.

L. B. D. (*mean*): $203\frac{1}{2} \times 29\frac{1}{2} \times 11$ feet = $62 \times 9 \times 3 \cdot 35$ metres.

Guns: 4F* (8·8 c/m.) + 8*. Torpedo tubes 0. Armour 0.

Machinery: 2 sets 3-cylinder triple expansion. Thornycroft boilers. 2 screws. I.H.P., Iltis 1500 = 15 kts. (*trial*). Others 1300 = 13 kts. (*est.*).

Coal: 160 tons.

(4.) GAZELLE (1898) & *NYMPHE* (1899) & *NIOBE* (bldg).

2645 tons. Complement 210.

L. B. D.: $328 \times 39 \times 16\frac{1}{2}$ feet = $100 \times 11 \cdot 80 \times 5$ metres.

Guns: 10E* (10·5 c/m.) + 18*. Torpedo tubes: 1, *submerged* (bow) and 2, *above water* (broadside).

Armour-deck (50 m/m.) = *d* amidships; but at ends (25 m/m.) = *e*. C. T. *d*.

Machinery: 2 sets triple expansion, Niclausse boilers (in Gazelle). 2 screws. I.H.P. (*estimated*) 6600 = 20 kts. Coal, *normal* 300 tons, *maximum* 500 tons.

Trials.

(1899) Gazelle 6 hours I.H.P. 6400 = 20 kts.
Nymphe „ „ = „
Niobe „ „ = „

(4 others, *C, D, E & F*, building or projected).

GAZELLE silhouette

GAZELLE

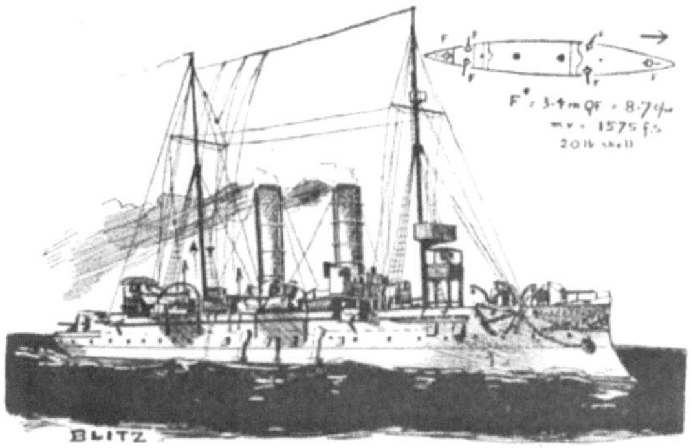

BLITZ & PFEIL (1882).

L.B.D. 246 × 33 × 14½ ft. = 75 × 10 × 4·2 m.

1382 Tonnen gross. Besatzung 130.

Guns: 6F* (8·7 c/m) + 8*. 1 torpedo tube (submerged).

Machinery: 2 sets compound. 2 screws. Locomotive boilers I.H.P. 2700 = 16 kts. (*trial*).

Coal: 250 tons.

(Pfeil has no top on foremast).

(IV.) ÆGIR (1895).

L.: 253 ft. = 77 m.

3600 tons. Complement 276.

B. 51 ft. = 15·5 m. D. 18 ft. = 5·5 m. (*mean*).

Guns: 3B (24 c/m, 35 calibres) + 10F* (8·7 c/m) + 8*.
Torpedo tubes: 4 (2 submerged).
Armour: deck = *d*. Belt *a*. Barbettes *b*. Hoods *c*. Hoists *f*. C. T. *b*.
Machinery: 2 sets triple expansion. Thornycroft boilers. 2 screws. I.H.P. 5600 = 16 kts. (*trial*).
Coal: 270 tons + petroleum.
Armour (Krupp *nickel steel*): deck 50 m/m Belt 38 m. (125 ft. long) 230 m/m (9 in.). Barbettes 200 m/m. Hoods 140 m/m (5½ in.). C. T. 200 m/m.

ÆGIR. (Plan, see p. 53). *Photo by A. Renard, Kiel.*

(II.) KAISER FRIEDRICH III. (1896) & KAISER WILHELM II. (1897) & *KAISER WILHELM DER GROSSE* (1899) & *K. KARL DER GROSSE* (1899) & *K. BARBAROSSA* (1900) & *C, D, E, F* & *G* (building).

11,130 tons. Complement 655.

L. (*p.p.*) × B. × D.: $377\frac{1}{2} \times 67 \times 27\frac{1}{2}$ feet = 115 × 20·50 × 8·50 metres.

Guns:

4B* (24 c/m 40 calibres).
18D* (15 c/m 40 calibres).
12F* (8·7 c/m 30 calibres).
12* (37 m/m revolver).
8 machine guns.

Torpedo tubes:

1 in bow *submerged*.
1 in stern *above water*.
4 on broadsides *submerged*

Armour (Harvey nickel in K. F. III).

12" Belt (amidships) ***aaa*** (300 m/m).
(8"-6") Belt (forward) ***a*** - *b* (200-152).
Prot. to vitals = ***aaa***.
10" B gun turrets and hoists *aa* (250).
6" D gun turrets, casemates *b* (152).
6" gun turrets, hoists *d*.
3" Deck aft = *b* (76 m/m).
12" Bulkhead aft *aaa* (300).
Fore C. T. ***aa*** (250); after C. T. *d* (100).

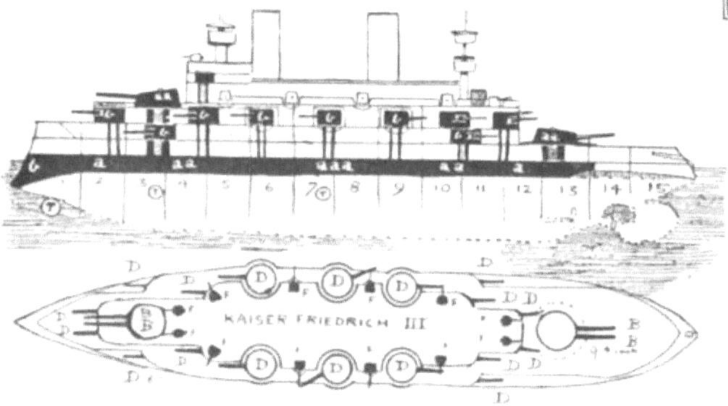

KAISER FRIEDRICH III. (1899). *Photo by Symc*

The K. W. II. and all later ships have Krupp armour. In consequence of its superior value, the armour of these ships is: Belt ***aaaa*** - *a*; B gun turrets and hoists *aaa*. D gun turrets and casemates ***a***, hoists ***c***. Deck aft = ***a***. Bulkhead aft ***aaaa***. Prot. to vitals = ***aaaa***. Fore C. T. ***aaa***; after C. T. ***c***.

Machinery: 3 sets triple expansion. 3 screws. Boilers, 12 cylindrical and 6 Thornycroft. I.H.P. 13,000 ***forced draught***.

Trials.

K. F. III. *hours nat. d.* I.H.P. = kts. 4 hours f. d. 13,500 I.H.P. = 18·2 kts.
K. W. II. " " " "
K. W. der G. " " " "

In 1899 K. F. III. made a mean of 18 kts. at sea on her way to Portsmouth.

Coal: 653 tons. ***maximum*** 1000 tons.

Note.—These ships have only a flat deck, 3 in. (75 m/m) on top of belt amidships. Forward and aft it is curved.

In 1899 the rig of K. F. III. was altered and the funnels heightened, as shown in the accompanying photograph.

C, D, etc. (building), 12,000 tons, similar to K. Wilhelm II., but with curved protective decks throughout, and I.H.P. 15,000 = 19 kts.

(II.) FÜRST BISMARK (1897).

"Armoured Cruiser."

10,650 tons. Complement 611.

L. (*p.p.*) × B. × D.: 394 × 65½ × 27 feet = 120 × 20 × 8·25 in.

(Length *over all* 410 feet = 125 metres.)

(Sheathed and Muntz-metalled.)

(Completing for sea, April 1900.)

ans (Krupp):
4B* (24 c/m 40 cals.).
12D* (15 c/m 40 cals.).
10F* (8·7 c/m 30 cals.).
10* (37 m/m).
8 Maxims.

rpedo tubes:
1 in bow, ***submerged.***
1 stern, *above water.*
4 broadside, ***submerged.***

Armour: Krupp.
8" Belt amidships *aa* (200 m/m).
4" Belt at ends *c* (100 m/m).
Prot. to vitals = *aa*.
8" B gun turrets and hoists *aa* (200).
4" Casemates and hoists *c* (100).
8" Fore C.T. *aa* (200).
4" After C.T. *c* (100).

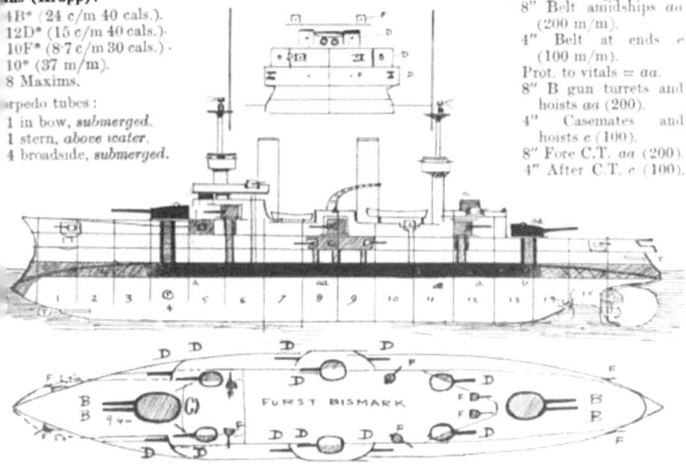

Machinery: 3 sets triple expansion. 3 screws. Boilers: 6 cylindrical + 8 Thornycroft. I.H.P. 14,000 = 19 kts., *natural draft.*
Trials.
Coal, 1000 tons. Bunker capacity, 1100 tons.
***Note.*—The armour deck, 2" (50 m/m), is flat on top of the belt amidships, but is curved and reinforces the belt at the ends.**

(III.) *PRINZ HEINRICH* (1900) & *b* (building).

"Armoured Cruisers."

8868 tons. Complement .

L. : B. : *D.* (mean): 394 × 65½ × 23¾ feet = 120 × 20 × 7·40 metres.

Guns:
2B* (24 c/m 40 cals.).
10D* (15 c/m 40 cals.).
10* (37 m/m).
4 Maxims.

Torpedo tubes:
1 in bow, ***submerged.***
2 broadside, ***submerged.***
1 stern, *above water.*

Armour: Krupp.
4-3" Belt *c - d* (150-75).
2¾" Deck on slopes = *a* (70).
Prot. to vitals = *aaa*.
6" B gun turrets *a* (152).
6" Casemates *a* (152).
Hoists ?

Machinery: I.H.P. 15,000 = 20 kts. (*est.*) Boilers, Dürr. 3 screws.

Note.—These ships are reduced Bismarks. The deck will re-inforce the belt.

FÜRST BISMARK. ***Photo, Renard, Kiel.***

(II.) KURFÜRST FRIEDRICH WILHELM (1891).
BRANDENBURG (1891). WEISSENBOURG (1891). WOERTH (1892).

10,040 tons. Besatzung, 567. L. B. D.: 361 × 64 × 26 ft. = 110 × 19·5 × 8 m.

Guns: 4A (28 c/m 40 cal.) + 2B (28 c/m 35 cal.) + 6E* (10·5 c/m) + 8F* (8·7 c/m) + 8*. 6 Torpedo tubes.

Armour: Belt (110 × 2 m.) *aa* - *a*. Barbettes and bases to amidships and after barbettes *a*. Floor to fore barbette 4" (10 c/m). Hoods: front *c*, sides *ε*. C. T. *a*. Casemate E, guns *f* (*ε* in Woerth). Flat deck = *d*. No bulkheads.

Machinery: 2 sets triple expansion. 12 cylindrical boilers. 2 screws. I.H.P. *circa* 10,000 = 16·5 - 17·2 kts. (6 hours' trial.) Coal capacity, 800. Funnels heightened. 1898. Sea speed: 14·5 kts.

(An armoured battery is to be substituted for the B guns amidships.)

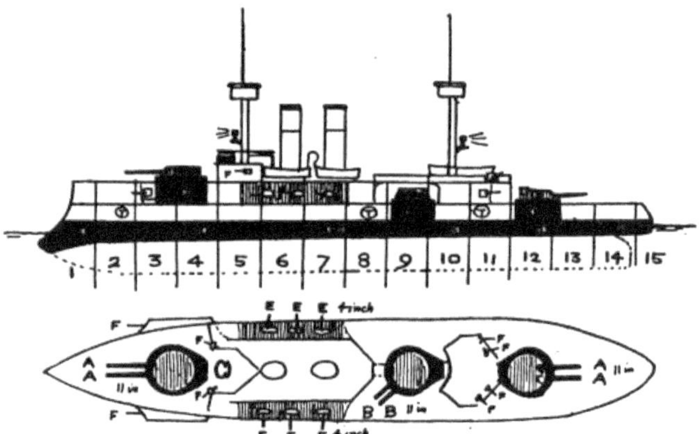

Note. In the Wörth all armour here marked *f* is ε

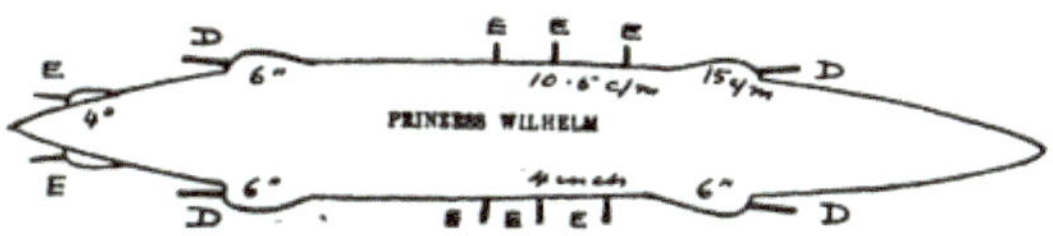

(3.) PRINZESS WILHELM: IRENE (— 1887).

L.: 340 ft. = 103·5 m.

4400 tons. Complement 365.
B. 45 ft. = 14 m. D. 23 ft. = 7 m.
Guns: 4D (15 c/m) + 8E* (10·5 c/m) + 6F* (8·7 c/m) + 16*.
5 torpedo tubes (1 submerged, 4 above water).
Armour: deck = *e*.
Machinery: 2 sets 4 cylinder comp. 4 double ended boilers. 2 screws. I.H.P. 8000 = *on trial* nearly 20 kts.
Coal capacity: 900 tons = 4500 miles at 10 kts.
Sea speed: 17 kts.

IRENE AND P. WILHELM. (*Identical*). *Photo by Symonds & Co.*

(IV.) *KONIG WILHELM* (1868 Blackwall) (1897) (1898).

9757 tons. Besatzung 732.

L. B. D.: 355 × 59 × (*max.*) 30 ft. = 108 × 18 × 9 m.

Guns: 22C (24 c/m) + 1D* (15 c/m) + 18F* (8·7 c/m) + 8*. 5 torpedo tubes.

Armour: Belt *c* - *e*. Deck = *e*. Battery and bulkheads *e*. C.T. *d*.

Machinery (*being re-engined*). 1 screw.

Coal capacity: 800 tons.

DEUTSCHLAND.

(IV.) KAISER: DEUTSCHLAND (1874 Poplar) (18

7531 tons: 7319 tons. Complement 650.

L: 280 ft. = 85·5 m. B.: 62½ ft. = 19 m. D.: 26 ft. = 8 m.

Guns (Deutschland): 8C (26 c/m) + 8D* (15 c/m) + 8F* (8·7 c/m) 5 torpedo tubes.

Guns (Kaiser): 8C (26 c/m) + 1D (15 c/m) + 6E* (10·5 c/m) + 9F* (+ 12*. 5 torpedo tubes.

Armour: Belt (280 × 11⅝ ft. = 85·5 × 3·5 m.) *d*—*e*. Battery and b

Machinery: Horizontal trunk. Penn. 1 screw. New boilers (K Deutschland 1896). I.H.P. = 14 kts., *trial* 1895–7. Sea speeds, *circa* 10 k

Coal capacity: 700 tons = *circa* 1200 miles at 8 kts.

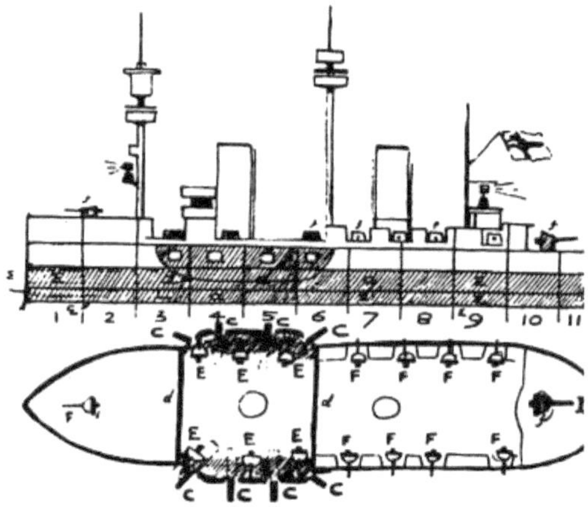

(6.) ARCONA: ALEXANDRINA (—1885).

L.: 235 ft. = 71·5 m.

Guns: 10D (15 c/m) + 4E* (10·5 c/m) + 6*. Armour: 0. Sea speed: *circa* 10 kts.

2373 tons. Complement: 263. 2 torpedo tubes.

Coal capacity: 300 tons = 2000 miles at 10 kts. 1 screw.

(6.) OLGA (1880)
(= Alexandrina).

L.: 226 ft. = 69 m.

Guns: 10D + 2F* + 10. Armour: 0. Sea speed: 11 kts.

2169 tons. Complement 269. 1 screw. Coal capacity: 250 tons.

(T.) GREIF (1886).

L.: 318 ft. = 97 m.

Guns: 8F* (8·7 c/m) + 12*. Armour: 0. Sea speed: 19 kts.

B.: 32 ft. = 9·7 m. D.: 15 ft. = 4·5 m.

Machinery: 2 sets compound, 2 cylinder with 6 cylindrical boilers. 2 screws. I.H.P. 5400 = 22·5 kts. (*trial.*)

Coal capacity: 180 tons.

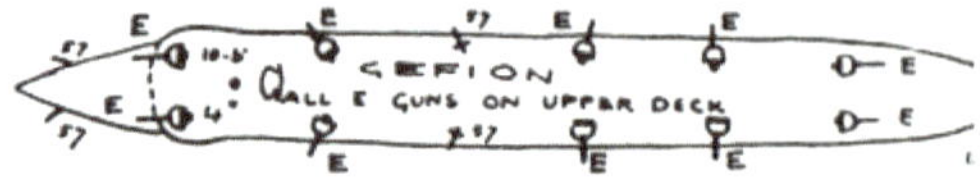

(4.) GEFION (1893).

L.: 334 ft. = 102 m. B.: 42½ ft. = 13 m. D.: 21½ ft. =

Guns: 10E* (10·5 c/m) + 14*. Sea speed: 18 kts.

4200 tons. Complement 302. 2 torpedo tubes.

Armour: Deck (over machinery only) = *ds* (75 m/m). Celluloe
Machinery: 2 sets triple expansion. 2 screws. 6 2-ended boiler
9828 = 20·5 kts (4 *hours' trial*).
Coal: (*max.*) 900 tons = *circa* 5000 miles at 8—9 kts.

(2.) KAISERIN-AUGUSTA (1892).

L.: 388 ft. = 118·5 m. B.: 49 ft = 15 m. D. 25½ ft. =

Guns: 12D* + 8F* (8·8 c/m) + 16*. Armour: *prot.* Sea sp

6050 tons. Complement 436. 5 Torpedo tubes (1 of them in bov

Armour deck = *d* (88 m/m [3½"]). Splinter deck = *f*.
Machinery: vertical triple expansion. Boilers 8 2-ended. 3
12,000 = 22·5 kts. (*short trial*).
Coal: 900 tons = *circa* 4000 miles at 10 kts.

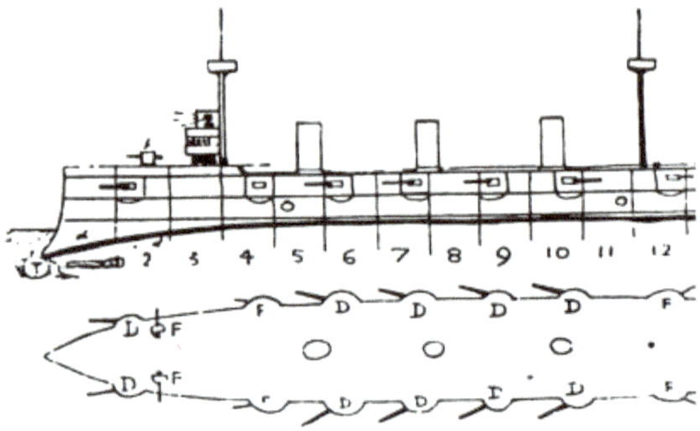

(2.) FREYA & HERTHA & VICTORIA LUISE (— 1897) : HANSA & VINETA (1898).

5650 tons. Complement 447.

L. B. D.: 344½ × 57 × 23 feet = 105 × 17·4 × 7 metres.

Guns:
- 2C* (21 c/m).
- 8D* (15 c/m).
- 10F* (8·8 c/m).
- 14*.

Torpedo tubes:
- 3 *submerged* (1 in bow).

Armour (Krupp):
- 4" Deck = *aa* (100 m/m).
- Cork belt amidships.
- Protection to vitals = *aaa*.
- 4" Turrets and casemates *c* (100 m/m).
- 5" Glacis to funnels = *aaa* (125 m/m).
- 3½" Hoists *c* (80 m/m).
- 8" Fore C.T. *aa*. After C.T. *c* (100 m/m).

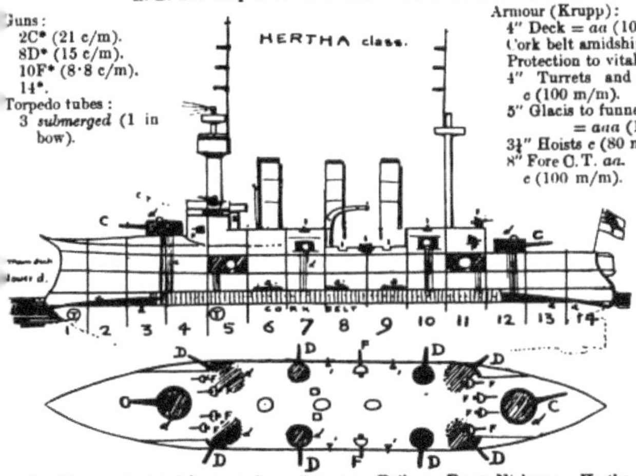

Machinery: 3 sets triple expansion. 3 screws. Boilers: *Freya*, Niclausse; *Hertha*, Belleville; *Victoria Luise*, Dürr; *Hansa*, Belleville; *Vineta*, Dürr. I.H.P. 10,000 = 19–20 kts. Coal: *normal* 500 tons, *maximum* 950 tons.

Trials.	*hours.*	I.H.P.	
Freya	?		= 20 kts.
Hertha			= 20 ,,
V. Luise			= 20 ,,
Hansa			= 19 ,,
Vineta			= 19·6 kts.

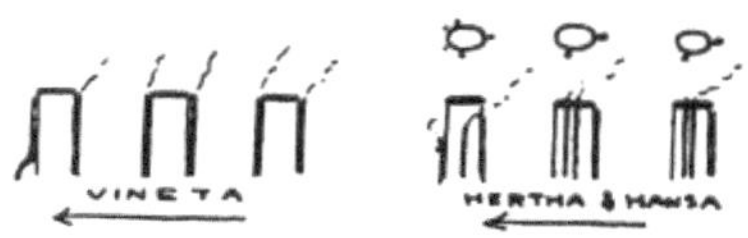

ENGLAND.

NAVAL GUNS.

Notation.	Nominal Calibre.		Length in Calibres.	Muzzle Velocity.		Muzzle Energy. *circa.*		Weight of Projectile.		Weight of Guns.
	in.	c.m.	*circa*	ft. secs.	m. secs.	ft. tons.	met. tnx. *circa*	lbs.	kgs.	tons.
A	16·25	41	30	2087	636	54400	16848	1800	816·4	110
A	13·5	34	30	2016	614	35230	10911	1250	566	67
A* (V)	12	30·5	...	2750	838	44500	13781	850	385·5	50
A (W)	12	30·5	35	2367	721	33000	10220	850	385·5	46
A (M)	16	40·5	18	1540	469	27960	9280	1700	771·1	80
A* (E)	12	30·5	40	2625	800	40600	12573	850	385·5	50
B	10	25·4	30	2040	621	14430	4470	500	226·8	29
B (W)	9·2	23	40	2347	715	14500	4486	380	172·3	25
B	12	30·5	25	1914	583	18100	5575	850	385·5	45
B O (M)	12	30·5	15	1570	478	14000	4336	818	370	38
O (M)	12	30·5	12	1292	393	7100	2199	600	272·0	25
O (M)	11	28	13	1360	414	7030	2177	550	249·4	25
O	9·2	23·5	31	2065	629	10900	3375	380	172·3	24
O	9·2	23·5	25	1780	542	8356	2587	380	172·3	21
D	8	20·3	30	2150	655	6730	2085	210	95·2	15
D	8	20·3	25	1953	595	5554	1720	210	95·2	14
D (M)	10	25·4	15	1379	420	5410	1675	400	181·4	18
D (M)	9	23	14	1440	439	3680	1139	256	116	12
D	6	15	26	1960	597	2665	825	100	45·3	5
E	5	12.6	25	1770	540	1124	348	50	22·7	2
Z	4	10	27	1900	579	625	193	25	11·3	1½

Quick Firers.

Notation.	Nominal Calibre.		Length in Calibres.	Muzzle Velocity.		Muzzle Energy.		Weight of
	in.	c.m.	*circa.*	ft. secs.	m. secs.	ft. tons.	metre tnx. *circa.*	lbs.
D* (V)	6	15	45	2784	848·5	...	...	
D* (W)	6	15	40	2200	670·5	3356	1039	100
D*	6	15	40	1882	573·5	2457	761	
D* (c)	6	15	26	1960	597·3	2665	825	
E* (c)	6	15	25	1880	572·4	1960	607	80
E* (W)	4·7	12	40	1786	544·3	995	308	45
E* (W)	4·7	12	40	2188	666·8	1491	462	
E* (W)	4	10	40	2300	701·0	1046	324	25
F*	3	7·6	40	2200	670·5	419	130	12

There are a few other guns, old B.L. or M.L. and various "Marks" of above.
(V) = Vickers. (W) = wire. (M) = muzzle loader. (C) = converted into Q.F. (
The Vickers 12 in. and 6 in. Q.F. are appropriated for new ships.
Most of the Q.F. and all the wire guns use smokeless powder.
Lyddite shell used by 6 in. and 4·7 in. Q.F.
The **A*** guns (Mark IX.) are rapid loaders (R.L.), and can be loaded at any elev
Canopus and Albion, and in all ships after and including Formidable.
The newest 6-in. guns are not officially known as "quick firers" as they have no me

2a

ock (1893), 240 tons, 3500 I.H.P. (*re-boilered* 1899, *'arrow*). 27 kts.

2b

ng (1893), Decoy (1894), 265 tons, *circa* 4500 I.H.P. *Thornycroft boilers, Thornycroft*). 28 kts.

2c

ent (1894). Bruizer (1895), 265 tons, 4500 I.H.P. *Thornycroft boilers, Thornycroft*). 28 kts. and 30 kts.

er (1894), 250 tons, 4200 I.H.P. (*Thornycroft boilers, hornycroft*). 29 kts.

ler, Ariel, Desperate, Fame, Mallard (1896), 272 tons, '00 I.H.P. (*Thornycroft boilers, Thornycroft*). 30 kts.

net, Cynthia, Coquette, Foam (1898), 310 tons, 5400 H.P. (*Thornycroft boilers, Thornycroft*). 30 kts.

dy, Hart, Hunter, 275 tons, 4000 I.H.P. (*Thornycroft ilers, Fairfield*). 27 kts.

2d

atross (1898), 395 tons, 7500 I.H.P. (*Thornycroft boilers, hornycroft*). 32 kts.

(1899), 312 tons, 5800 I.H.P. (*Thornycroft boilers, hornycroft*). 30 kts.

2e

rger (1894), 250 tons, 3100 I.H.P. (*Thornycroft boilers*). kts.

ra (1895), 300 tons, 3850 I.H.P. (*White boilers, Thames ron Works*). 27 kts.

2f

ty (1894), 250 tons, 3100 I.H.P. (*Thornycroft boilers, 'arrow*), re-boilering.

3a

9 Bat, Chamois, Crane, Flying Fish, Star, Whiting (—1896), Fawn, Flirt (—1897), Spiteful (1898), 360 tons, 5900 I.H.P. (*Reed boilers, Palmer*). 30 kts.

3 Lee (1898), Sylvia, Violet (1897), 335 tons, 5500 I.H.P. (*Yarrow boilers, Doxford*). 30 kts.

+ *1* *Myridom* (bldg.), 360 tons, 6900 I.H.P. (*Reed boilers, Palmer*). 30 kts.

3b

2 Bullfinch, Dove (1898), 340 tons, 5800 I.H.P. (*Yarrow boilers, Earle*). 30 kts.

2 Fairy, Osprey (1897), 355 tons, 6000 I.H.P. (*Thornycroft boilers, Fairfield*). 30 kts.

2 Cheerful, Mermaid (1898), 355 tons, 5800 I.H.P. (*Thornycroft boilers, Hawthorn Leslie*). 30 kts.

1 Dasher (1894), 260 tons, 3800 I.H.P. (*Thornycroft boilers*; re-boilered 1898). 26·2 kts.

+ *8* *Arab, Falcon, Greyhound, Ostrich, Racehorse, Roebuck, Success, Vixen* (building), 360 tons, 6900 I.H.P. 30 kts.

3c

4 Avon, Bittern (1896), Leopard, Otter (1897), 3000 tons, 6000 I.H.P. (*Thornycroft boilers, Vickers*). 30 kts.

3 Starfish, Sturgeon, Skate (1895), 270 tons, 4000 I.H.P. (*Blechynden boilers, Vickers*). 27·8 kts.

3 Opossum, Ranger, Sunfish (1895), 295 tons, 4000 I.H.P. (*Yarrow boilers, Hawthorn Leslie*). 27·5 kts.

2 Rocket, Shark, Surly (1894), 280 tons, 4100 I.H.P. (*Normand boilers, Clydebank*). 27·5 kts.

(Masts of some of these as **3d.**)

2 Hardy, Haughty (1895), 290 tons, 4200 I.H.P. (*Yarrow boilers, Doxford*). 27 kts.

3d

3 Janus, Lightning, Porcupine, 280 tons, 3900 I.H.P. (*Reed boilers, Palmer*). 28 kts.

3e

4 Brazen, Electra, Recruit, Vulture (1896), 355 tons, 5800 I.H.P. (*Normand boilers, Clydebank*). 30·5 kts.

3f

1 Gipsy (1897), 340 tons, 6000 I.H.P. (*Thornycroft boilers, Fairfield*).

3g

3 Conflict, Teazer, Wizard (1895), 270 tons, 4500 I.H.P. (*White boilers, White*). 27 kts.

3h

2 Viper, 325 tons, 10,000 I.H.P. (*Parsons' turbine, Hawthorn Leslie*). 36 kts. & Cobra, *ditto, built at Elswick.*

72+9 building.

4a

2 Spitfire, Swordfish (1895), 295 tons, 4500 I.H.P. (*Yarrow boilers, Elswick*).

2 Fervent, Zephyr (1895), 280 tons, 3850 I.H.P. (*re-boilered* 1899, *Thornycroft boilers, Hanna, Donald & Wilson*). 27 kts.

1 Kestrel (1898), 355 tons, 5800 I.H.P. (*Normand boilers, Clydebank*).

1 + *1* Peterel (1899), *Syren* (bldg.), 360 tons, 6000 I.H.P. (*Normand boilers, Palmer*). 30 kts.

4b

1 Hornet (1893), 240 tons, 3800 I.H.P. (*re-boilering, Yarrow*). 27 kts.

4c

1 Lynx (1894), 290 tons, 4,400 I.H.P. (*Normand boilers, Laird*). 27·3 kts.

4d

2 Salmon, Snapper, 280 tons, 3600 I.H.P. (*Yarrow boilers, Earle*). 27 kts.

4e

4 Banshee, Contest, Dragon, 295 tons, 4400 I.H.P. (*Normand boilers, Laird*). 28 kts. & Ferret, 280 tons. 27·6 kts.

12 + *2* Earnest, Griffon, Hawk, Locust, Panther, Thrasher, Quail, Seal, Virago, Wolf (1895–6), Orwell (1898), *Lively, Sprightly* (bldg.), 360 tons, 6300 I.H.P. (*Normand boilers, Laird*). 30–30·4 kts., *except* Thrasher, *which, owing to damages received, is now less.* Sparrowhawk, 360 tons.

4f

1 Express (1897), 430 tons, 9250 I.H.P. (*Normand boilers, Laird*). 33 kts. Length, 235 feet = 71·60 m.

99+*12 building.*

Guns: Daring, 2F* + 3 6 prs.* Decoy, Ferret, Havock, Hornet, Lynx 1F* + 3 6 prs.* *All the others* 1F* + 5 6 prs.*

Torpedo tubes: Daring, Decoy, Ferret, Havock, Hornet, Lynx 3 *tubes* (= 1 *bow tube* + 1 *pair*). *All the others*, 2 *single tubes* (18 inch) (46 c/m).

All the older 3 and 4-funnelled boats were designed to carry the mast abaft fore funnel, but in most it has now been shifted to before the funnel.

ENGLAND.

DESTROYERS.

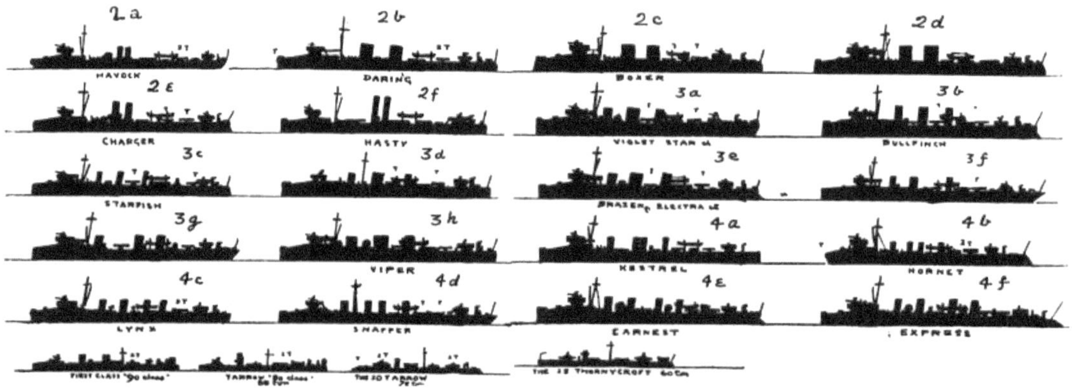

TORPEDO BOATS.

First Class. *Trial, max. speed*

1 (Nos. 97),	(1893), Birkenhead boat,	130 tons,	140 feet long,	23·5 kts.,	2690 H.P.	2 Torpedo tubes.
3 (,, 96–94),	(1894), White ,,	130 ,,	140 ,,	23·2 ,,	2000 ,,	3 ,, ,,
3 (,, 93–1),	(1894), Thornycroft,	130 ,,	140 ,,	23 24 ,,	2400 ,,	3 ,, ,,
3 (,, 90–89),	(1895), Yarrow,	110 ,,	140 ,,	22 ,,	1600 ,,	3 ,, ,,
1 (,, 80),	(1887), Yarrow,	105 ,,	135 ,,	23 ,,	1500 ,,	5 ,, ,,
11						

Second Class. *Trial, max. speed*

6 (Nos. 87–82),	(1889), Yarrow,	85 tons,	130 feet long,	23 kts.	1100 H.P.,	3 Torpedo tubes.
1 (,, 81),	(1885), White,	125 ,,	150 ,,	19 ,,	? ,,	3 ,, ,,
1 (,, [illegible]),	(1886), Yarrow,	75 ,,	125 ,,	22·4 ,,	1000 ,,	5 ,, ,,
20 { (,, 61, 63–74, 76–78), (30–33) }	(1886), Yarrow,	75 ,,	125 ,,	20 ,,	700 ,,	5 ,, ,,
25 (60–41) (29–25),	(1886), Thornycroft,	60 ,,	127 ,,	21 ,,	700 ,,	4 ,, ,,
5 (38–34)	(1886), White,	65 ,,	125 ,,	19 ,,	950 ,,	5 ,, ,,
2 (23, 24)	(1885), Yarrow,	67 ,,	113 ,,	19·5 ,,	600 ,,	3 ,, ,,
2 (22, 21)	(1885), Thornycroft	63 ,,	113 ,,	20 ,,	730 ,,	3 ,, ,,
62						

2a

1 Havock (1893), 240 tons, 3500 I.H.P. (*re-boilered* 1899, *Yarrow*). 27 kts.

2b

2 Daring (1893), Decoy (1894), 265 tons, *circa* 4500 I.H.P. (*Thornycroft boilers, Thornycroft*). 28 kts.

2 Ardent (18

3a

9 Bat, Chamois, Crane, Flying Fish, Sta
Fawn, Flirt (—1897), Spitef
I.H.P. (*Reed boilers.*

3 Lee (1898), Syl
(*Yarr*

+ 1

1

1

1 h

25

0 tons. Complement 67. Guns: 1Z (4 in.) + 6*.
in stern, 2 training on broadside by funnel). I.H.P.
a speed: 18 kts. 2 screws. Coal: 100 tons = 4 to

1886). 1580 tons. Complement 147. Guns:
tubes: 1 in bow *submerged*, 6 *above water* (1
mour-deck = *f*. I.H.P. 3370 = 17·6 kts. (*trial*,
speed: 13·5 kts. Coal: 450 tons.

ement 160. Guns: 6E* (4·7 inch) + 4*.
I.P. 3000. On *trial, forced draught* 4 *hours*
Sea speed: *circa* 14 kts. Coal: 160 tons

MES (1885) & FORTH (1886).
nt 325.
onverted Q.F.) + 11*. Torpedo tubes:
r, 2 *above water*; Mersey, 2 *submerged*,

lirect tube boilers. 2 screws. Sea

d d. 6335 = 17·3 kts.
„ 6628 = 17·5 „
„ 5886 = 16·8 „
„ 5736 = 16·6 „
The Severn has shorter steam

t 240. Guns: 2C (12 in.,
tubes. Armour: belt *d–e*.
service only.)

4C (12 in., 25 ton M.L.)
peed: 8 kts. (*Harbour*

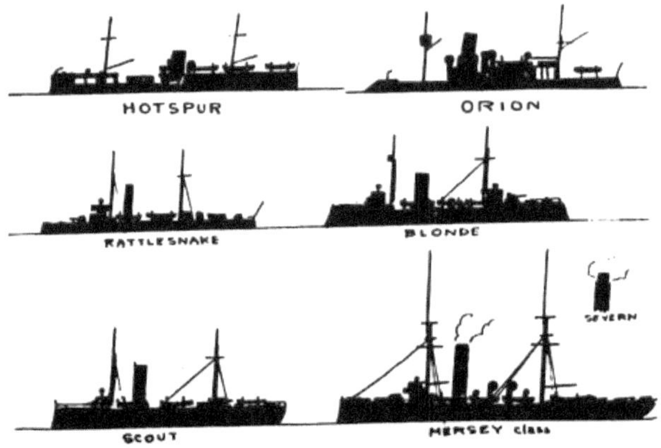

Mersey. *Photo by W. Nye,*

COLONIAL.

ia (1888),	Thornycroft,	96 tons,	135 feet long,		Speed	23	kts.,	5	torpedo	tubes.
ia (1889),	White,	95 ,,	130	,,	,,	20	,,	3	,,	,,
ia (1888),	?	92 ,,	130	,,	,,	21	,,	3	,,	,,
toria (1891),	Yarrow,	82 ,,	130	,,	,,	23	,,	3	,,	,,
toria (1883),	Thornycroft	65 ,,	113	,,	,,	20	,,	2	,,	,,

THIRD CLASS.

						Max. trial.			
. 39–40)	(1895),	Yarrow,	40 tons,	100 feet long,	Speed	20 kts.	1	Torpedo	tube.
19)	(1878),	White,	28 ,,	87 ,,	,,	21 ,,	2	,,	tubes.
20)	(1880),	—	28 ,,	87 ,,	,,	17 ,,	2	,,	,,
18–14)	(1877-8),	Yarrow,	33 ,,	86 ,,	,,	21 ,,	2	,,	,,
13)	(1878),	Maudslay,	28 ,,	87 ,,	,,	21 ,,	2	,,	,,
12–1)	(1880–77),	Thornycroft,	28 ,,	87–90 ,,	,,	21·19 kts.,	1	Torpedo	tube.
101)	?	?	? ,,	60 ,,	,,	16 ,,	2	,,	tubes.
-76, 73–64)	(1886–3),	Thornycroft,	15 ,,	60 ,,	,,	16·5 ,,	2	,,	,,
96, 75, 74)	(1883),	Yarrow,	12 ,,	62 ,,	,,	16 ,,	2	,,	,,
38)	(1889-7),	Yarrow,	16–15 ,,	60 ,,	,,	16·5 ,,	1	,,	,,
51)	(1879-8),	Thornycroft,	12 ,,	60 ,,	,,	16·5 ,,	2	,,	,,

) ( ), White, 12 tons, 56 ft. long, Speed (*max. trial*) 14·5 kts. Spar torpedo.

COLONIAL.

oria (1884),	Thornycroft,	12 tons,	63 ft. long,	Speed (*max. trial*)	17·5 kts.	1	Torpedo	tube.
. Wales (1879),	Thornycroft,	16 ,,	63 ,,	,, ,,	17 ,,	1	,,	,,
mania (1884),	Thornycroft,	12 ,,	63 ,,	,, ,,	17 ,,	1	,,	,,
nsland (1884),	Thornycroft,	12 ,,	63 ,,	,, ,,	17 ,,	1	,,	,,

Zealand (1884), Thornycroft, 12 tons, 63 ft. long, Speed (*max. trial*) 17 kts. Spar torpedo.

SUBVENTED CRUISERS.

2 Cunarders (Campania and Lucania), 12,950 tons gross, 30,000 I.H.P., 21 knots sea speed, 610 feet long.
2 White Star (Majestic and Teutonic), 9800 ,, 16,000 ,, 20 ,, 565 ,,
2 P. & O. (Australia and Himalaya) 6900 ,, 10,000 ,, 17 ,, 465 ,,
2 P. & O. (Victoria and Arcadia) 6900 ,, 7,000 ,, 16 ,, 465 ,,
3 Canadian Pacific Railway Co. (Empress of India, Japan, and China), 5900 tons gross, 10,000 I.H.P., 16 knots sea speed, 440 feet long.

11

also

2 Cunarders (Etruria and Umbria), 8000 tons gross, I.H.P. 14,500 tons, 19½ knots sea speed, 500 feet long.
2 Cunarders (Servia, Aurania), *circa* 7300 tons gross, I.H.P. 9500 tons, 17 knots sea speed, 500 and 470 feet long.
3 1 Cunarder (Gallia), and 2 White Star (Britannic and Germanic), *circa* 5000 tons gross, *circa* 16 knots sea speed.
2 P. & O. (Britannia and Oceana), *circa* 6000 tons gross, 16 knots sea speed.
6 P. & O. (Peninsular, Oriental, Massila, Rome, Carthage, Valetta), *circa* 5000 tons gross, 15 knots speed.
2 P. & O. (Ballarat and Parramatta), *circa* 4700 tons gross, 14 knots sea speed.

17

River Gunboats (Nile).
SULTAN & SHEIKH (Yarrow).

150 tons. L.B.D.: 145 × 24½ × 2 ft. = 44 × 7 × 4 × 0·60 m.

Guns: 2 F* + 8 *Maxims.*
Armour: 1" (25 m/m) on sides.
Machinery: I.H.P. 500 = *circa* 12 knots. 2 Screws. 2 Yarrow boilers.
"Raised propellers" in tunnels.

&

6 Gunboats for Niger River (Yarrow).

80 tons. L.B.D.; 100 × 20 × 2 ft. = 30·5 × 6 × 0·60 m.

Guns: 2 6-pdr. (57 m/m) + 6 *Maxims.*
Machinery: 2 sets triple expansion. 2 Screws. 1 Yarrow boiler.

(TX.) POLYPHEMUS (1881).

L.: 240 ft. = 73 m. B.: 40 ft. = 12·2 m. D.: 21½ ft. = 9·6 m.

Guns: 6* Armour: *prot.* Sea speed: 17 kts.(?).

2640 tons. Complement 142. 5 submerged torpedo tubes.

Armour: deck = *d* (75 m/m). C. T. *d.* 2 screws. 8 boilers. I.H.P. *forced* 5500 = 17·8 kts. *trial.* Coal: 300 tons.

POLYPHEMUS (E

CYCLOPS class. ABYSSINIA class GLATTON

(V.) GLATTON (1871).

4900 tons. Complement 191.

L. B. D.: 245 × 54 × 23 ft. = 75 × 16·4 × 7 m.

Guns: 2C (12-in. m.l.) + 3 6-pdrs.* Armour: Belt (2-75 × 2 m.) *c–d.* Turret *c* (300 m/m). Redoubt *c.* Bulkhea 75–33 m/m (3–1½"). Iron armour.

Machinery: I.H.P. 2000. Speed: 9 kts. (*trial* 189–).

(V.) CYCLOPS & GORGON & HECATE & HYDR

3560 tons. Complement, 175.

Guns: 4D (10-in. m.l.) + 4*. Armour: Belt, redoubt and tu
Sea speed, *circa* 6 kts.

(V.) ABYSSINIA & CERBERUS & MAGDALA

2900 tons. Complement, 200.

Guns: 4D (8-in.) + 7 machine. Armour: Belt and redoubt,
Sea speed, 6-8 kts.

(V.) CONQUEROR (1881): HERO (1885).

L.: 270 ft. = 82 m. B.: 58 ft. = 17·7 m. D.: 24⅝ ft. = 7·5 m.

6200 tons. Complement 330.

Guns: 2B (12 in.) + $4D_{\bullet}$ (6 in. [26 cals.]) + 6 6-pdrs.* (57 m/m) + 5 3-pdr.* (47 m/m). 6 torpedo tubes (14 in. = 355 m/m).

Armour: Belt, from bow (250 ft. = 76 metres long) *a* - *c*. Redoubt and turret *a*. Bulkhead *a*. Redoubt bulkheads *b*. C.T. *a*. Prot. deck aft = *de* (63 m/m). Flat deck on armour 43 m/m (1·7").

Machinery: 2 sets compound. 2 screws. 8 1-ended boilers. I.H.P. Conqueror 4500 = 14·75 kts. (4 *hours' trial* 1883) *forced* 6000 = 15·5 kts. (4 *hours* 1883). Hero, 4351 I.H.P. *Forced* 6162 = 15·6 kts. (4 *hours' trial* 1887). Sea speed: 1897 *circa* 11 kts. Coal 650 tons = *circa* 3000 miles.

(Bad sea boats.)

(V.) RUPERT (1874) (1893).

L.: 250 ft. = 76 m. B.: 53 ft. = 16·2 m. D.: 24½ ft. = 7·4 m.

Guns: 2B. (10-in.)+ $2D_{\bullet}$ (6-in., 26 cal.) + 4 6-pdr.* + 6 3-pdr. 4 torpedo tubes (14 in.).

Armour: Belt (250 × 7 ft. = 76 × 2 m.) *c* - *e*. Redoubt *c*. Turret *bc*. Bulkheads to redoubt *c*. C.T. *d*. Deck 75 m/m.

Machinery: I.H.P. 6000 *forced* = 14·4 kts. (*trial*). Sea speed: *circa* 10 kts. Coal (1894) *max.* = *circa* 200 tons = *circa* 500 miles at 8 kts. 2 screws. Reboilered 1892–3.

(Bad sea boat.)

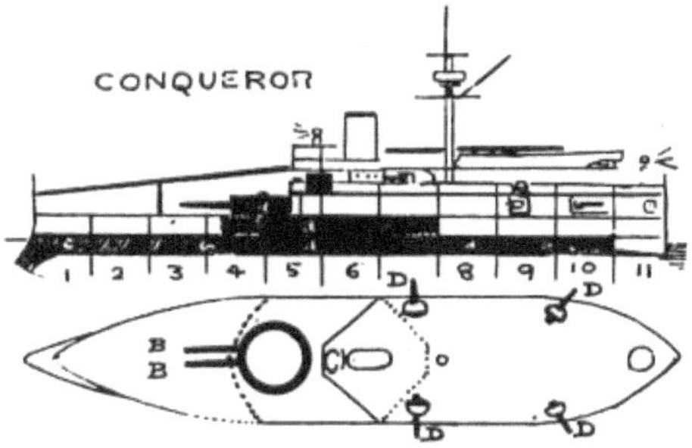

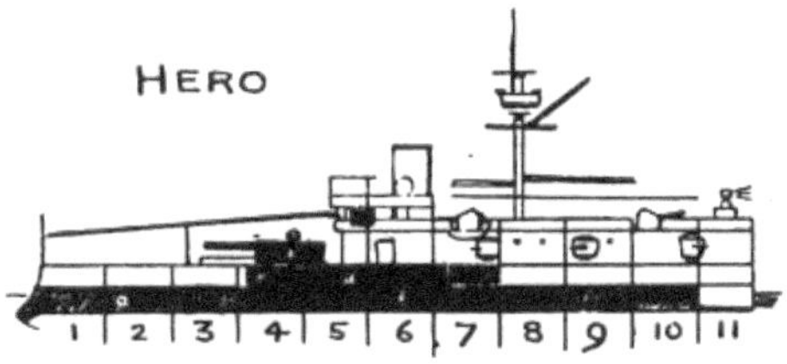

(T.) **RATTLESNAKE** (1886). 550 tons. Complement 67. Guns: 1Z (4 in.) + 6*. Torpedo tubes: 4 (one in bow, one in stern, 2 training on broadside by funnel). I.H.P. *forced* 2700 = 18·5 kts. (*trial*). Sea speed: 18 kts. 2 screws Coal: 100 tons = 4 to 5 days at sea.

(5.) **SCOUT** (1885) & **FEARLESS** (1886). 1580 tons. Complement 147. Guns: 4E* (5 in., converted) + 8*. Torpedo tubes: 1 in bow *submerged*, 6 *above water* (1 in bow, 1 in stern, 4 on broadside). Armour-deck = *f*. I.H.P. 3370 = 17·6 kts. (*trial, 4 hours forced draught*). 2 screws. Sea speed: 13·5 kts. Coal: 450 tons.

(5.) **BLONDE** (1889). 1580 tons. Complement 160. Guns: 6E* (4·7 inch) + 4*. Torpedo tubes: 2. Armour deck = *f*. I.H.P. 3000. On *trial, forced draught 4 hours* I.H.P. 2762 = 16·5 kts. (1891). 2 screws. Sea speed: *circa* 14 kts. Coal: 160 tons = *circa* 4 to 5 days at sea.

(4.) **SEVERN** & **MERSEY** (1885): **THAMES** (1885) & **FORTH** (1886).
4500 tons. Complement 325.

Guns: 2D (8 inch, 25 cals.) + 10E* (6 in., 25 cals., converted Q.F.) + 11*. Torpedo tubes: Forth & Thames, 2 *above water*; Severn, 1 *submerged* in bow, 2 *above water*; Mersey, 2 *submerged*, 2 *above water*.

Machinery: 2 sets horizontal compound. 6 circular direct tube boilers. 2 screws. Sea speed: 14·5 kts.

First trials.

Severn	*natural d.* 4603 I.H.P.	= 15·6 kts.	*forced d.* 6335	= 17·3 kts.
Mersey	„ „ 4515 „	= 15·3 „	„ „ 6628	= 17·5 „
Thames	„ „ 4162 „	= 15·2 „	„ „ 5886	= 16·8 „
Forth	„ „ 3590 „	= 14·7 „	„ „ 5736	= 16·6 „

Note.—Thames and Forth have their 6 inch guns sponsoned. The Severn has shorter steam pipes than the Mersey, etc.

(V.) **HOTSPUR** (1870). *Circa* 4500 tons. Complement 240. Guns: 2C (12 in., 25 ton M.L.) + 2D* (6 in.) + 6F* + 6*. 2 torpedo tubes. Armour: belt *d—e*. Turret *c*. Base of turret *d*. Sea speed: 8 kts. (*Harbour service only.*)

(V.) **ORION** (1879). 4870 tons. Complement 280. Guns: 4C (12 in., 25 ton M.L.) + 6*. Armour: Belt *c—e*. Battery and bulkheads *d*. Speed: 8 kts. (*Harbour service only*)

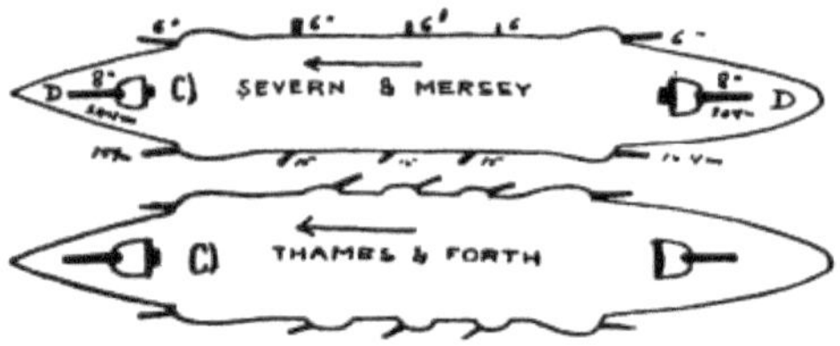

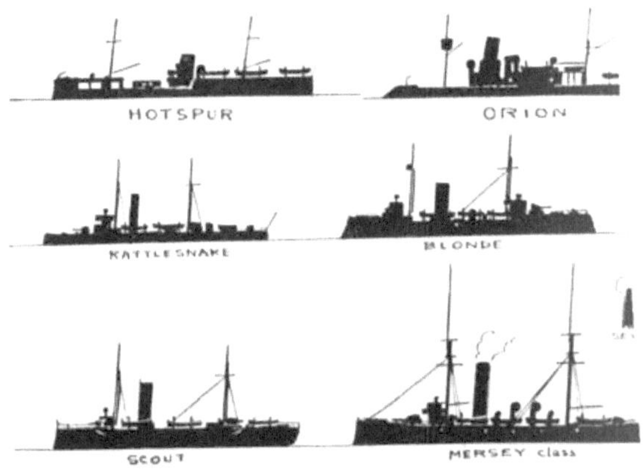

MERSEY. *Photo by W. N*

(III). COLOSSUS (1882) (1898) : EDINBURGH (1882) (1897–8).

9420 tons. Complement 396.

L.B.D.: 325 × 68 × 29 ft. = 99 × 20·7 × 8·8 m.

Guns: 4B (12-in.) + 5D* (6-in. Q.F. converted, 26 calibres) + 4 6-pdrs.* (57 m/m) + 10 3-pdrs.* (47 m/m). 2 torpedo tubes.

Armour (compound): Deck = *d* (76 m/m = 3"). Belt (457–350 m/m = 18–14" *compound*) *aa*–*a*. Bulkheads *aa* (450–350 m/m). Turrets *aa* (406 m/m = 16). C. T. *a* (14" = 350 m/m). Flat deck amidship (75 m/m = 3").

Machinery: Colossus, I.H.P. *forced* = 16·5 kts. (5 *hours trial*, 1884). Edinburgh: I.H.P. *forced* 6809 = 16 kts. (6 *hours trial*, 1884). Engines: 2 sets 3-cylinder compound. Boilers (replacing by water tube boilers). 2 screws. Coal: 970 tons = *circa* 3000 miles at 10 kts.

(V.) AGAMEMNON (1879) & AJAX (1880).

8600 tons. Complement 400.

L.B.D.: 280 × 66 × $24\frac{2}{3}$ ft. = 85 × 17·7 × 7·5 m.

Guns: 4BC (12 in. M.L.) + 2E* (6 in., 25 cals., converted Q.F.) + 6* (57 m/m) + 8* (47 m/m) + 5 machine. Torpedo tubes, 2.

Armour (iron): Belt amidships (100 ft. long) *a*–*b*. Bulkheads *b*. Protection to vitals = *a*. Turrets (compound) *aa*. C.T. *a*. Cork belt at ends.

Machinery: 2 sets compound. 2 screws. I.H.P. 4500 = 12·2 kts. (*trial*).

Sea speed, *circa* 8 kts., maximum speed at sea, 9·5 kts. Coal, 950 tons = *circa* 2500 miles at 8 kts.

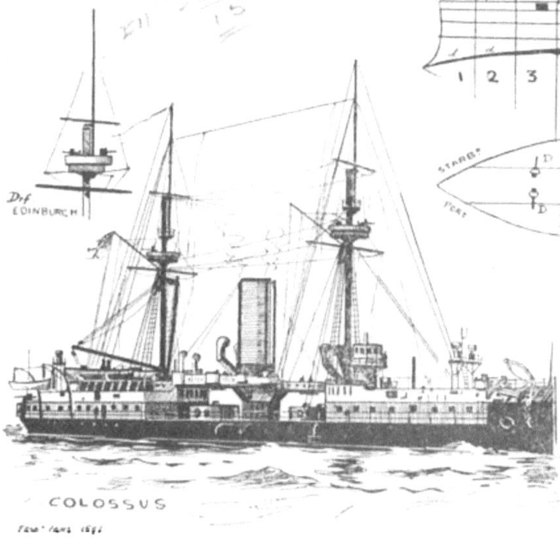

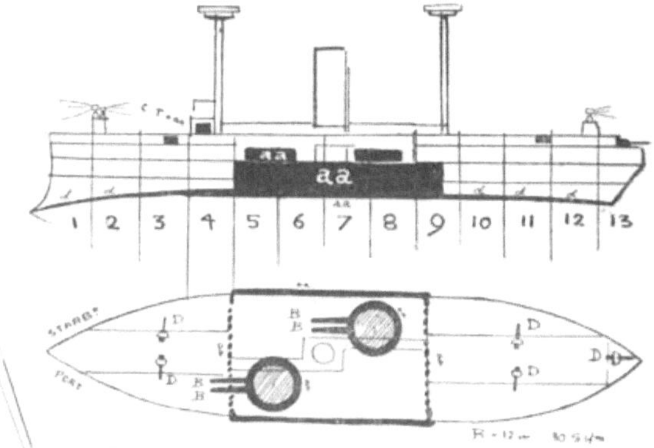

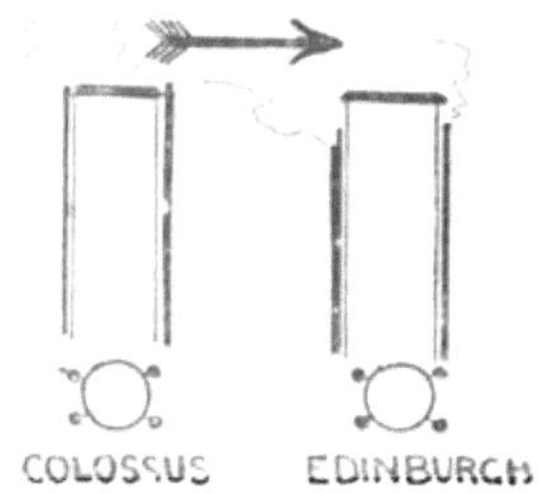

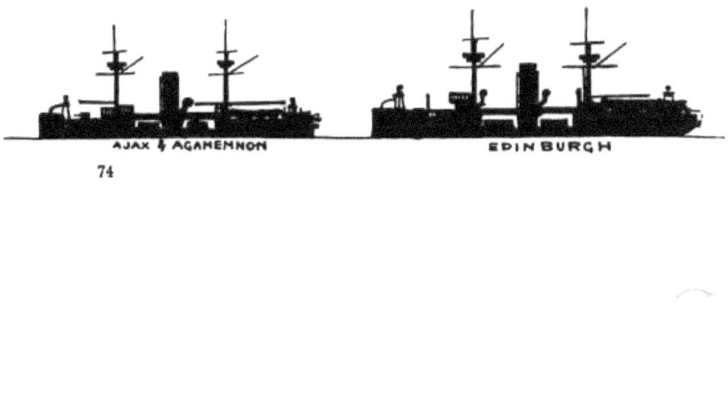

(6.) COCKATRICE & RATTLER & LIZARD (1886–87). 715 tons. Guns: 6Z (4 inch). Armour to engines $\frac{3}{4}$ inch deck. *Trial* speeds, 11·5 kts. *natural*, 13 kts. *forced*.

(6.) GOLDFINCH & LAPWING & MAGPIE & REDBREAST & REDPOLE & RINGDOVE & SPARROW & THRUSH & WIDGEON (—1889), 805 tons: & PLOVER & PEACOCK & PHEASANT & PIGMY & PIGEON (1888), 755 tons. Guns: 6Z (4 inch) + 2* (47 m/m, 3 pdrs.) for the 805 ton ships—the 755 do not carry any 3 pounders. Speeds (*trial*), 11 kts. *natural*, 13 kts. *forced*.

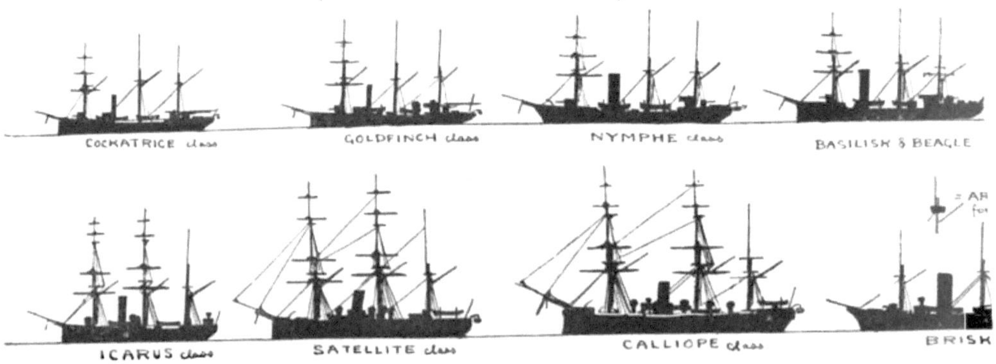

(6.) NYMPHE & BUZZARD & DAPHNE (1887–88), 1140 tons. Guns: 8E (5 in.) [in Buzzard E* (4·7 inch)]. Speeds: 14·5 kts. *forced draught trial*: & SWALLOW (1885), 1130 tons. Guns: 8E (5 in.) + 4* (3 pdrs.). Speed, *forced d. trial* 13·5 kts.

(6.) BASILISK & BEAGLE (1889), 1170 tons. Guns = Nymphe. Speed *trial*, 13·5 kts. *forced*: (6) BRAMBLE & BRITOMART & DWARF & THISTLE (1898–99), 710 tons. Guns: 2E* (4 inch) + 4F* (3 in.). Belleville boilers. Nominal speeds, 13·5 kts. *forced draught*.

(6.) ICARUS (1885) & MELITA & RACER (1883), 970 tons. Guns: 8E (5 in.) + 4* (3 pdrs.). *Nominal* speeds, *circa* 12–13 kts.: also ALERT & TORCH (1894), 960 tons: & ALGERINE & PHŒNIX (1895), 1050 tons. Nominal speeds, 13 knots: & CONDOR & ROSARIO (1898) & *SHEERWATER* & *VESTAL* & *MUTINE* & *RINALDO* (1900), 980 tons. Belleville boilers. Speeds, 13·25 kts. Guns: 6E* (4 in.) + 4* (3 pdr.): also *ESPIEGLE* & FANTOME (bldg.) & *2 others* (to be built), 1075 tons.

(5.) SATELLITE (1881) & RAPID & ROYALIST (1883) (1884). 1420 tons. Guns: 2E (6 in.) + 6 to 10E (5 inch). [Pyl
Armour, over engines = *f*. Sea speeds: *circa* 9–11 kts.

(5.) CALLIOPE & CALYPSO & CHAMPION & CLE
COMUS & CORDELIA (1878–84). 2380 tons. Guns: 10D,
Sea speeds: 11–8 kts. (Several of these will soon be removed fron
Single screw.

(5.) BRISK & ARCHER & RACOON & COSSACK &
PORPOISE & TARTAR (1886–87). 1770 tons. Guns:
converted Q.F.) + 8*. 5 torpedo tubes. I.H.P. *circa* 4000 *forced*
18–16 kts. Sea speeds: *circa* 13–14 kts. Coal: 325 tons = 2500 k
economical steamers).

ENGLAND (BRITISH EMPIRE).

MONARCH. *Photo, W. Nye, Esq.*

(IV.) MONARCH (1869) (1897).

8320 tons. Complement 600.

. D.: 330 × $57\frac{1}{2}$ × 28 feet = 100·50 × 17·50 × 8·50 m

uns: 4C (25 ton 12 in. M.L.) + 2D (9 in. M.L.) + 1E M.L.)+4F* (3 in. Q.F.)+10* (3 pdr. 47 m/m) + 10 machine. lo tubes, 2.

rmour (iron): Belt *e*. Protection to vitals *e*. Redoubt *e*. s *d*. Bulkheads and battery fore and aft *f*.

achinery: Triple expansion. Boilers, 8 cylindrical. 1 screw. gined, etc., 1895.) I.H.P. 8216 = 15·8 kts., 4 *hours forced* *at trial* (1897). Coal, 630 tons.

) AUDACIOUS (1869): IRON DUKE (1870) INVINCIBLE (1869): SWIFTSURE (1870) & TRIUMPH (1870).

6100 tons, *except* Swiftsure 6910, and Triumph 6600.

uns: 10D (9 in. M.L.) + 4 to 8Z (4 in.) + 10-12*. Torpedo 4.

rmour (iron): Belt *d*-*e*. Protection to vitals *d*. Battery and ads *e*.

achinery: I.H.P. *circa* 3300. Sea speeds, 8-10 kts. Coal, s = *circa* 1200 miles at 8 kts.

(Most of these are to be removed from the Navy list.)

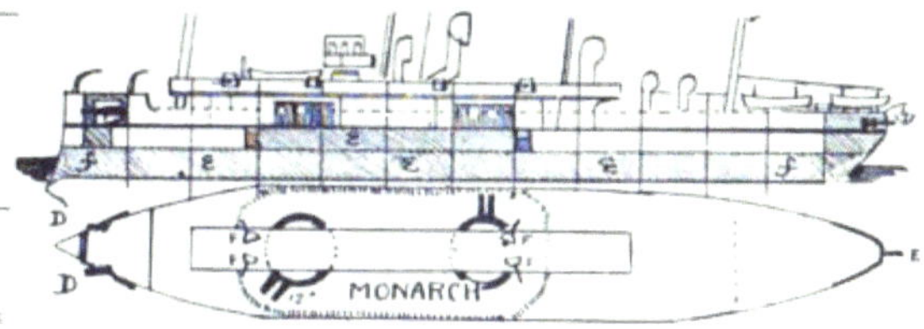

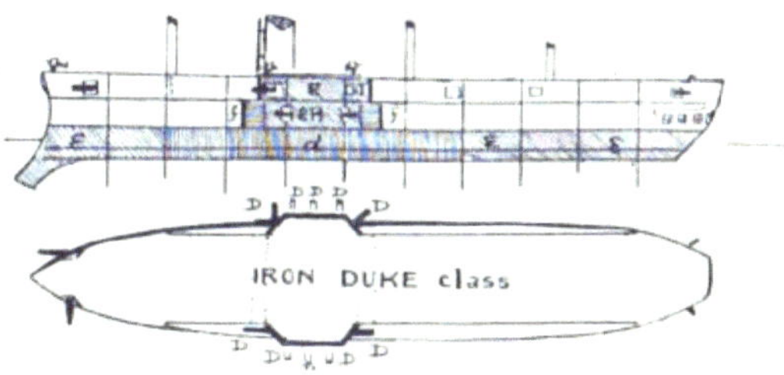

NILE. *Photo, R. Ellis, Malta.*

(II.) TRAFALGAR (1887) : NILE (1888).

12,530 tons. Complement. 570.

L. : 345 ft. = 105 m. B. : 73 ft. = 22 3 m. D. : 30 ft. = 9 m.

Guns : 4A (13·5 in.) + 6D* (6 in., 40 cals.) + 8* (6 pdrs.) + 11* (3 pdrs.) 4 tubes (2 of them *submerged*, 14 in.) Sea speed : 13·5 kts.

Armour (*compound*) : Deck = *d* (75 m/m). Belt {8½ ft. / 2·5 m.} wide *aa* (450 18 in.) Bulkheads *aa* (450 m/m). Redoubt above belt *aa*. Turret *aa*. B (100 m/m). Screens in battery *f*. Armour on torpedo tubes *a*. C T. *aa*.

Machinery : 2 sets vertical triple expansion. 6 cylindrical boilers. Speed, *forced draught trials*, 1890–91. Nile, 12,109 H.P. = 16·9 kts. Trafalgar, = 17·2 kts. Sea speeds (1897–98) : *circa* 13·5 kts. 2 screws. Coal *nor* tons, *max.* 1200 tons = *circa* 3000 miles at 10 kts.

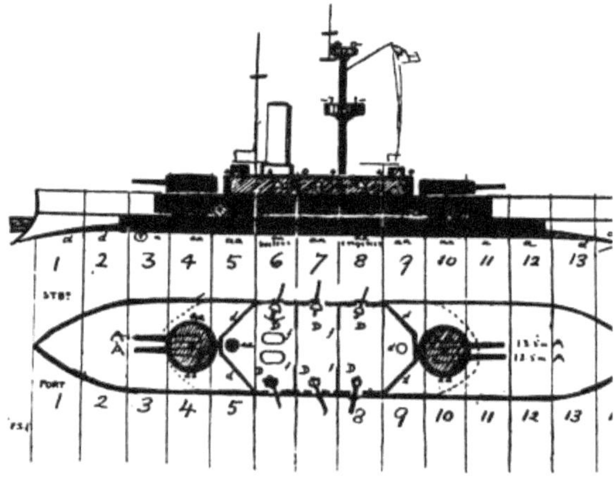

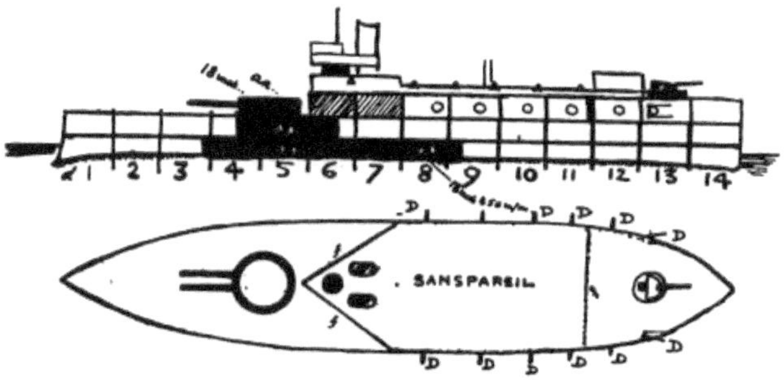

(II.) SANSPAREIL (1887).

11,000 tons. Complement, 630.

L.: 340 ft. = 103·5 m. B.: 70 ft. = 21·3 m. D.: 29½ = 8·8 m.

Guns: 2A (110 tons) + 1B (10 in.) + 12D* (6 in. 26 cals. converted) + 12* (6 pdrs.) + 12* (3 pdrs.) Sea speed: 14·5 kts.

Torpedo tubes, 6 (2 of them *submerged*). Armour: Deck = *d* (75 m/m = 3") Belt $\left\{\begin{matrix}162 \times 8\frac{1}{3}\text{ ft.}\\ 50 \times 2\cdot5\text{ m.}\end{matrix}\right\}$ *aa* (450 m/m) Bulkheads *aa* (450 m/m) Redoubt and turret *aa* (450 m/m) Bulkheads to battery *f*. Shield of B gun *d*. C. T. *a*.

Machinery: Triple expansion. 2 screws. 8 cylindrical boilers. I.H.P. 8 *hours trial* (1889) *natural draught* 8000 = 16 kts.; 4 *hours forced draught* 14,483 = 17·75 kts.

Photo. by W. Nye, Esq.

(5.) BARRACOUTA (1889) & BLANCHE (1889) & BARROSA (1889).

1580 tons. Complement, 160.

L: 220 ft. = 67 m. B.: 35 ft. = 10·7 m. D.: 15 ft. = 4·6 m.

Guns: 6E* + 4*. 2 Torpedo tubes. Sea speed: *circa* 14 kts.

Armour: Deck = *fe* $\left\{\begin{matrix}50\text{ m/m}\\ 2''\text{ steel}\end{matrix}\right\}$. Machinery, triple expansion vertical. 2 screws.

Boilers: Blanche, 2 Laird water tube; Barracouta and Barrosa, 2 2-ended cylindrical.

Coal: 160 tons = *circa* 2000–2500 miles. *Trials:* Barrosa, 2133 H.P *natural* 8 hours.; kts. *forced* 3111 H.P. 4 *hours*. Barracouta, 8 *hours natural*, 1920. Blanche, 8 *hours natural*, 1832 H.P. = 16·5 kts. *forced* 4 hours, 2849 H.P.

Centurion. (Barfleur identical). *Photo. by West, Southsea.*

(III.) BARFLEUR & CENTURION (189

Circa 10500 tons. Complement, .

L. B. D.; 360 × 70 × 27 ft. = 110 × 21·3 × 8·2 n

Guns: 4B (10 in.) + 10E* (4·7 in.) + 8* (6 pdr.) + 9* (3 pd
Torpedo tubes: 2 *submerged* + 5 *above water*.

Armour (*compound*): Deck = *d* (63 m/m = 2¼"). Belt, *a* (30
heads, *c* (200 m/m). Casemates on main deck *ed* (100 m/m *nickel*
m/m). Barbettes, *bc* (325 m/m = 9"). Hoods (slanting) = *a* (150 n
Lower deck, *ed* (100 m/m *nickel*).

Machinery: Greenock Foundry Co. Vertical triple expans
Boilers, cylindrical. Coal, 750 tons, *max.* = 1125 = *circa* 6000
nominal, 9000 = 17·5 kts. *natural draught*, 13000 = 18·5 *forced dr*

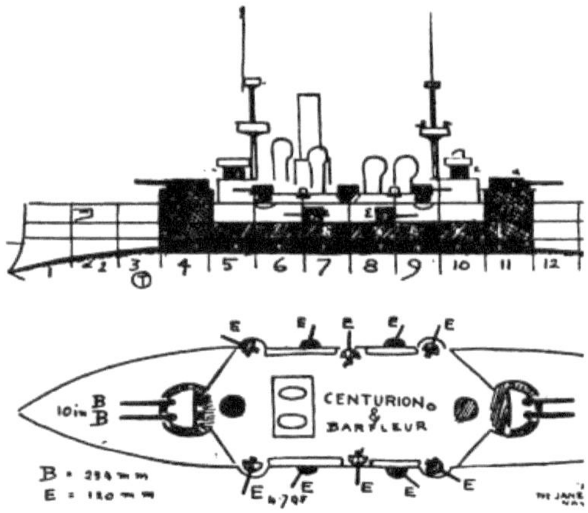

(II.) ROYAL SOVEREIGN & EMPRESS OF INDIA (–1891) & RAMILLIES & REPULSE & RESOLUTION & REVENGE: ROYAL OAK (–1892).

Circa 14200 tons. Complement, 712.

L. B. D.: 380 × 75 × 30 ft. = 115 × 22·8 × 9·14 m.

Guns: 4A (13·5 in.) + 10D* (6 in. 40 cals.) + 16* (6 pdr., 57 m/m) + 12* (3 pdr., 47 m/m) + 8 Maxims. Torpedo tubes, 2 *submerged* + 5 *above water*.

Armour (*compound*): Deck = *d* (75 m/m). Belt (8½ ft. = 2·6 m. *wide*), *aa* - *c* (450–200 m/m). 4 casemates, main deck for D* guns, *d* (150 m/m). Bulkheads, *aa* (400 m/m). Lower deck amidships, *c* (100 m/m), (in *Ramillies*, *Repulse*, *Revenge*, and *Royal Oak*, lower deck armour is *dc* (100 m/m *nickel steel*) and casemates, *dc*). Barbettes, *aa* (425 m/m). Hoists to upper deck guns, *dc* (200 m/m). Fore C.T., *aa* (350 m/m = 14"). After C.T., *f* (75 m/m).

Machinery: *Empress of India*, *Repulse* and *Royal Sovereign*, Humphrys and Tennant. *Ramillies*, Thomson. *Resolution* and *Revenge*, Palmer. *Royal Oak*, Laird. Vertical triple expansion. 2 screws. 8 cylindrical boilers each with 4 furnaces. I.H.P. *nominal*, 9000 *natural draught* = 15 kts. 13000 *forced* = 17 kts. Sea speed: 15 kts. Coal *normal* 900 tons, *max.* 1450.

Royal Sovereign (1894) 72 hours, Plymouth to Gibraltar = 15 kts. *average* speed.

(II.) HOOD (1891).

Circa 15400 tons *actual*. Complement, 654.

L. B. D.: 380 × 75 × 31 ft. = 115 × 22·8 × 9·40 m.

Guns: = *Royal Sovereign*, *except* 10 6-pdr.* instead of 16.

Armour: = *Ramillies*, *except* bulkheads, 425 m/m (17"). Turrets instead of barbettes.

Machinery: Humphrys and Tennant. I.H.P. *nominal* 9000 *natural* = 15 kts. 13000 *forced* = 17·5 kts. Sea speed: 15 kts. Coal = *Royal Sovereign class*.

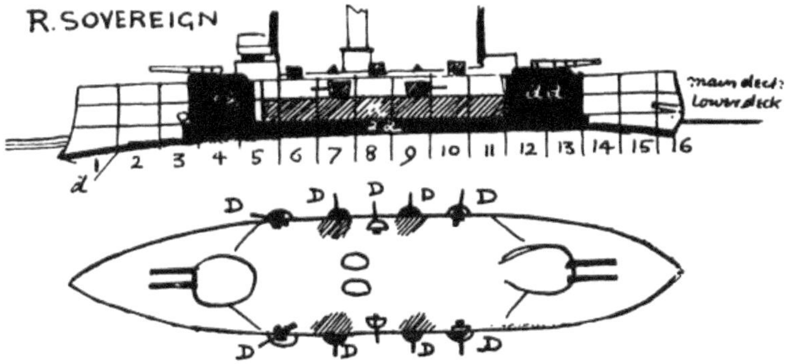

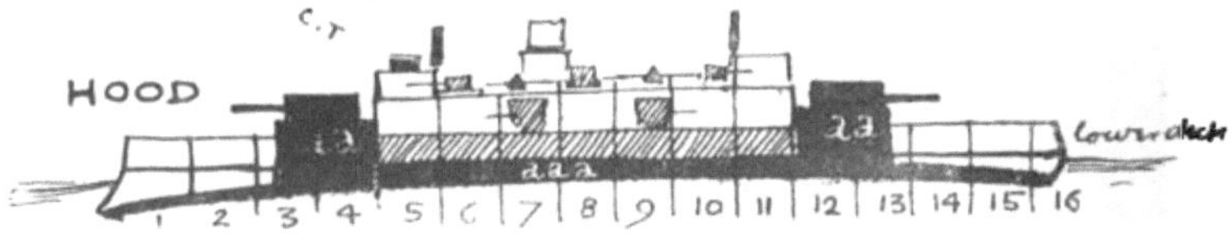

Royal Oak

Photo, Symonds & Co., Portsmouth.

(II.) RENOWN (1895).

12,350 tons. Complement 764.

L. B *D.* (*mean*) 380 × $72\frac{1}{3}$ × $26\frac{3}{4}$ ft. = 115·80 × 22 × 8·10 metres

Guns:
- 4B (10 inch, 40 cals.).
- 10D* (6 inch, 40 cals.).
- 8F* (3 in. 12 pdrs.).
- 12* (47 m/m, 3 pdrs.).
- 8 Maxims.

Torpedo tubes:
- 4 *submerged.*
- 1 *above water* (in stern).

Armour:
- 8"–6" Belt amidships
- 3" Armour deck = *c* (
- Protection to vitals =
- 6" Lower deck amidsl
- 10" Barbettes and bul
- Shields to B guns = *a*
- 6" Casemates (main de
- 4" Casemates (upper de
- Fore C.T. *a*, After C.T

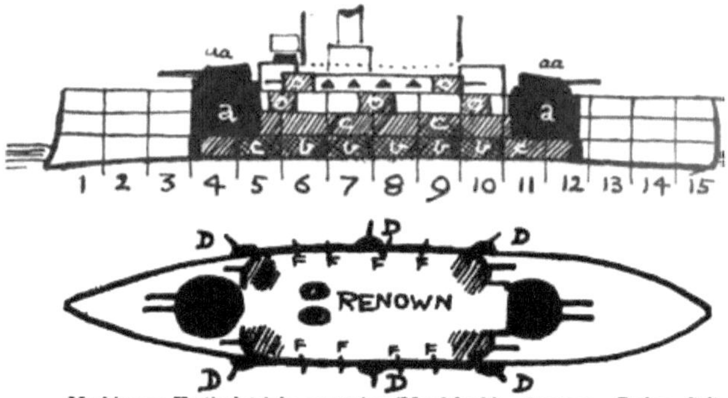

Machinery: Vertical triple expansion (Maudslay's). 2 screws. Boilers, Cyli
normal 800 tons, *maximum* 1600 tons = *circa* 8500 miles at 15 kts.

Trials (1897).

8 *hours natural draft* I.H.P. 10,708 = 17·9 kts. *mean.*
4 *hours forced draft* I.H.P. 12,901 = 18·75 kts. *mean.*
4 *days* at sea with squadron in station 17 kts. *mean* (1898).

ENGLAND.

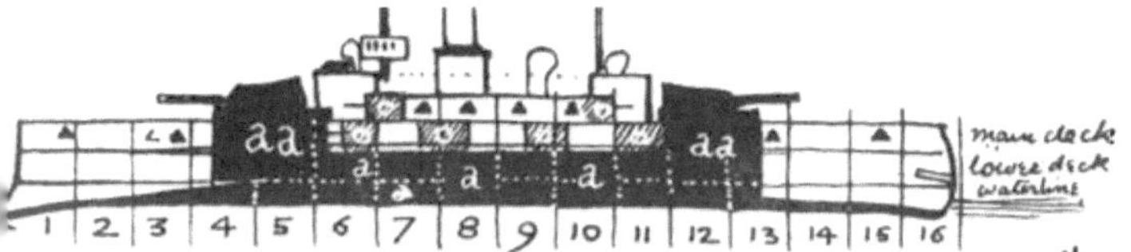

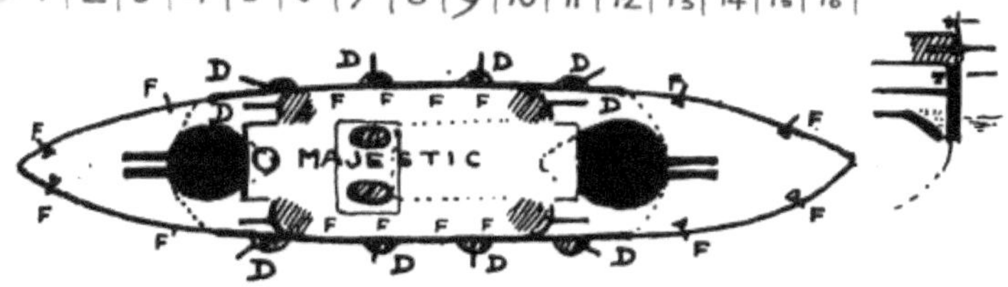

(I.) MAJESTIC (1895) & MAGNIFICENT (1894): PRINCE GEORGE & MARS & JUPITER: HANNIBAL & ILLUSTRIOUS & CÆSAR & VICTORIOUS (1895–96.)

14,900 tons. Complement, L. B. D.: 390 × 75 × 30 ft. = 119 × 22·8 × 9·14 m.

Guns: 4A (12 in. wire) + 12D* (6 in. 40 cals.) + 16F* (3 in.) + 12* (3 pdrs., 47 m/m) + 8 *Maxims* Torpedo tubes: 4 *submerged* + 1 in stern *above water*.

Armour (*Harvey*): Deck = *b* (100 m/m, *i.e.*, 4", on slopes). Belt {220 × 16 ft. / 67 × 4·9 m.} *a* (225 m/m = 9'.) Protection to vitals, belt + inclined deck + coal = *aaaa*. Bulkheads and barbettes *aa* (350 m/m = 14'.) A gun shields (inclined) = *aa* (250 m/m = 10'.) Casemates for D guns *c* (150 m/m.) Fore C. T. *aa*. After C. T. *e*. Shields to F guns = *d*.

Machinery: 3-cylinder inverted triple expansion. *Cæsar*, Maudslay; *Hannibal*, Harland & Wolff; *Illustrious* and *Magnificent*, Penn; *Jupiter*, Thompson; *Majestic*, Vickers; *P. George*, Humphrys & Tennant; *Victorious*, Hawthorn Leslie. 2 screws. Boilers: 8 cylindrical, Scotch.

Sea speeds: *circa* 15 kts. Coal, *normal circa* 1200 tons, *maximum* 2200 = *circa* 5000 miles at 12 kts. Consumption for auxiliary purposes in harbour, *circa* 12 tons *per diem*. On cruises, average 30–35 tons *per diem* (*sea and harbour*) for all purposes.

Trials (1896–97).

	Displacement circa.		I.H.P.	kts.		I.H.P.	kts.
Majestic	13,000 tons,	8 *hours natural*,	10,418	= 16·9	4 *hours forced*,	12,097	= 17·9
Majestic	15,000 „	8 „ „	10,083	= 16·0	(3 runs) „	11,795	= 16·85
Magnificent	13,000 „	8 „ „	10,301	= 16·5	4 *hours* „	12,157	= 17·6
Victorious	13,500 „	8 „ „	10,316	= 16·9	4 „ „	12,201	= 18·7
Jupiter	15,000 „	8 „ „	10,258	= 15·8	4 „ „	12,457	= 18·4 (at 14,000 tons).
P. George	13,000 „	8 „ „	10,464	= 16·5	4 „ „	12,253	= 18·3
Mars	13,000 „	8 „ „	10,159	= 15·96	4 „ „	12,434	= 17·7
Hannibal	13,000 „	8 „ „	10,357	= 16·3	4 „ „	12,138	= 18·0
Illustrious	13,500 „	8 „ „	10,241	= 15·96	4 „ „	12,112	= 16·5 (rough weather).
Cæsar	13,500 „	8 „ „	10,630	= 16·7	4 „ „	12,652	= 18·7

(30 hours consumption trials speed averaged 14–15 kts.—the ships light at the time).

...nce George. *Photo, W. Nye, Esq.*

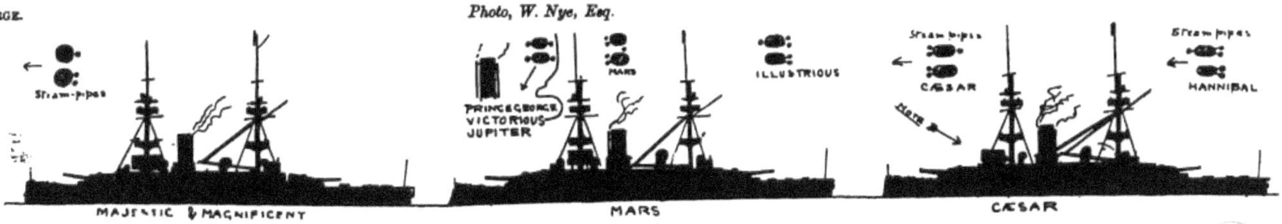

Photo, West, Southsea.

(IV.) IMPERIEUSE (1883): WARSPITE (1884).

Circa 9000 tons. Complement, 544.

L.: 315 ft = 96 m. B.: 62¾ ft. = 19·1 m. D.: 28 ft. = 8·5 m.

Guns: 4C (9·2 in.) + 10D, (6 in. 26 cals.) + 13*. Sea speed: *circa* 14

Armour (*compound*): Deck (75 m/m) 3″ = *ds*. Belt, $\frac{140 \times 8}{42 \cdot 7 \times 2}$ (250 m/m). Bulkheads, *cb* (225 m/m = 9″). Barbettes (225 m/m), *cb*. C (9″).

Machinery: vertical compound. 2 screws. Boilers, cylindrical.

Coal, 900 tons. *Max.* 1130 tons = *circa* 4000 miles at 8–10 kts.

Trials (1886)—Imperieuse (*light*) 7500 *natural draught* = 16 kts. 1(
17·2 kts. *forced draught.*
Warspite (*light*) 7451 *natural draught* = 15·5 kts. 1(
17·3 kts. *forced draught.*

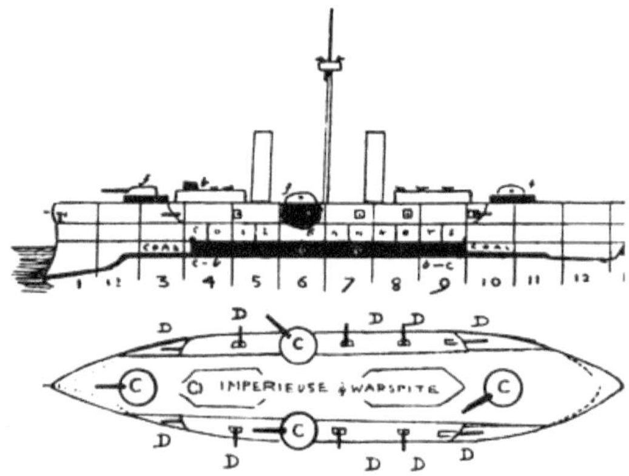

F 2

V.) THUNDERER (1872) (1890): DEVASTATION (1871) (1891).

9400 tons. Complement, 420.

L.: 285 ft. = 87 m. B.: 63 ft. = 19·2 m. D.: 28 ft. = 8·5 m.

uns: 4B (10 in.) + 14*. Devastation, 2 *submerged* torpedo tubes. Sea speed: ı 12 kts.

Thunderer, 2 torpedo tubes, lower deck behind *c* armour.

Armour (*iron*): Belt *c* - *d* (300-200 m/m). Bulkheads *d*. Turrets *c*.

Machinery: 2 inverted compound. 2 screws. Cylindrical boilers.

Speed, Devastation (1893): 8 *hours natural draught trial*, 6000 H.P. = 13·3 kts.; urs *forced*, 144 = 14·5. Coal, 1600 tons.

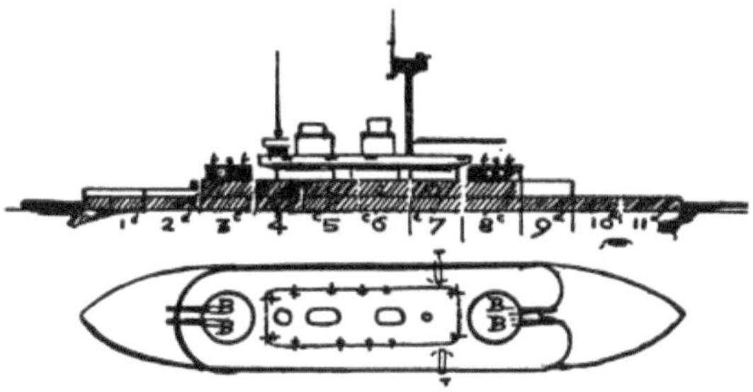

(IV). *DREADNOUGHT* (1875) (1879-9).

11,000 tons. Complement, 453.

L.: 320 ft. = 97·5 m. B.: 64 ft. = 19·5 m. D.: 29 ft. = 8·8 m.

Guns: 4B (12·5 in. M.L.) + 18*. 2 submerged torpedo tubes. Sea speed: kts.

Armour: = Thunderer. Coal, 1200 tons. Re-boilered 1895.

CAMPERDOWN. *Photo, West, Southsea.*

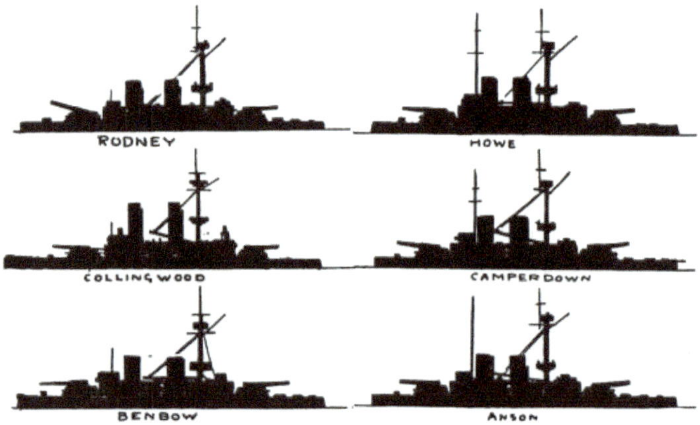

(III.) HOWE (1885) & RODNEY (1884) (10,300 to
(1886): CAMPERDOWN (1885) (10,600 to

Complement, 545.

L.: 325 ft. = 99 m. B. 68 ft. = 20·7 m. D. *circa* 29 ft. :
Guns: 4A (13·5 in.) + 6D$_*$ (6 in., 26 cals.) + 22*. Torpedo tubes

Armour: (*compound*). Deck 75 m/m (3″) on slopes = *d*. Belt: *aa* - *c* $\left(\frac{450 - 200 \text{ m/m}}{18 - 8''.}\right)$. Bulkheads: *aa* (400 m/m). Barbettes (
(350 m/m). Bulkheads to battery *f*. Hoists: *a*. C.T. *a* (350 m/
barbettes.

Machinery: Comp. inverted 3 cylinder, 2 screws. Cylindrical
natural draught 7500, *forced draught* 11,500. Sea speeds: (18
Coal *normal*, 900 tons. *Max*. 1200 = *circa* 5000 miles at 8 kts. *Ho*
(*Rodney* & *Howe* = *aaaa* (300 m/m)).

(III.) COLLINGWOOD (1882): BENBOW (

9500 tons. 10,600 tons.

L. B. D. = Howe *nearly*.

Guns: Collingwood: 4B (12 in.) + 6D$_*$ (6 in., 26 cal.) + 20
(110 ton) + 10D$_*$ + 26*. Benbow: Sea speed: 13 kts.
Armour & Machinery: *Collingwood* = Howe: *Benbow* = Anson
(Colling
1897.)
Trials:
speeds.
Six
draught 8
Thirty hou
= 16·1 kts

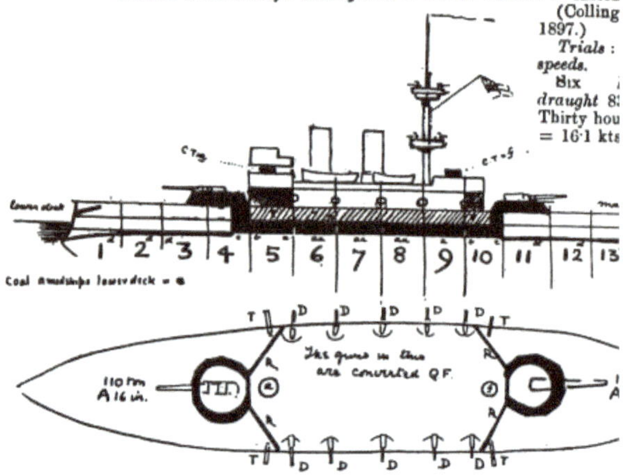

(T.) VULCAN (1889).

6620 tons. Complement, 433.

L.: 350 ft. = 107 m. B.: 58 ft. = 17·7 m. D.: 24⅝ ft. = 7·5 m.

Guns: 8E* (4·7 in.) + 12* (3 pdrs.) + 12N. Torpedo tubes: 2 *submerged* + 4 above water.

Armour: Deck 5" (125 m/m), on slopes = *c*.

Machinery: Vertical triple expansion; Humphrys & Tennant. 2 screws. 4 2-ended boilers + 1 1-ended *auxiliary* boiler.

Coal: 1000 tons, *max.* capacity 1300 tons.

Trials (in 1895).	8	*hours natural draught*,	I.H.P	8167 =	*circa* 18	kts. *mean*.
	4	,, *forced* ,,	,,	12,062 =	,, 20·2	,, ,,
	48	,, *natural* ,,	,,	7230 =	,, 17·75	,, ,,

(Carries 6 torpedo boats, 3rd class.)

Photo, R. Ellis, Malta.

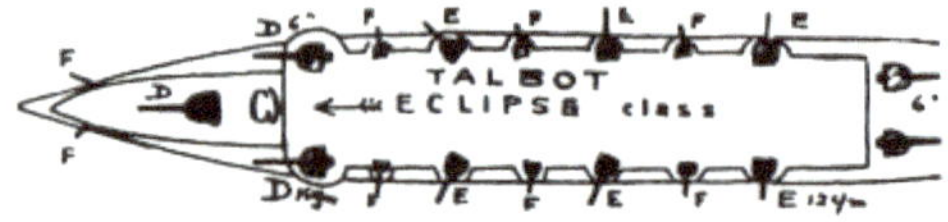

(2.) TALBOT (1895).

5600 tons. Complement, 437.

L.: 350 ft. = 106·5 m. B.: 54 ft. = 16·5 m. D.: 22¾ ft

Guns: 5D* (6 in., 40 cal.) + 6E* (4·7 in.) + 8F* (3 in.) · *Maxims.* Torpedo tubes, 2 *submerged* + 1 *above water* in stern.

Armour (*Harvey*): Deck (3" (75 m/m) on slopes) = *c* (150 m/m) *inclined* = *aa.* C.T. (150 m/m) *c.*

Machinery: inverted triple expansion. 2 screws. 8 cylindr *normal* 550 tons. *Max.* capacity of bunkers = 1000 tons.

(1896) *Trials* 8 *hours natural draught* 8462 H.P. = 19·2 kts.
4 „ *forced* „ 9766 „ = 20 „
(1898) 4 days at sea ? „ = 17 „

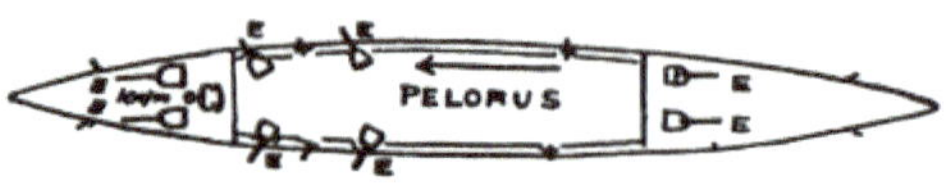

(4.) PELORUS & PROSPERINE (1896) & PA PEGASUS & POMONE (1897) & PROMETHEUS & PYRAMUS (1898) & PANDORA & PIONEER & PS

2135 tons, *except* last 3, which are 2200 tons. Compleme

L. B. D. (*mean*): 300 × 36½ × 17 feet = 91·50 × 11·10 × 5

[Mean draught of Pandora, Pioneer, and Psyche is 17½ feet (5

Guns: 8E* (4 in.) + 8* (3 pdr., 47 m/m). Torpedo tubes: 2

2" armour deck = *d.* Protection to vitals, deck + coal = *d.*

Machinery: inverted 3 cylinder triple expansion. 2 screws. except *Pomone*, Blechynden; *Prosperine*, Thornycroft. Coal, *norm*

Trials.		8 *hours natural draught.*		"60 *hours natural.*"	
Pelorus	I.H.P.	5075 = 19·7 kts.	...	4277 = 17·3 kts.	(60 hrs.) ..
Prosperine	„	5534 = 19·7 „	...	3615 = 17 „	(60 „) ..
Pactoclus	„	5428 = 19·1 „	...	3631 = 16·6 „	(30 „) ..
Pegasus	„	5400 = 20·0 „	...	3698 = 17·3 „	(30 „) ..
Perseus	„	5239 = 19·1 „	...	——	..
Pomone	„	5500 = 20·2 „	...	3604 = 16·5 „	„ ..
Prometheus	„	5183 = 19·8 „	...	3556 = 17·5 „	„ ..
Pyramus	„	5424 = 19·9 „	...	3605 = 17·5 „	„ ..
Pioneer (*foul bottom*)	„	5265 = 17·7 „	...	3665 = 16·3 „	„ ..
Psyche	„	5000 = 19·5 „	...	3647 = 16·8 „	„ ..
Pandora		[*not yet tried*]		=	

(4.) PALLAS, PEARL, PHOEBE, PHILOMEL (1890). KATOOMBA, MILDURA, RINGAROOMA, TAURANGA, WALLAROO (1889).

2575 tons. Complement 217.

L.: 266 ft. = 81 m. B.: 41 ft. = 12·5 m. D.: 17 ft. = 5·2 m.

Guns: 8E* (4·7 in.) + 8* (3 pdr.). Sea speed: *circa* 15 kts.

Armour: Deck 65 m/m (2½") on slopes = *e*.

Machinery: Vertical triple expansion. 2 Screws. Boilers 4, 2-ended. I.H.P. *nominal*, 4500 *natural*, and 7500 forced = 19 kts. *nominal*.

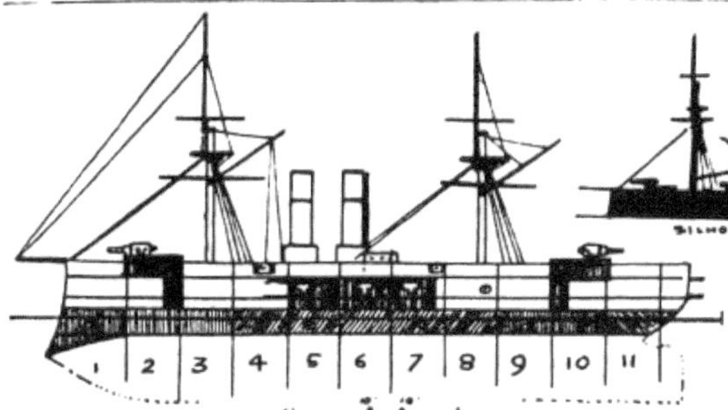

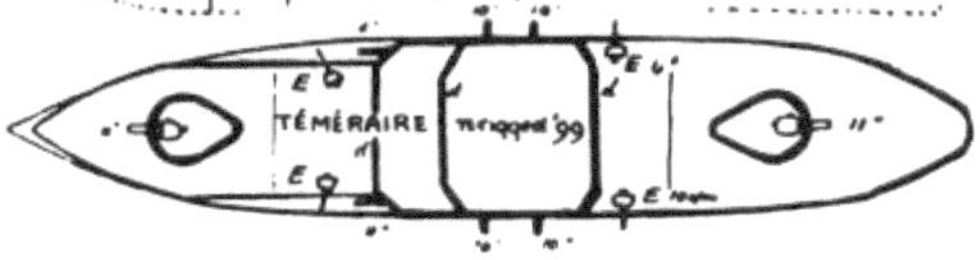

(IV.) TÉMÉRAIRE (1876).

8540 tons. Complement 590.

L.: 285 ft. = 87 m. B.: 62 ft. 18·9 m. *D.: mean* 27 *ft.* = 8·2 *m.*

Guns: 4C (11 in. *m.l.*) + 4C (10 in. *m.l.*) + 6E* (4 in.) + 14*. Torpedo tubes: 2.

Armour (*wrought iron*): Belt: $\left(\frac{275-250 \text{ m/m}}{11 - 10''}\right)$ *dc*. Battery (200 m/m) *dc*. Barbettes (250 m/m) *d*. Deck, 50 m/m (2").

Machinery: I.H.P. *nominal* 6500 *natural draught* = 13·6 *trial*. Sea speed: 11·5 kts.

ENGLAND.

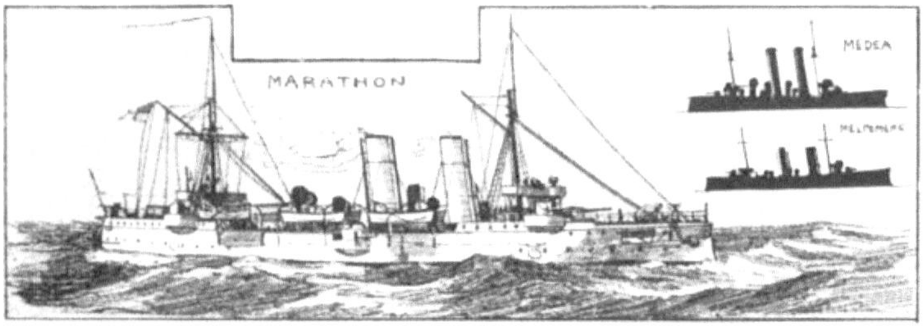

(4.) MEDEA & MEDUSA (1888).

2800 tons. Complement, 218.

L. B. *D.* (*mean*): 265 × 41 × 16½ ft. = 81 × 12·5 × 5 m.

MELPOMENE: MAGICIENNE & MARATHON (1888)

2950 tons. Complement, 218.

L. B. *D.* (*mean*): 265 × 41¾ × 17½ ft. = 81 × 12·7 × 5·4 m.

Guns: 6D* (6 in. 25 cals.) + 10*. Torpedo tubes, 4.

Armour (deck): *Medea* class, 50 m/m (2") = *e*; *Melpomene* class, 35 m/m

Machinery: *Medea* class, vertical; *Melpomene* class, horizontal triple exp 2 screws. Boilers, 4 2-ended. Coal, 400 tons. *Trials*, see below.

Melpomene class sheathed and coppered.

Dif. = Melpomene has less conspicuous ventilator cowls than Marathon. Se

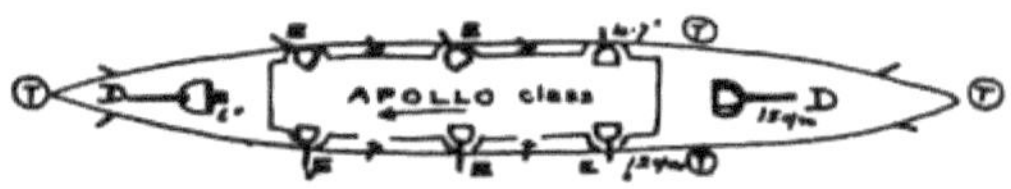

(3.) APOLLO (1891)

& ANDROMACHE & LATONA & MELAMPUS & NAIA SYBILLE & TERPSICHORE & THETIS (1890): . . . & SA & SCYLLA & TRIBUNE . . . (1891):

3400 tons. Complement, 273.

L. B. D.: 300 × 43 × 17½ ft. = 91·5 × 13·1 × 5·3 m.

Sea speed: *circa* 19 kts.

&

PIQUE & SIRIUS (1890): . . . & ÆOLUS & BRILLIA INDEFATIGABLE & INTREPID & IPHIGENIA & RAIN & RETRIBUTION & SPARTAN—(1891):

3600 tons. Complement, 273.

L.: 300 ft. = 91·5 m. B.: 43¾ ft. = 13·3 m. D.: 18½ ft. = 5·6 m.

Guns: 2D* (6 in. cals.) + 6E* (4·7 in.) + 9*. Armour (deck) = *e* (50 inches on slopes). Engine hatches = *d* (125 m/m = 5 in.)

Machinery: Vertical triple expansion. 2 screws. Boilers: 3 2-ended + 2 Coal, *normal* 400 tons: *max.* 535 tons.

Sea speed: *circa* 18·5 kts.

Note.—Apollo and Andromache have higher ventilator cowls than the ot of this class.

Trials.—	*8 hours natural draught.*		*4 hours forced draught.*	
Magicienne	5408 =	kts.	9280 = *circa* 19	kts.
Marathon	5530 =	,,	9173 = 18·8	,,
Melpomene	6215 =	,,	9653 =	,,
Apollo	7488 =	,,	9226 =	,,
Andromache	7234 =	,,	9044 =	,,
Latona	7261 =	,,	9455 = 20	,,
Melampus	7684 =	,,	9135 =	,,
Naiad	7547 = 18·6	,,	9261 =	,,
Sappho	7301 =	,,	9618 =	,,
Scylla	7614 =	,,	9280 =	,,
Sybille	7598 =	,,	9524 =	,,
Terpsichore	7133 =	,,	8825 =	,,
Thetis	7034 =	,,	9496 —	,,

Trials.—	*8 hours natural draught.*		*4 hours forced draught.*	
Tribune	7523 =	kts.	9101 =	kts.
Medea (189–)	6027 = 18	,,	9185 = 19·9	,,
Medusa	6144 = 18	,,	9435 = 19·8	,,
Æolus	7504 =	,,	9315 =	,,
Brilliant	7522 = 19·2	,,	9180 =	,,
Indefatigable	7349 =	,,	9047 =	,,
Intrepid	7522 =	,,	9489 = 19·8	,,
Iphigenia	7251 =	,,	9337 =	,,
Pique	7193 = 18·5	,,	9258 =	,,
Rainbow	7879 = 19·5	,,	9741 = 20·2	,,
Retribution	7645 =	,,	9367 =	,,
Sirius	7491 =	,,	9281 =	,,
Spartan	7832 =	,,	9254 = 20·4	,,

(2.) HERMIONE: ASTRÆA: CAMBRIAN: CHARYBDIS: FLORA: FORTE: FOX (—— 1893): BONAVENTURE (1892).

4360 tons. Complement 318.

L.: 320 ft. = 97·5 m. B.: 49½ ft. = 15·1 m. D.: 21 ft. = 6·4 m.

Guns: 2D* (6 in. 40 cal.) + 8E* (4·7 in.) + 8* (6 prs.) + 1* (3 pdr.).

Armour: Deck = *e* (50 m/m, 2″ on slopes). Engine hatches = *c* (25 m/m). Gun shields *e*. C.T. *f* (75 m/m).

Machinery: Vertical triple expansion. 2 screws. 8 1-ended boilers. Coal: *normal* 400 tons; *max. capacity* 1000 tons = *circa* 4000 miles.

		8 *hours natural d.*	4 *hours forced d.*
Trials:—	Astræa	7603 = 19 kts.	9151 = 19·75 kts.
	Bonaventure	7423 = 19 2 „	9365 = 20 „
	Cambrian	7164 = 19·4 „	9259 = 20·4 „
	Charybdis	7125 = 19·3 „	9136 = 20·5 „
	Flora	7211 = 19·1 „	9008 = 20·1 „
	Forte	7427 = 18·7 „	9063 = 19·9 „
	Fox	7049 = 19 „	9382 = 19·9 „
	Hermione	7393 = 19 „	9264 = 19·5 „

Note. Hermione, Astræa, Bonaventure, Fox, have stern walks (*balcons*) and carry an extra pair of boats aft.

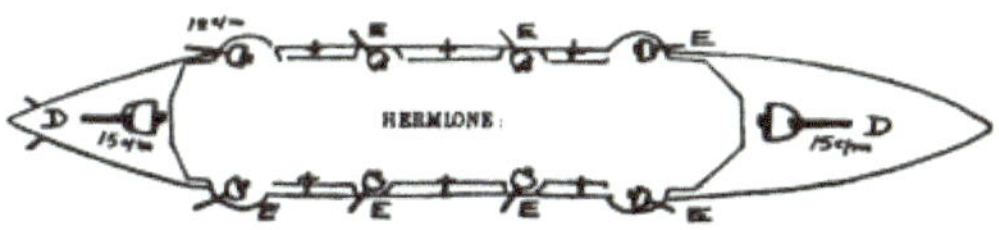

Photos, W. Nye, Esq.

(1.) BLAKE (1891): BLENHEIM (1890).

9000 tons. Complement 590.

L.: 375 ft. = 114 m. B.: 65 ft. = 19·8 m. D.: 27½ ft. = 8·4 m.

Guns: 2C (9·2 in. 30 cals.). + 10D* (6 in. 26 cals. [?]) + 16* (3-pdrs.). To tubes: 2 *submerged* + 2 *above water*. Sea speed: Blake 17·5 kts. Blenheim 18·

Armour: Deck = c (125 m/m = 5") *max*. on slopes. Engine hatche (150 m/m). 4 casemates on main deck *d* (150 m/m compound). C. T. *a*. Ho (175 m/m).

Machinery: 4 sets vertical triple expansion (2 coupled on each shaft). 2 sc 6 2-ended boilers. Coal *normal*: 1800 tons = *circa* 10,000 miles at 10 kts.

Trials (*mean*): Blake, 8 *hours natural draught*, 14525 = 19·1 kts.; 4 *hours f draught* (1897), 19579 = 21·5 kts. Blenheim, 8 *hours natural draught*, 14? 20·4 kts.; 4 *hours forced draught* (1893), 21411 = 21·8 kts. Blake was reboil etc., 1899.

(1.) CRESCENT (1892): ROYAL ARTHUR (1891).

7700 tons. Complement 567.

L.: 360 ft. = 110 m. B.: 60¾ ft. = 18·5 m. D.: 25 ft. = 7·6 m.

Guns: 1C (9·2 in. 30 cals.) + 12D* + 12* (6-pdrs.) + 5* (3-pdrs.) + 6 Ma Sea speed: 18·5 kts.

Torpedo tubes, Armour, Machinery, Coal, etc. = *Gibraltar* and *St. George*.

1st *Trials*: Crescent, 8 *hours natural draught*, 10378 = 18·6 kts. Royal A 8 *hours natural draught*, 10086 = kts. 48 *hours full power I.H.P.* ? 18 kts. (*mean*).

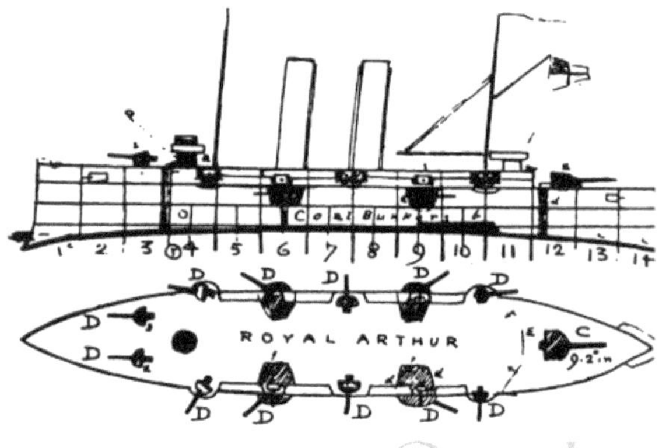

) HAWKE (1891): GRAFTON (1892): GIBRALTAR (1892): T. GEORGE (1892): EDGAR (1890): & ENDYMION (1891): & THESEUS (1892).

7700 - 8100 tons. Complement, 544.

L.: 360 ft = 110 m. B.: 60 ft. = 18·3 m. D.: *circa* 26 ft. = *circa* 7·90 m.

Guns: 2C (9·2-in. 30 cal.) + 10D* (6-in. Q.F. 26 cal.) + 12* (6-pdrs. 57 m/m) 5* (3-pdrs. 47 m/m) + 6 Maxims. Sea speed: 18·5 kts. 2 *submerged* torpedo)es, and 2 *above water*.

Armour: (5" = 125 m/m) Deck = *c*. Engine hatches = *b* (6" = 150 m/m). semates on main deck to 4 of the D guns *d* (6 in. = 150 m.m. compound). Hoists *d*. T *a* (12" = 305 m/m).

Machinery: Vertical inverted triple expansion. 4 2-ended and 1 1-ended lindrical boilers. 2 screws. (Theseus and Grafton, 8 1-ended boilers). Engines: gar and Hawke by Elder; St. George and Theseus by Maudslay; Gibraltar by pier; Endymion by Earle; Grafton by Humphrys & Tennant. Coal (*normal*):) tons; *maximum capacity* = *circa* 1200 tons = *circa* 10000 miles

First trials (4 *hours natural d.*): Edgar, 10172 H.P. = kts.; Endymion, 662 = 20·9 kts.; Grafton, 10956 = kts.; Hawke, 10761 = 19·5 kts.; braltar, 10536 = 20·4 kts.; St. George, 10585 = 20·25 kts.; Theseus, 10608 = ·6 kts. (all *mean*). 8 *hours forced d.*: Edgar, 12550 = 20 kts.; Grafton, 13483 = kts.; Hawke, 12521 = 20 kts. (all *mean*).

Endymion, St. George and Gibraltar were light on trials.

(T.) GRASSHOPPER, SPIDER (1887).

L.: 200 ft. = 76 m.

SANDFLY (1887).

525 tons. Complement 70.

L.: 200 ft. = 76 m. B.: 23 ft. = 7 m. D.: 9 ft. = 2·7 m.

Guns: 1F (4 in.) + 6* (3 prs.). Sea speed: 13—13·5 kts. Spider: 1E* (4 in.). 4 Torpedo tubes (1 of them in bow, 1 in stern). Coal 80 tons = *circa* 5 days.

Speeds *on trial*, 1888, 19·5 kts. 3 *hours forced draught*.

Grasshopper with 3 in. (75 m/m) pressure = 13·5 kts. at sea (1893) 12 *hours on cruise*.

(T.) ASSAYE (1891), BOOMERANG (1892), GOSSAMER (1890), GLEANER (1890), KARRAKATTA (1889), PLASSEY (1890), SALAMANDER, SHELDRAKE, SKIPJACK, SPANKER, SPEEDWELL (1889)

(T.) SHARPSHOOTER (1888)

SEAGULL (1889) (

735 tons. Complement 90.

L. B. D.: 230 × 27 × 12 ft. = 70 × 8·2 × 3·60 m.

Guns: 1E* (4·7 in.) + 4* (3 prs. = 47 m/m). 5 Torpedo tubes (1 of them . *Assaye & Plassey* have only 3 tubes.

Machinery: Vertical triple expansion. 2 Screws.

Boilers: 4 Locomotive. Except—*Seagull*, Niclausse; *Sharpshooter*, Belleville; Temple; *Salamander*, Thornycroft; *Sheldrake*, Babcock & Wilcox. (Re-boilered 18

Mean H.P. developed on First *Trials*.

	8 *hours natural draught.*		4 *hours forced draught.*	
Assaye	2774 =	kts.	=	kts.
Boomerang	2612 =	kts.	3509 =	kts.
Gossamer	2634 =	kts.	3632 =	kts.
Gleaner	2606 =	kts.	3654 =	kts.
Karrakatta	2598 =	kts.	3840 =	kts.
Plassey	2895 =	kts.	=	kts.
Speedwell	2601 =	kts.	3588 =	kts.

	8 *hours natural draught.*		4 *hours forced draught.*	
Salamander	=	kts.	=	kts.
Seagull	2664 =	17·5 kts.	=	kts.
Sharpshooter	2620 =	kts.	3238 =	19 kts.
Sheldrake	=	kts.	4050 =	20·5 kts.
Skipjack	=	kts.	=	kts.
Spanker	=	kts.	3920 =	20 kts.

ALARM, ANTELOPE
CIRCE, JASEUR etc

DRYAD, HALCYON etc

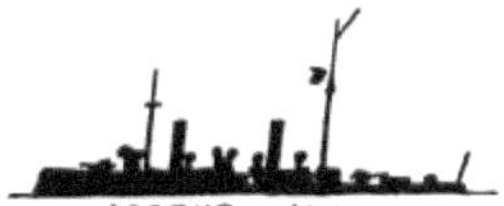
JASEUR — 1900.

Photos, W. Nye, Esq.

(T.)	*Trials.* 8 *hours natural draught* (*mean*).	4 *hours forced draught* (*mean*).
ALARM	I.H.P. 2593 = 18 kts.	I.H.P. 3886 = 19·2 kts.
ANTELOPE	„ 2653 = 17 „	„ 3621 = 19 „
CIRCE	„ 2621 = 18·25 „	„ 3508 = 19·3 „
JASEUR	„ 2546 = 18 „	„ 3711 = 19 „
JASON	„ 2676 = 19 „	„ 3552 = 20 „
HEBE	„ 2702 = 17·8 „	„ 3566 = 19 „
LEDA	„ 2696 = 17·75 „	„ 3601 = 18·3 „
NIGER	„ 2710 = 17·5 „	„ 3785 = 19·3 „
ONYX	„ 2526 = 17·5 „	„ 3548 = 19 „
RENARD	„ 2609 = 17·6 „	„ 3962 = 19·4 „

810 tons. Complement, 85.

L.: 230 ft. = 70 m. B.: 27 ft. = 8·2 m. D.: *circa* 12 ft. = 3·6 m.

Guns: 2E* + 4*. Sea speed: *circa* 17 kts.

3 torpedo tubes (1 of them fixed in bow).

Machinery: vertical triple expansion. 2 screws. Locomotive boilers, 4. Coal, 100 tons = *a* 5 days.

Launched 1892.

(T.) DRYAD, HALCYON, HARRIER, HAZARD, & HUSSAR (1894).

1070 tons. Complement, 120.

L.: 250 ft. = 76 m. B.: 30½ ft. = 9·3 m. D.: *circa* 13 ft. = 4 m.

Guns: 2E* (4·7 m., 12 c/m) + 4*. Armour: 0.

Sea speed: 16·5 kts.

Machinery: Vertical triple expansion. 2 screws. Boilers, 4 locomotive. Coal, 100 tons.

Trials.	8 *hours natural draught* (*mean*).	4 *hours forced draught* (*mean*).
Dryad	I.H.P. 2696 = 16·6 kts.	I.H.P. 3709 = 18·2 kts.
Halcyon	„ 2590 = 16·9 „	„ 3546 = 17·7 „
Harrier	„ 2709 = 18 „	„ 3608 = 19 „
Hazard	„ 2621 = 17·1 „	„ 3734 = 19 „
Hussar	„ 2548 = 17·5 „	„ 3553 = 19·7 „

Orlando Class.

(I.) AURORA & GALATEA (—1887): AUSTRALIA: NARCISSUS: ORLANDO & UNDAUNTED (—1886) & IMMORTALITÉ (—1887).

Circa, 5700 tons. Complement, 497.

L.: 300 ft. = 91·5 m. B.: 56 ft. = 17·1 m. D.: 26 ft. = 7·92 m.

Guns: 2C (9·2 in. 25 cals.) + 10DE* (6 in. 5 ton, 25 cals.) + 10* (3 pdrs.) + 6 Nordenfelt. **Sea speed**: 15·5 - 15 kts. **Torpedo tubes**: 4 - 2 *above water*.

Armour (*compound*): Deck (steel) = *d* (75 m/m = 3″.) Belt, below water line {200 × 5½ ft. / 61 × 1·7 m.} *b* (250 m/m = 10″.) Bulkh ads *aa* (400 m/m = 16″.) C. T. *a* (300 m/m = 12″.)

Machinery: Horizontal triple expansion. 2 screws. 4 2-ended boilers. Coal *max.* 900 tons = *circa* 5000 miles.

1st trials.	*8 hours natural.*		*4 hours forced draught.*
Aurora,	I.H.P. 5706 =	kts.	I.H.P. 9013 = 18·5 kts.
Australia,	,, 5810 =	,,	,, 8876 = 18·8 ,,
Galatea,	,, 5871 =	,,	,, 9220 = 19 ,,
Immortalité,	,, 6090 =	,,	,, 8738 = 19·5 ,,
Narcissus,	,, 5353 =	,,	,, 8589 = 18·5 ,,
Orlando,	,, 5628 = 17·2	,,	,, 8739 = 19·3 ,,
Undaunted	,, 5682 = 17	,,	,, 8670 = 19·4 ,,

All the ships (except Aurora) were very light on trials; only displacing *circa* 5000 tons.

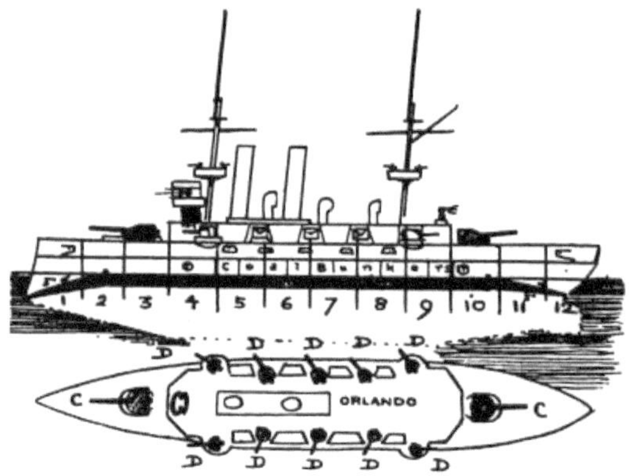

(2.) MINERVA: DIANA & VENUS & JUNO & DORIS (1895): ECLIPSE (1894): DIDO & ISIS (1896).

5600 tons. Complement 437.

L.: 360 ft. = 110 m. B.: 54 ft. = 16·5 m. D.: 22¾ ft. = 6·9 m.

Guns: 5D* (6-in. 40 cals.) + 6E* (4·7-in.) + 8F* (3-in.) + 7* (3-pdrs.). Torpedo tubes: 2 *submerged* + 1 *above water* fixed in stern.

Armour (*Harvey*): Deck = *c* (3″ = 75 m/m on slopes). Engine hatches = *aa*. C. T. *c* (6″ 150 m/m).

Machinery: Inverted triple expansion. 2 screws. 8 cylindrical boilers. Coal: *normal* 550 tons; *max. circa* 1000 tons.

Trials (Ships Light).

	8 hours natural d.	*4 hours forced d.*	*30 hours natural d.*
Diana,	8252 = 19·7 kts.	9875 = 20·2 kts.	4916 = 17·2 kts.
Doris,	8391 = 19·1 ,,	9851 = 20·1 ,,	4938 = 16·5 ,,
Eclipse,	8220 = 19·2 ,,	9853 = 20·1 ,,	4838 = 16·8 ,,
Isis,	8208 = 19·8 ,,	9840 = 21 ,,	4925 = 17·5 ,,
Juno,	8272 = 18·8 ,,	9771 = 20 ,,	4863 = 16·1 ,,
Minerva,	8221 = 19·6 ,,	9891 = 20·3 ,,	4919 = 17·5 ,,
Venus,	8290 = 19·3 ,,	9774 = 20·2 ,,	4876 = 16·8 ,,

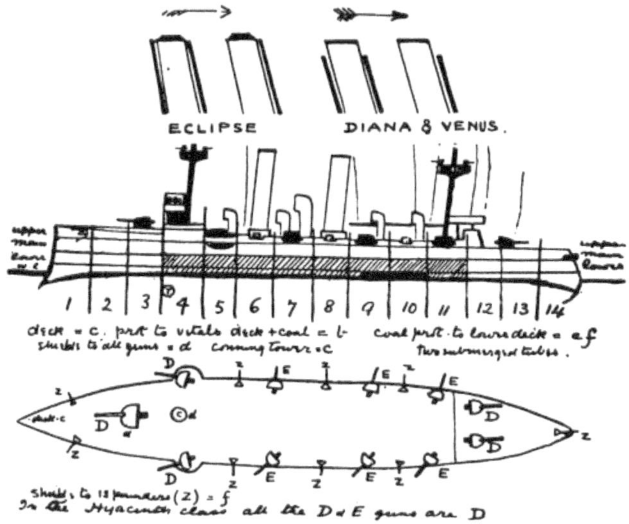

DIDO & ISIS.

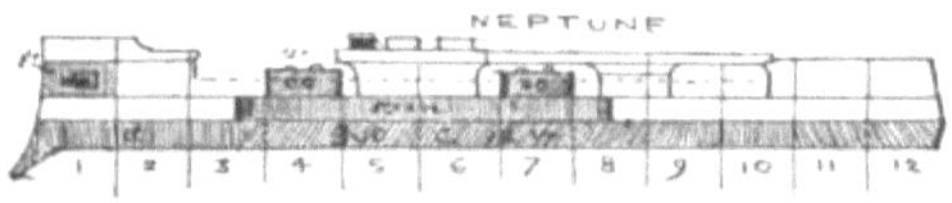

(IV.) NEPTUNE (1874).

9310 tons. Complement,

L.: 300 ft. = 91·5 m. B.: 63 ft. = 19·2 m. D.: 27 ft. = 8·2 m.

Guns: 4B (12·5 in. *muzzle loaders*) + 2D (9 in. *m. l.*) + 14*. 2 torpedo tubes. Sea speed: 8 kts.

Armour (iron): Belt (300 × 8½ ft. = 91·5 × 2·60 m.) *c* - *d* (300 - 225 m/m = 12 - 9"). Turrets *c*. Redoubt *d* (250 m/m). Bulkheads to redoubt *de* (200 m/m). Deck 2 - 3" (75 - 50 m/m). Battery forward (D guns) *de* (200 m/m = 10").

Machinery: 1 screw. Horizontal trunk boilers. I.H.P. *nominal* 6000 = 13·4 kts. (in 1880).

Coal, 670 tons = *circa* 900 miles at 8 kts.

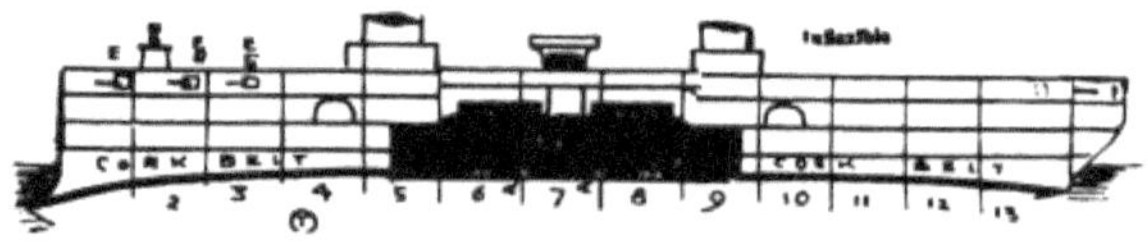

(III.) INFLEXIBLE (1876).

11,880 tons. Complement 485.

L.: 320 ft. = 97·5 m. B.: 75 ft. = 22·9 m. D.: 29 ft. = 8·8 m.

Guns: 4A (80 ton *m. l.* 16 in.: 40·6 c/m) + 8E* (4·7 in.) + 4* (6-pdrs. 57 m + 2* (3-pdrs.) + 15 machine. Torpedo tubes: 2 *submerged* + 2 *above water*.

Armour: Deck (*steel*) *d* (75 m/m = 3"). Citadel amidships (*iron*) *aa* (600 - 400 m/m = 24 - 16"). Turrets (*compound*) *aa* (450 m/m = 18"). Deck citadel 75 m/m (3"). Cork ends.

Machinery: Compound 3-cylinder. 2 screws. 12 cylindrical boilers. I.H. 1892, 6264 = 13·8 kts. *natural draught*. Sea speed: 10 kts. Coal: 1300 to *circa* 3000 miles at best speed.

(I.) CANOPUS (1897) & GOLIATH (1898) & OCEAN (1898) & GLORY (1898) & ALBION (1898) & VENGEANCE (1899).

12,950 tons. Complement 750.

Dimensions: L. (p.p.) × B. × D.: 390 × 74 × 26 ft. = 119 × 22·55 × 7·90 metres.
(Length over all 410 feet.)

Guns:
- 4A (12 in. Vickers).
- 12D* (6 in. 45 cal.).
- 10F* (3 in.).
- 6* (3 pdr., 47 m/m).
- 8 Maxims.

Torpedo tubes:
- 4 *submerged* (18 inch).

Armour (Harvey nickel): m/m.
- 6" Belt amidships . *b* (152).
- 2" Belt bow . . . *f* (50).
- 2½" Deck on slopes . = *c* (63).
- 12" Barbettes . . *aaa* (300).
- 12" Bulkheads . . *aaa* (300).
- 8" Hoods to A guns = *aaa* (200).
- 5" Casemates . . *b* (125).
- 12" Fore C.T. *aaa*,
- After C.T. . . . *e* (75).

(Albion alone has a 2 inches belt aft.)

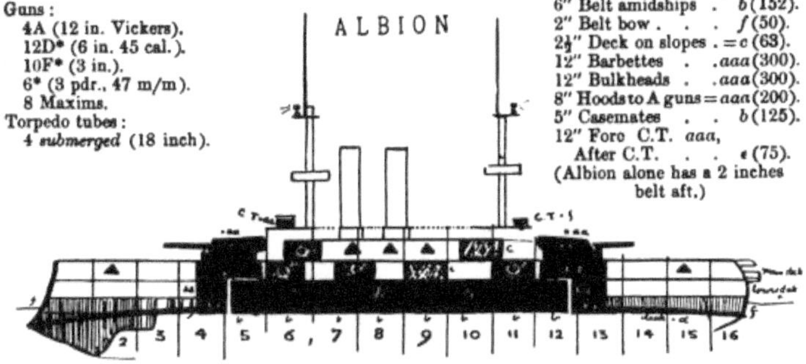

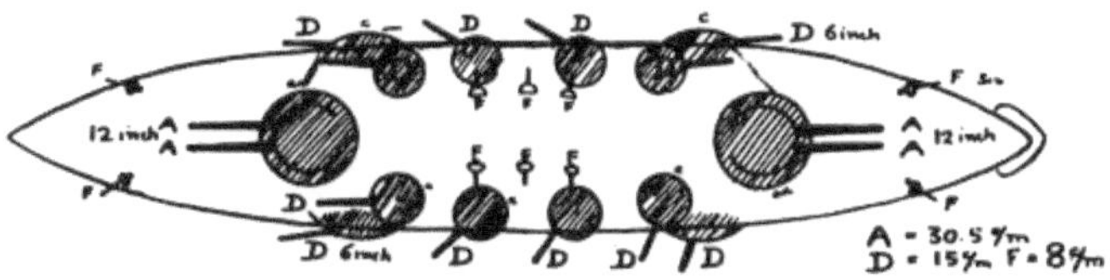

Machinery: Vertical triple expansion. 2 screws. Boilers, 20 Belleville. I.H.P., *nominal* 13,500, *natural draft* = 18·25 knots. Sea speed, 17 kts. Coal, 1000 tons, maximum capacity 2300 = *nominal* 8000 miles at 10 kts.

Vengeance has Krupp armour, hence Belt *a–c*, Deck = *b*, Prot. to vitals = *aaa*, Barbettes and bulkheads *aaaa*, Hoods = *aaaa*, Casemates *a*.

Notes.—Canopus and Albion big guns on central pivot mountings: they can be loaded at any elevation or direction. The others have their guns mounted on the same system as the Majestic class. All have central loading positions only. The ships are on an even keel.

The belt is 195 × 13 feet (59·43 × 3·96 metres). The ram is only 8 feet under water. 9 inch wood sheathing is carried aft of the belt. Bulkheads are at right angles to the belt. Barbettes, in these and all later ships, are circular. The main deck is of steel 1 inch (25 m/m) thick. The deadwood aft is cut away. All these ships are fitted for wireless telegraphy. Guns, mountings, ammunition, etc., weigh 1000 tons. The machinery weighs 1200 tons.

CANOPUS. *Photo by West.*

(Ocean has steam pipes both sides of each funnel. Glory = Canopus.)

Trials.	*30 hrs. 4/5 power.*	*8 hrs. full power.*
Canopus (1899)	I.H.P. 10,454 = 17·2 kts.	I.H.P. 13,763 = 18·5 kts.
Goliath (1899)	„ 10,413 = 17·3 „	„ 13,918 = 18·4 „
Ocean (1899)	„ 10,314 = 16·2 „	„ 13,728 = 18·5 „
Glory (1900)	„ = „	„ = „
Albion (1900)	„ = „	„ = „
Vengeance (19..-)	„ = „	„ = „

(I.) FORMIDABLE (1898) & IRRESISTIBLE (1898) & IMPLACABLE (1898).

15,000 tons. Complement 750. (Completing for sea April 1900.)
Dimensions: L. × B. × D.: 400 × 75 × 28½ feet = 122 × 22·85 × 8·68 metres.
[Length over all 430 feet.]

Guns:
4A (12 in. Mark IX.).
12D* (6 in. Vickers).
16F* (3 in.).
6* (3 pdr. 47 m/m).
8 Maxims.
Torpedo tubes:
4 *submerged* (18 in.).

Armour (Harvey nickel):
9" Belt *a* (225 m/m).
2" Belt at bow *f* (50 m/m).
3" Armour deck = *b* (76 m/m).
Prot. to vitals = *aaaaa*.
12" Bulkheads *aaa* (300 m/m).
9" Lower deck amidships *a* (225 m/m).
12" Barbettes *aaa* (300 m/m).
8" Hoods to big guns = *aaaa* (200 m/m).
6" Casemates *b* (152 m/m).
12" Fore C. T. *aaa*; after C. T. *d* (75 m/m).

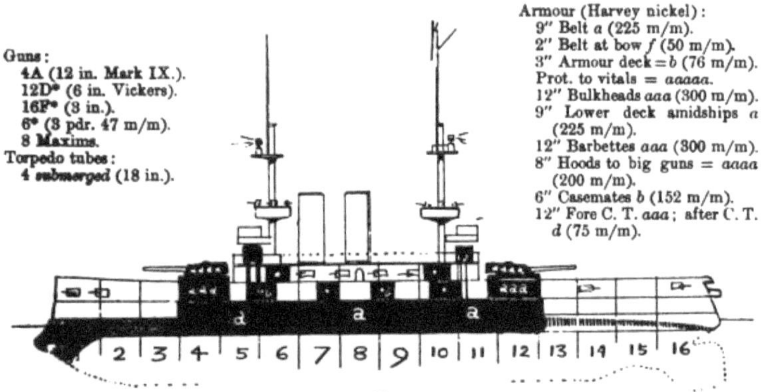

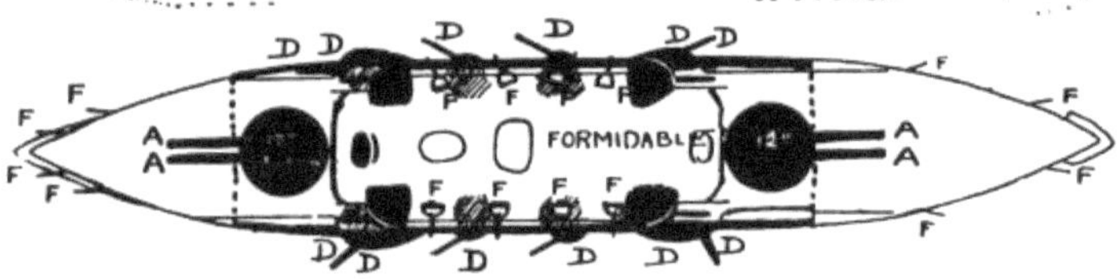

Machinery: 3 cylinder triple expansion. 2 screws. Boilers. 20 Belleville (with economisers). I.H.P. 15,000, *natural draught* = 18 kts. *estimated.* Sea speed, 16·5 kts. *est.* Coal: 900 tons. Maximum capacity 2200 tons = *nominal* 8000 miles at 10 kts.

Notes.—The big guns load at any elevation. They weigh 50 tons each. Belt amidships, 216 feet long by 15 feet deep (66 × 4·57 metres). This belt reaches to the main deck. It is continued to the bow, at a thickness of 2 in. (50 m/m), but is not carried down over the ram so much as in the Canopus class. The hoods to the big guns are inclined at a 40° angle. Main deck 1 in. steel (25 m/m). Casemates are 6 in. (152 m/m) in front, 2½ in. (65 m/m) at the backs.

There are 6 searchlights. Three launches fitted to work 14 in. torpedoes are carried.

The lower bunkers in this and the London type can be coaled independently of the upper. The boilers are able to stand a pressure of 300 lbs. to the square inch (21 kgs. per *centimetre car.*). The heating surface is 37,000 square feet (3437 *métres car.*).

The ships have the deadwood aft cut away.

I.) *LONDON* (1899) & *BULWARK* (1900) & *VENERABLE* (1900).

15,000 tons. Complement 750.

Dimensions: L. × B. × D.: 400 × 75 × 29 ft. = 122 × 22·85 × 8·37 metres.

(12 in. Mark IX.).
)* (6 in. Vickers).
* (3 in. Q.F.).
(3 pdr. 47 m/m).
laxims.
do tubes:
ibmerged (18 inch).

Armour (Krupp):
9" Belt amidships *aa* (225 m/m).
6"-2" Belt forward *a* to *c* (152-50 m/m).
3" Armour deck = *a* (76 m/m).
Prot. to vitals = *aaaaaa*.
12" Bulkheads (aft) *aaaa* (300m/m).
9" Lower deck amidships *aa* (225 m/m).
6"-2" Lower deck forward *a-c* (152-50 m/m).
12" Barbettes *aaaa*.
8" Hoods to big guns = *aaaaa* (200 m/m).
6" Casemates *a* (152 m/m).
12" Fore C.T. *aaaa*; after *d* (75 m/m).

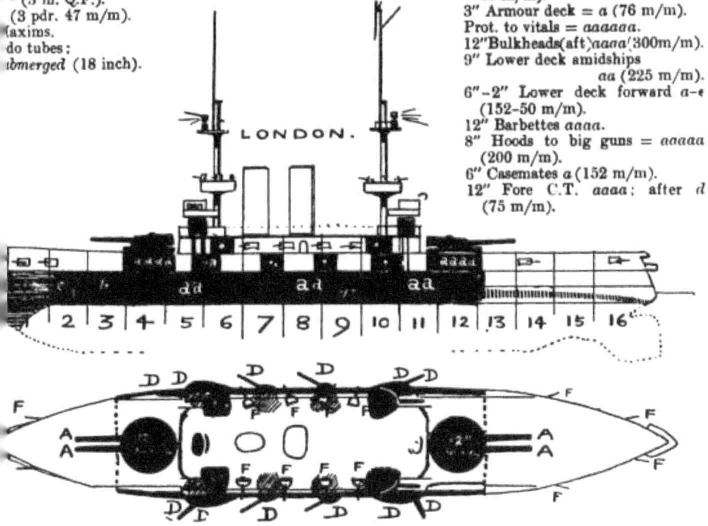

Machinery: 3 cylinder triple expansion. 2 screws. Boilers, 20 Belleville. I.H.P., *nominal* ,000, *natural draft* = 18 kts. (*estimated*). Coal: 900 tons. Maximum capacity 2200 tons = *minal* 8000 miles at 10 kts.

Note.—The belt tapers from the amidships 9 inches to 4 inches (*c*) when 30 feet from the w. The rest is 2 inches thick. All other armour, etc., is identical in thickness with the rmidable class, but being Krupp instead of Harvey nickel it has a higher resisting value. In is class the forward bulkhead is dispensed with—being disposed on the sides instead. *London* ilding at Portsmouth; *Bulwark*, Devonport; *Venerable*, Chatham.

(I.) *Duncan* & *Cornwallis* & *Exmouth* & *Russell* & *Albemarle* & *Montagu* & *two new ships to be laid down* 1900.

14,000 tons. Complement 750.

Dimensions: L. × B. × D.: 405 × 75½ × 26½ ft. = 123·50 × 23 × 8·07 metres.

Guns:
4A (12 in. Mark IX.).
12D* (6 in., 45 cal.).
12F* (3 in.).
6* (3 pdr., 47 m/m).
8 Maxims.
Torpedo tubes:
4 *submerged* (18 inch).

Armour (Krupp):
7" Belt amidships *a* (175 m/m).
5"-3" Belt forward *b-d* (125-75 m/m).
2½" Armour deck = *b* (63 m/m).
Prot. to vitals = *aaaaa*.
11" Bulkhead (aft) *aaa* (278 m/m).
7" Lower deck amidships *a* (175 m/m).
5"-3" Lower deck forward *b-d* (125-75 m/m).
11" Barbettes *aaa* (278 m/m).
6" Hoods to big guns = *aaaa* (152 m/m).
6" Casemates *a* (152 m/m).
12" Fore C.T. *aaaa*; after C.T. *d* (75 m/m).

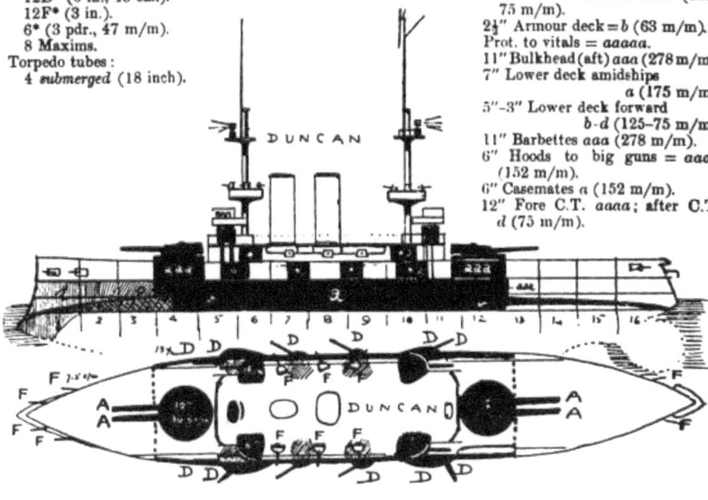

Machinery: 4 cylinder inverted triple expansion. 2 screws. Boilers, 24 Belleville (with economisers). I.H.P., *nominal* 18,000, *natural draft* = 19 kts. (*estimated*). Sea speed, 18 kts. (*est.*). Coal: 900 tons. Maximum capacity 2000 tons = *nominal* 7200 miles at 10 kts.

Note.—These ships are practically Londons, with less thick armour and slightly reduced armament. The armour belt tapers to the bow. The big guns load at any elevation, and are generally identical with those of the London and Formidable. The ships are of light draught to enable them to go through the Suez Canal. The deadwood aft is cut away, and they will be especially handy craft. A feature of these ships is the open upper deck, which leaves the 3 inch guns there less liable to be put out of action by the first shell.

The *Duncan* and *Cornwallis* are building at the Thames Ironworks; the *Russell*, at Hull; the *Exmouth*, at Laird's, of Birkenhead; *Albemarle*, Chatham; *Montagu*, Devonport.

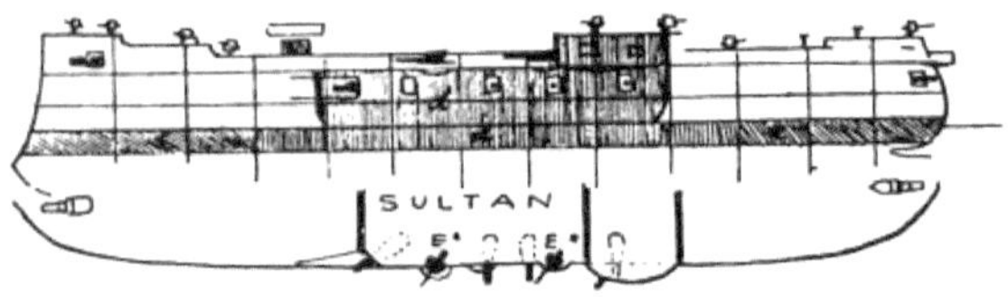

(IV.) SULTAN (1870) (1890).

9290 tons. Complement 661.

L : 325 ft. = 99 m. B.: 59 ft. = 18 m. D.: 29 ft. = 8·8 m.

Guns: 8D (10 in. *m. l.*) + 4D* (6 in. 40 cals.) + 4E (4 in.) + 9* (6-pdrs. 57 m/m) + 13* (3-pdrs. 47 m/m). Torpedo tubes: 4 *above water*.

Armour (iron): Belt *d* - *e* (225 - 150 m/m = 9 - 6"). Battery *d* (225 m/m). Bulkheads to battery *ef* (125 m/m).

Machinery (re-engined and re-boilered 1891–6) by Thomson. Vertical triple expansion. 1 screw. Boilers: 8 cylindrical. Coal: 810 tons = *circa* 1200 miles at 8 kts.

Trials (*circa* 1896): 8 *hours natural draught* = 6531 H.P. = 14·6 kts.; 30 *hours natural draught* = 4014 H.P. = 13·8 kts. (*mean*); 4 *hours forced draught* = 8244 H.P. = 15·3 kts. (*mean*).

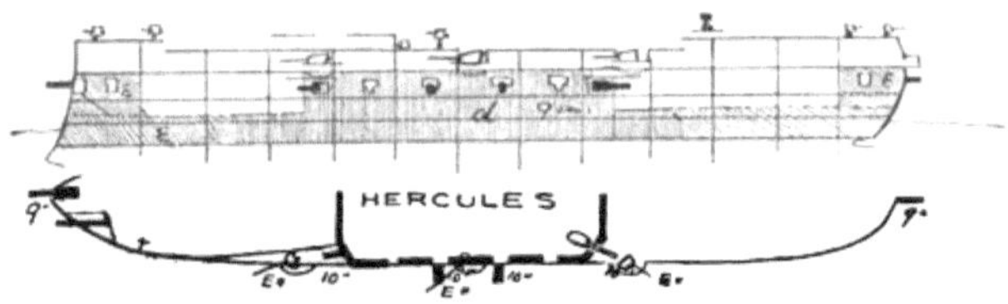

(IV.) HERCULES (1868) (1892).

8680 tons. Complement 638.

L.: 325 ft. = 99 m. B.: 59 ft. = 18 m. D.: 28 ft. = 8·5 m.

Guns: 8D (10 in. *muzzle loaders*) + 2 D (9 in. *m. l.*) + 2D* (6 in. 40 cals.) + 6E* (4·7 in.) + 9* (6 pdr. 57 m/m) + 13* (3 pdr. 47 m/m) + 7*. Torpedo tubes: 4 *above water*. Armour: = Sultan.

Machinery: (Re-engined and re-boilered 1891–2) by Penn. Vertical triple expansion. Boilers: ?. 1 screw. Coal: 610 tons = miles at 8 kts. I.H.P. *nominal* 7000 *natural* = 14 kts., *forced* 8500 = 15 kts.

Photo, Symon

AMPHION : ARETHUSA : LEANDER : PHAETON (1882–3).

circa 4300 tons. Complement, 309.

L. B. D : 300 × 46 × $22\frac{5}{8}$ ft. = 91·5 × 14 × 6·9 m.

Guns: 10D* (6 m., cals.) + 4* (3 pdr.). *Arethusa* 8*. Torpedo tubes: 4 *e water.*

Armour: partial steel deck *over machinery only*, $1\frac{1}{2}$" (40 m/m) = *f*.

Machinery: Horizontal compound. 2 screws. Cylindrical boilers, I.H.P. *nominal natural draught (circa 5500 was developed on trials). Forced draught Leander* . Coal, *normal* 725 tons; *max. capacity*, 1000 tons = *circa* 6000 miles at omical speed. Sea speed *circa* 17 kts.

(6.) IRIS (1877): MERCURY (1878).

3730 tons. Complement 290.

L.: 300 ft. = 91·5 m. B.: 46 ft. = 14 m. D.: 23 ft. = 7 m.

Guns: 13E (5 in. = 12·7 c/m) + 4*. Torpedo tubes, 4 *above water*. Armour: 0. speed: *circa* 15 kts.

Machinery: Maudslay. Horizontal tandem compound. 2 Screws. Boilers, cylin- al. Coal, 780 tons = *circa* 1500 miles *or less*, at 10 kts. I.H.P. *nominal*, 6000 *ral draught.*

Trials (1897–8). Iris, 17·9 kts.; Mercury, 18·9 kts. *forced draught*: Mercury 8), 1 *hour*, 5890 H.P. = 17 kts.; 4 *hours*, *max.* 6147 H.P. = mean speed kts.

(IV.) NORTHAMPTON (1878) : NELSON (1876).

7630 tons. Complement, 560.

L.: 280 ft. = 85·5 m. B.: 60 ft. = 18·2 m. D.: 29 ft. = 8·83 m.

Guns: 4D (10 in. *m.l.*) + 8D (9 in. *m.l.*) + 4E* + 20*.

our: Deck (steel) 3" = 75 m/m. Partial belt (iron), *d* (225 m/m 9"). heads, *d*. Bulkheads to battery and over 10 m/m guns, *d*.

Sea speeds: Nelson, *circa* 12 kts.; Northampton, *circa* 8 kts.

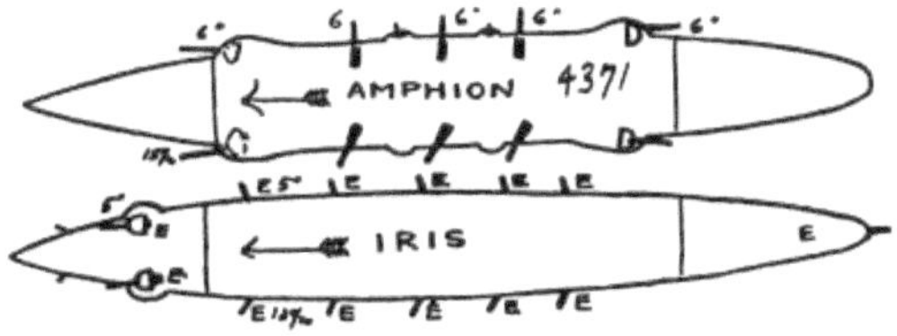

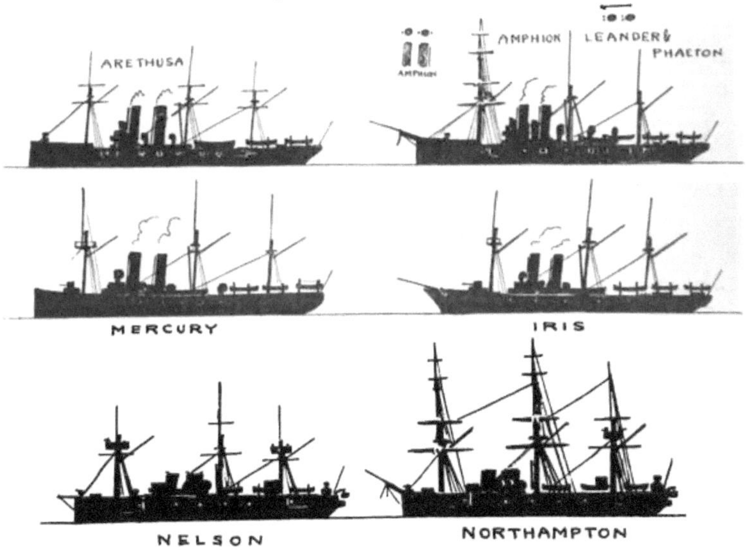

Alexandra, 1900 *Photo, Symonds*

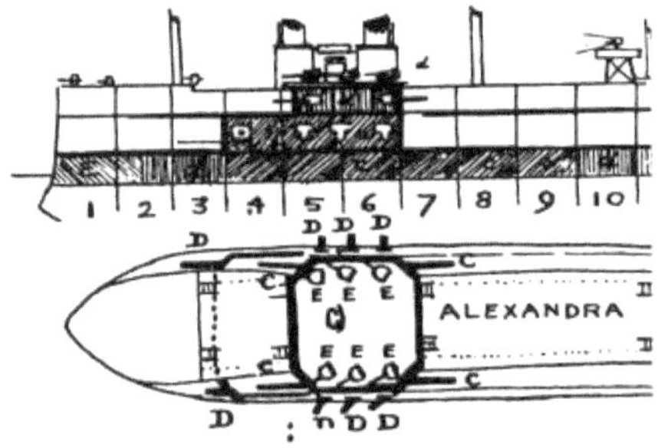

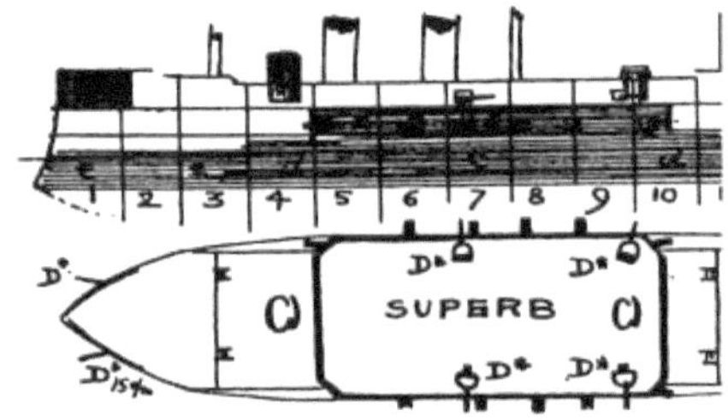

(IV.) ALEXANDRA (1875) (1890) (1898).

Circa 9500 tons. Complement 690

L.: 325 ft. = 99 m. B.: 63⅔ ft. = 19·4 m. D.: 29 ft. = 8·8 m.

Guns: 4C (9·2 in. 25 cals.) + 8D (10 in. *m. l.* in lower battery) + 6E* (4.7 in.) + 4* (6-pdrs.) + 6* (3-pdrs.). Torpedo tubes: 4.

Armour (*iron*): Belt (325 × 10½ ft. = 99 × 3·2 m.) *c - e* (300–150 m/m). D gun battery *c* C gun battery *d* (200 m/m = 8"). E gun shields (*Harvey*) *d* (110 m/m). Steel deck 2" (50 m/m).

Machinery (*re-boilered* 1890): Humphrys & Tennant 2 screws. I.H.P. *nominal* 7000 = 14·3 kts. Sea speed: 13 kts. *or more*. Coal: 680 tons = *circa* 1500 miles at 10 kts.

(IV.) SUPERB (1875) (1890) (1898).

Circa 9200 tons. Complement 654.

L. B. D.: 332 × 59 × 29 = 101·2 × 18 × 8·8 m.

Guns: 12D (10 in. *m. l.*) + 10D* (6 in. 35 cals.) + 6* (6-pdrs.) +
Torpedo tubes: 4.

Armour (*iron*): Belt (332 × 10½ ft. = 101·2 × 3·2 m.) *c - f* (300
Battery *d* (250 m/m = 10"). Bulkheads *d* (250 m/m). C. T. *d*
(50 m/m).

Machinery (*re-boilered and re-engined* 1890): Humphrys & Tenna
triple expansion. 1 screw. Boilers: 4 2-ended, 1 1-ended. I.H.P. *nomi*
13·5 kts. *natural d.*; 8500 = 14·5 kts. *forced d.* Coal: 970 tons = *circ*

(5.) BARHAM (1889) (1898) & *BELLONA* (1890) (1898-9).

1830 tons. Complement, 169.

L.: 280 ft. = 85·5 m. B.: 35 ft. = 10·6 m. D.: 16 ft. = 4·87 m.

Guns: 6E* (4 in. wire) + 4* (3 pdr.).

Armour: deck 2" 50 m/m on slopes.

Machinery: Vertical triple expansion. Thornycroft boilers (1897-8). Coal, tons.

Trials: Barham, 8 *hours natural draught*, 4490 = 18·6 kts.
4 *hours forced draught*, 5410 = 19·61 kts.
Bellona, 8 *hours natural draught*, 4280 = 18·8 kts.
4 *hours forced draught*, 6038 = 19·75 kts.

(Both alike.) *Photo, West.*

(T.) SPEEDY (1893).

810 tons. Complement, 91.

L.: 230 ft. = 70 m. B.: 27½ ft. = 8·2 m. D.: 13 ft. = 3·9 m.

Guns: 2E* (4·7 m.) + 4* (3 pdr.). Armour: 0. Sea speed: *circa* 17 kts.

Machinery: Thornycroft vertical triple expansion. 2 screws. 8 Thornycroft boilers. Coal, 100 tons = 5 days (*circa*).

Trials: 8 *hours natural draught*, 3046 = 18·5 kts.
4 *hours forced draught*, 4703 = 20·2 kts.

GLADIATOR. (All 4 are alike.) *Photo, West, Southsea.*

HIGHFLYER. *Photo, Symonds, Portsmouth.*

(2.) ARROGANT & FURIOUS & GLADIATOR (—1896) VINDICTIVE (1897).

"Ram-cruisers." 5750 tons. Complement 480.

L. B. D.: 320 × 57½ × 24 (?) feet = 97·50 × 17·50 × 7·30 metres.

Guns: 4D* (6 in.) + 6E* (4·7 in.) + 8F* (3 in.) + 3* (47 m/m, 3 pdrs.).
Torpedo tubes: 2 *submerged*.
3" Armour deck = c (75 m/m). Over ram and 40 feet (12·20 m.) of bow 2" arn
Gun shields d. 9" C. T. a (225 m/m).
Machinery: Vertical triple expansion. 2 screws. Boilers, 18 Belleville.
Coal: *normal* 500 tons: *maximum* 1200 tons.
Note.—These ships are fitted with double rudders, and have an exceedingly small turni

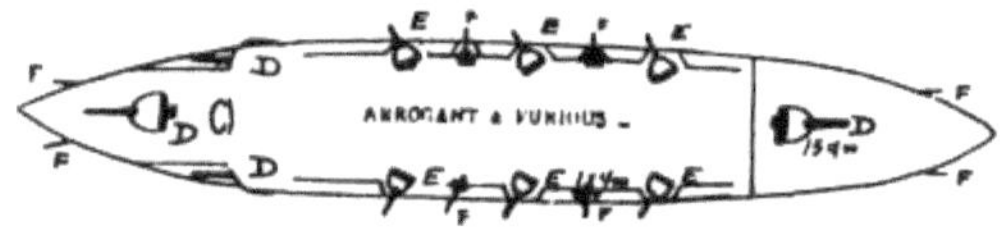

Trials.

Arrogant	8 *hours full power*	10.290 *H.P.*	= 19·6 kts.	30 *hours n. d.*	
Furious	8 ,, ,, ,,	10.272 ,,	= 20·1 ,,	30 ,, ,,	7155 *H.P.* =
Gladiator	8 ,, ,, ,,	10 018 ,,	= 19·1 ,,	30 ,, ,,	7149 ,, =
Vindictive	8 ,, ,, ,,	10,262 ,,	= 20·1 ,,	30 ,, ,,	7164 ,, =

(Gladiator had a foul bottom on trials.)

(2.) HERMES & HIGHFLYER & HYACINTH (—1898).

(Sheathed and coppered.) 5600 tons. Complement 450.

L. B. D. (mean): 350 × 54 × 22 feet = 106·70 × 16·50 × 6·70 m.

Guns: 11D* (6 in.) + 8F* (3 in.) + 6* (3 pdrs.). Torpedo tubes: 2 *submerged*.
3" Armour deck = b (75 m/m).
Machinery: 4 cylinder triple expansion. 2 screws. Boilers, 18 Belleville.
Coal: *normal* 500: *maximum* 1100 = *nominal* 9000 miles.

Trials.

Hermes	8 *hours full power*	10,224 *H.P.*	= 20·5 kts.	30 *hours ⅗ power*	7713 *H.P.* =
Highflyer	8 ,, ,, ,,	10,344 ,,	= 20·1 ,,	30 ,, ,,	7644 ,, =
Hyacinth	8 ,, ,, ,,	10,536 ,,	= 19·4 ,,	30 ,, ,,	7718 ,, =

(Hyacinth had a foul bottom on trials.)

Note.—These ships are improved Eclipses—a plan of which class is on page 96.

I.) *Drake* & *King Alfred* & *Leviathan* & *Good Hope* (*ex Africa*).

"Armoured cruisers." Building.

14,100 tons. Complement

ıs: L. × B. × *D.* (*mean*): 500 × 71 × 26 feet = 152·39 × 21·64 × 7·92 metres.

(The length given is between perpendiculars.)

ı.
).
).
).

Armour (Krupp):
6" Belt amidships *a* (152 m/m).
3" Belt bow *d* (75 m/m).
4" Deck on slopes = *aa* (100 m/m).
Protection to vitals = *aaaaa*.
8" Bulkheads *aa* (200 m/m).
6" Barbettes and bases *a* (152 m/m).
5" B gun hoods = *aaa* (125 m/m).
5" Casemates *b* (125 m/m).

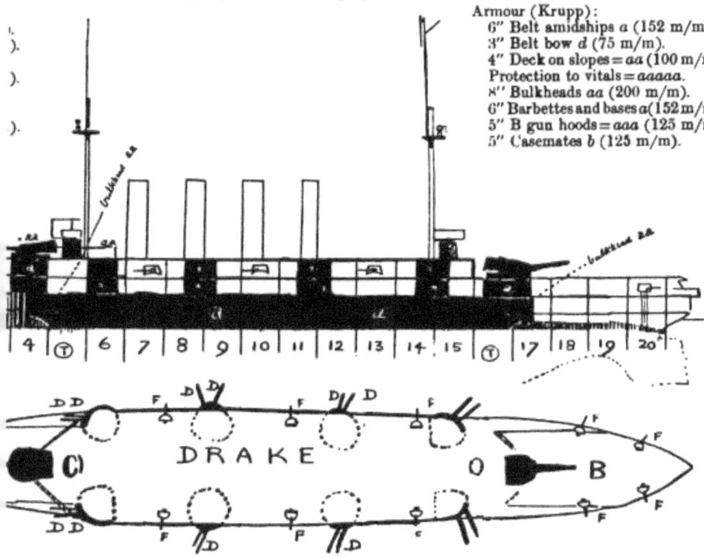

y: 2 screws. Boilers, 48 Belleville (with economisers). I.H.P. 30,000 = 23 kts. (*est.*)
;0 tons. Bunker capacity 2500 tons.
ıe Drake is building at Pembroke; King Alfred by Vickers; Leviathan at Clydebank; Good
ow (Fairfield Co.). This last ship was previously known as the Africa.
of 23 knots has to be made for 8 hours on contractors' trials, natural draught. A continuous
speed of 21 knots is expected. The machinery will weigh 2500 tons, and the boilers will
85 feet.

(III.) *Essex* & *Monmouth* & *Kent* & *Bedford.*

"Armoured cruisers." Building.

9800 tons. Complement .

Dimensions: L. × B. × *D.* (*mean*): 440 × 66 × 24½ feet = 134·10 × 20·11 × 7·46 metres.

Guns:
14D* (6 in. 45 cals.).
10F* (3 in.).
3* (3 pdrs.).
Maxims.
Torpedo tubes:
2 submerged (18 in.).

Armour (Krupp):
4" Belt *c* (100 m/m).
2" Belt at bow *e* (50 m/m).
Armour deck = ?
Protection to vitals = ?
4" Lower deck amidships *c* (100 m/m).
2" Lower deck bow *e* (50 m/m).
5" Bulkheads *b* (125 m/m).
5" Turrets and supports (Harvey Nickel) *c* (100 m/m).
4" Casemates *c* (100 m/m).

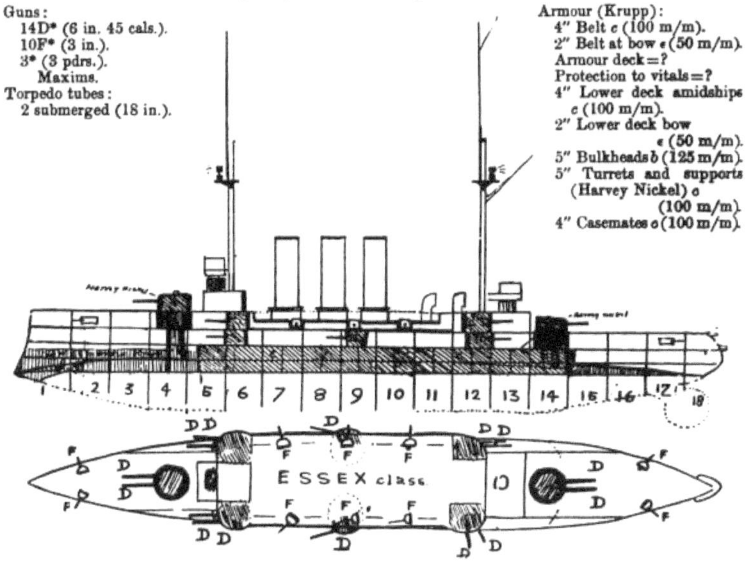

Machinery: 2 screws. Boilers, Belleville (with economisers). I.H.P. 22,000 = 23 kts. (*est.*)

Note.—The special feature of these ships is the system of mounting the turret guns, which first appears in the British Navy in these ships. The guns are in pairs, so mounted that while either can be used independently, both can be trained together and fired as one piece.

(II.) *CRESSY* (1900) & *ABOUKIR* (1900) & *SUTLEJ* (1899) & *HOGUE* (1899) & *EURYALUS* (1900) & *BACCHANTE* (1900).

("Armoured Cruisers" building.)

12,000 tons. Complement 700.

Dimensions: L. × B. × *D.* (*mean*): 440 × 69½ × 26¼ feet = 134 × 21·18 × 8 metres.

Guns:
- 2B (9·2 in. 40 cals.).
- 14D* (6 in. 45 cal. Q.F.).
- 12F* (3 in. Q.F.).

Torpedo tubes:
- 2 submerged.

Armour (mostly Krupp):
- 6" Belt *a* (152 m/m).
- 2" Belt at bow *e* (50 m/m).
- 3" Armour deck = *b* (76 m/m).
- Protection to vitals = *aaaa*.
- 5" Bulkheads *b* (125 m/m).
- 6" Lower deck amidships *a* (152 m/m).
- 2" Lower deck forward *e* (150 m/m).
- 6" Barbettes and hoists *a* (152 m/m).
- 6" Hoods to B guns = *aa* (152 m/m).
- 5" Casemates *b* (125 m/m).

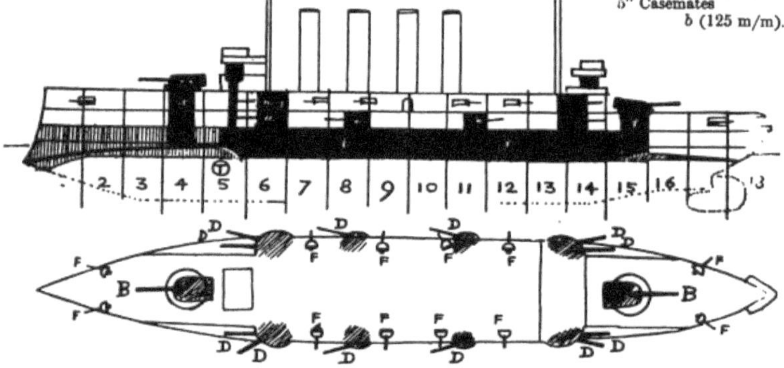

Machinery: 4 cylinder inverted triple expansion. 2 screws. Boilers, 30 Belleville, with economisers. I.H.P., 21,000 = 21 kts. (*estimated*). Coal: 800 tons. Maximum capacity 1600 tons.

Notes.—The armour for this ship weighs 1200 tons, of which about 800 tons This belt is 11½ feet (3·50 metres) deep, of which 6½ feet (1·97 m.) is designed water line. The barbettes, 6 in. (152 m/m), have supports of the same thicknes all the mechanism is contained. 130 feet are devoted to the 4 boiler compartmen pressure will be 300 lbs. to the square inch, reduced to 250 lbs. at the engines (2 le c/m²). Coal bunkers are arranged round the boilers' rooms, both above and be deck. 120 revolutions are to give the 21-knot speed on an eight-hour trial. Th weigh 1800 tons.

Where built: *Cressy* & *Aboukir*, Fairfield & Co., Glasgow; *Hogue* & *E*ı *Sutlej* & *Bacchante*, Clydebank (Thomson).

To be built: 6 armoured cruisers—probably Essex type.

(1.) DIADEM (1896) & ANDROMEDA (1897) & EUROPA (1897) & NIOBE (1897) & ARGONAUT (1898) & ARIADNE (1898) & SPARTIATE (1898) & AMPHITRITE (1898).

(Wood sheathed and coppered.)

11,000 tons. Complement 600.

Dimensions: L. × B. × D.: 435 × 69 × 27½ feet = 132·58 × 21·03 × 8·37 metres.

(The length over all is 462 feet = 140·80 metres.)

* (6 in. 45 cals.).
* (3 in.).
(3 pdr. 47 m/m).
axims.
do tubes:
bmerged (18 in.).

Armour (mostly Harvey nickel):
4" Deck = *b* (100).
Prot. to vitals = *a*.
4½" Casemates and shields *c* (114).
12" Fore C.T. *aaa*; after C.T. *c*.
6" Tube to fore conning tower *b* (152).
2" Hoists *c* (50).

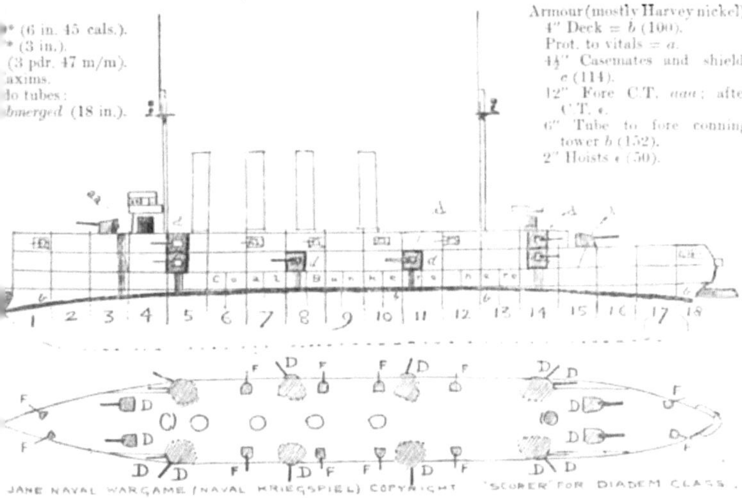

Machinery: 2 sets 4-cylinder triple expansion. 2 screws. Boilers, 30 Belleville. *Nominal* I.H.P. 16,500 *n. d.* in first four = 20·25 kts. Sea speed 19 kts. In last four I.H.P. *nominal* 18,000 *n. d.* = 20·75 kts. Coal, 1000 tons. Bunker capacity, 2000 tons = *circa* 2000 miles at 19 knots.

DIADEM. *Photo. West, Southsea.*

Trials.

	8 hours full power, 30 boilers *n.d.*	4 hours *forced d.* 24 boilers.	30 hours *n. d.* 30 boilers.
Diadem (1897)	I.H.P. 17,262 = 20·65 kts.	I.H.P. 15,861 = 19·9 kts.	I.H.P. 12,791 = 19·80 kts.
Andromeda (1898)	" 16,751 = 20·41 "	" = "	" 12,621 = 19·30 "
Europa (1898)	" 17,137 = 20·40 "	" 16,298 = 19·9 "	" 12,739 = 19·33 "
Niobe (1898)	" 16,834 = 20·59 "	" 16,834 = 20·5 "	" 12,961 = 19·27 "
Argonaut (1899)	" 18,894 = 21·17 "	" = "	" 13,815 = 19·86 "
Ariadne (1899)	" 19,156 = 21·50 "	" = "	" 14,046 = 20·10 "
Spartiate ( )	" = "	" = "	" = "
Amphitrite (1900)	" 18,229 = 20 78 "	" = "	" 13,635 = 19·50 "

Notes.—There are coal bunkers on the lower deck, and above and below the armour deck. The pair of 6 in. guns forward and aft are served by one hoist to the pair, each casemate has its own hoist: those to the double casemates being double. The machinery weighs 1630 tons: the boilers occupy a space of about 132 feet (40 metres). Working pressure is 300 − 250 lbs. (21 − 17·5 kgs.). The Diadem (Dec. 1898) did a mean of 19·27 kts. in a 69 hour run from Gibraltar to the Nore with 14,268 horse power. For 60 hours with 24 boilers with 12,791 I.H.P. she made 19·3 kts. on trial. Andromeda did a sea trip at 8000 I.H.P. = 16 kts.

Distinctions.—All the ships are almost absolutely identical in appearance. The Diadem and Andromeda have no steam pipes abaft the first 2 funnels, while all the others have steam pipes before and abaft each funnel.

ENGLAND.

(1.) POWERFUL & TERRIBLE (1895.)

14,200 tons. Complement, 894.

L. (*p.p.*) B. D.: 500 × 71 × 31 ft. = 152·3 × 21·6 × 9·45 m.

…s:
B (9·2 in., 40 cals.).
…2D* (6 in., 40 cals.).
…3F* (3 in., 12 pdrs.).
…2* (47 m/m, 3 pdrs.).
Maxims.

…pedo tubes:
submerged.

Armour (Harvey):
6" deck, on slopes = *aa* (152 m/m).
Protection to vitals, deck + coal = *aaa*.
6" Casemates and B gun shields *c* (152 m/m).
6" supports and hoists to B guns *c* (152 m/m).
Hoists to D guns, *c*.
Fore C.T., *aa*. After C.T., *d*.
Tube to fore C.T., *a*.

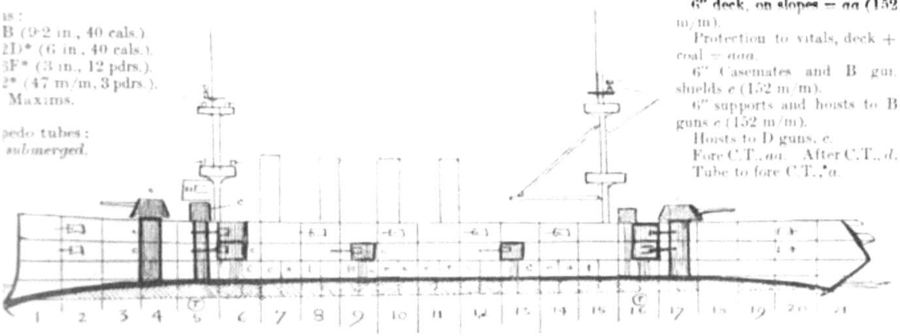

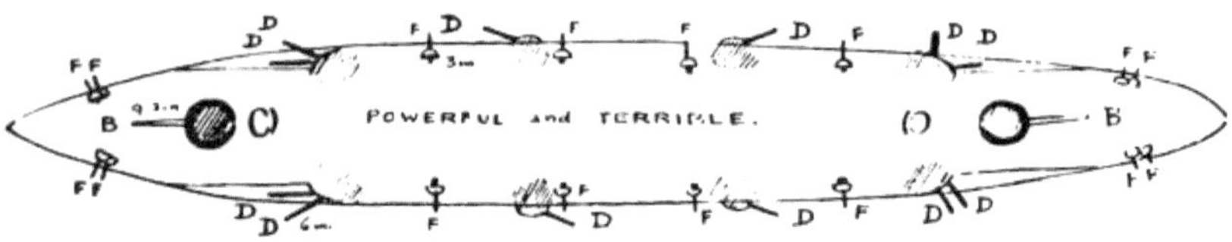

POWERFUL. Pho…

Machinery: 4 cylinder triple expansion engines; 48 Belleville boilers; 2 screws.
Speed trials: 30 hours natural draught. Powerful (1896): 18,433 I.H.P. = 20·6 kts. Terrible (1897): 18,493 H.P. = 20·96 kts.
4 hours forced draught. Powerful (1896): 25,866 I.H.P. = 21·8 kts. (in bad weather). Terrible (1897): 25,572 H.P. = 22·4 kts.
Terrible (1898): 60 *hours* with 20 boilers ($\frac{1}{5}$ power) 5084 = 12·8 kts. 68 *hours* at $\frac{2}{5}$ power 10,246 = 17 kts.
60 *hours* at $\frac{3}{5}$ power, 15,554 = 19·6 kts. 60 *hours* $\frac{3}{4}$-power, 18515 = 20·3 kts. 8 *hours* (22,000 I.H.P. *nominal.*) = 23,053 H.P. = 20·7 kts. 4 *hours full power*: 25512 H.P. = 21·9 kts.
Powerful (1898) ? *hours trial*, 22 kts. H.P. ?
Dec. 1898. Terrible *sea trip*, Portsmouth to Gibraltar, 12,500 H.P. = 18 kts. *average*. Gibraltar to Malta, *average* 21·5 kts.

ESPAGNE.

ARTILLERIE.

Nominal Calibre.			Longeur en Calibres.	Initial Velocite.		Initial Energie.		A.P. Shell. Obus de rupture.	
m/m	in.			met. secs. *circa.*	ft secs. *circa.*	foot tons. *circa*	metre tx. *circa.*	lbs.	kgs.
	12·5	Hontoria (1) .	35	620	2034	29000	8980	1036	470
	11	" .	35	628	2034	24000	7430	837	380
	10·0	Canet (2) .	45	800	2625	22750	7045	528	240
	9·4	Canet (3) .	45	800	2625	15000	4650	317·4	144
	9·4	Hontoria .	35	620	2034	12600	3900	438·6	199
	7·9	" (4) .	35	620	2034	6800	2105	253·3	115
	7	" .	35	620	2034	5400	1670	187·4	85
	6·4	...	35	618	2008	3700	1145	130	59
	5·5	converted .	35	620	2034	2500	775	86	39
	6·4	'79 model converted .	25	500	1640	1700	525	94	42·5
	4·7	converted .	35	606	1988	1500	465	53	24·1
	4·7	'79 model .	25	530	1740	?	?	?	?
·5	3	Canet . .	50	810	2658	650	200	13·2	6

a and Numancia retain some Armstrong muzzle loaders (see 'England' for details).
(old armament). (2) Carlos V. (3) Pelayo (new armament). (4) Alphonso XIII. and

to the ships illustrated the Spanish navy contains 2 wooden frigates, Navarra and Aragon, ing 6 old Krupp guns. 6 gunboats of 500 tons, 3 old guns; and about 80 small gunboats, for police duties, etc., all of no fighting value.

DESTROYERS.

TERROR (Clydebank 1896). L. B. D.: $= \frac{220 \times 22 \times 5\frac{1}{2} \text{ ft.}}{67 \times 6{\cdot}7 \times 1{\cdot}6 \text{ m.}}$ 300 tons. 6000 I.H.P.

Trial speed: 28 kts. 2 screws.

Guns: 2F* (14-pdrs.) + 2 6-pdrs.* + 2*. 2 torpedo tubes. *Boilers*: Normand.

AUDAZ & OSADA & PROSPERINE (Clydebank 1897). L. B. D.: $= \frac{226 \times 25\frac{1}{4} \times 5\frac{3}{4} \text{ ft.}}{69 \times 7{\cdot}7 \times 1{\cdot}7 \text{ m.}}$

400 tons. 7500 I.H.P. 2 screws. *Trial speed*: 30 kts.

Guns: 2F* (14-pdrs.) + 2 6-pdrs.* (57 m/m) + 2*. 2 torpedo tubes. *Boilers*: Normand.

(Furor 300 tons and Pluton 400 tons sunk in war 1898.)

TORPEDO BOATS.

SECOND CLASS.

2 Yarrow . . .	(1887),	108 tons,	134½ ft.	(41 m.) long.	*Max. trial* speed:	24 kts.	3 torpedo tubes.
2 Thornycroft (Ariéte and Rayo)	(1887),	97 ,,	147½ ,,	(45 m.) ,,	,, ,,	26 ,,	2 ,,
2 Normand . . .	(1887),	66 ,,	126 ,,	(38·5 m.) ,,	,, ,,	19·5 ,,	2 ,,
2 Thornycroft . .	(1885),	65 ,,	118 ,,	(36 m.) ,,	,, ,,	20 ,,	3 ,,
1 Schichau . . .	(1887),	60 ,,	111 ,,	(34 m.) ,,	,, ,,	25 ,,	2 ,,
1 Yarrow . . .	(1886),	70 ,,	118 ,,	(36 m.) ,,	,, ,,	20·5 ,,	2 ,,
3 Various . . .	(1885–3),	85 ,,	? ,,	,,	,, ,,	19–14 ,,	? ,,
13							

THIRD CLASS.

3 boats (1883–78), 25 tons, and 3 spar torpedo boats. 1 submarine boat (Peral), 87 tons, 70 ft. (21·5 m.) long. Speed: 10 kts.

ESPAGNE.

(T.) NUEVA ESPAÑA (1889) & MARTIN ALONZO PINZON (1889) & YANEZ PINZON (1891).

570 tons. Complement 100.

L.: 190 × 23 × 13 ft. = 58 × 7 × 4 m.

Guns: 2E + 4*. **Armour**: 0. **Sea speed**: *circa* 12 kts.
2 Torpedo tubes. Coal: 105 tons.

(5.) MARQUES DE LA ENSEÑADA (1890).

1040 tons. Complement 164.

L.: 185 × 30 × 15 ft. = 56·5 × 9 × 4·5 m.

Guns: 4E + 2*. **Armour**: Deck *ε*. **Sea speed**: *circa* 12 kts.
3 Torpedo tubes. Coal: 100 tons.

(T.) FILIPINAS (1892).

747 tons. Complement 110.

L.: 233 ft. = 71 m.

Guns: 2E* + 4*. **Armour**: 0. **Sea speed**: 15 kts.
4 Torpedo tubes (1 *submerged ?*) Coal: 100 tons.

(6.) INFANTA ISABEL (1885) & ISABEL II. (1886) & CONDE DE VENADITO (1888).

1152 tons. Complement, 150.

L.: 210 ft. = 64 m.

Guns: 4E + 4*. **Armour**: 0. **Sea speed**: *circa* 8 kts.
2 torpedo tubes. 1 screw.

(IV.) NUMANCIA (1863) (transformé, La Seyne, 1898).

7035 tons. Complement, 600.

L.: 314 ft. = 95·5 m.

Guns: 4D* (16 c/m (?)) + 11E* (12 c/m Canet) + 8*. 2 torpedo tubes.
Armour (*iron*): Belt and lower deck *ℓ* (120 m/m = 4¾".)
Machinery: 1 screw.

ESPAGNE.

(T.) MARIA DE MOLINA (1896).

L. : 235 ft. = 71·5 m.

Guns : 2E* + 4*. Armour : 0. 830 ns. Complement 90. 3 Torpedo tubes of them submerged). I.H.P. 3500 = ? s. Coal, 120 tons. 2 screws.

(T.) GALICIA (1891) & MARQUES DE MOLINS (1891) : TEMERARIO (1889).

L. : 190 ft. = 58 m

Guns : 2E + 4*. Armour : 0. Sea speed : *circa* 13·5 kts. 570 tons. Complement 90. I.H.P. 2600 = 20 kts. *Trial* : Boilers, 2 locomotive : 2 cylindrical. 2 Torpedo tubes. 2 screws.

(T.) DESTRUCTOR (1886 Clydebank).

L. : 200 ft. = 61 m.

Guns : 1E + 4*. Armour : 0. 386 tons. Complement 110. I.H.P. 3800 = 20 kts. *Trial* : Boilers (1897-8) 4 Normand. 2 screws. Coal, 45 tons.

(IV.) CARDENAL CISNEROS (1896) & PRINCESA DE ASTURIAS (1896) & CATALUÑA (1898.)

700 tons. Complement 500.

L.: 348 ft. = 106 m. B. 60 ft. = 18·3 m. D. 25 ft. = 7·6 m.

Guns : 2B (27 c/m) + 10D* + 7·3-pdr.* + 8*. Sea speed (?) 17 kts.

Armour : Deck = *ed*. (50 m/m = 2"). Belt *a* – *b*. (400 = 250 m/m) Glacis to cylinders = *aaa*. Barbettes *a*, C. T. *a*. Hoists, c.

Machinery : I.H.P. 15,000 = 20 kts. (*trial*). 8 Torpedo tubes (2 of them submerged ?) 2 screws. Coal, 1000 tons.

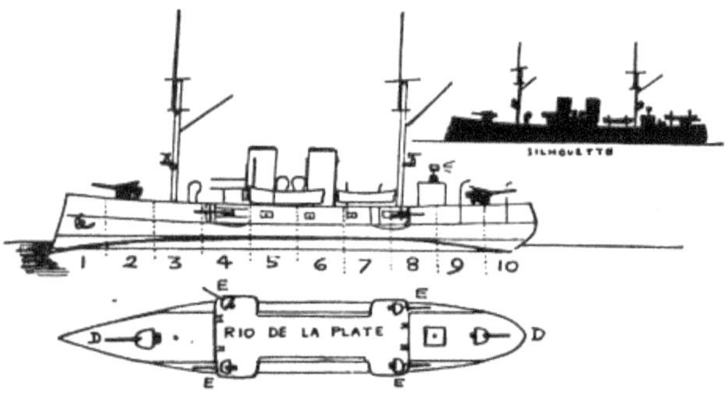

(4.) RIO DE LA PLATE (Havre 1898).

1775 tons. Complement .

L. B. D. : 246 × 35½ × 16 feet = 75 × 10·70 × 4·90 m.

Guns: 2D* (14 c/m Canet) + 4E* (10 c/m) + 4* (57 m/m) + 6*. Torpe
Armour deck 20 m/m (·8 inch) = *f*.

Machinery: 2 sets triple expansion. Boilers, Normand-Sigaudy. 2 screws. = 18 kts. (*trial*).

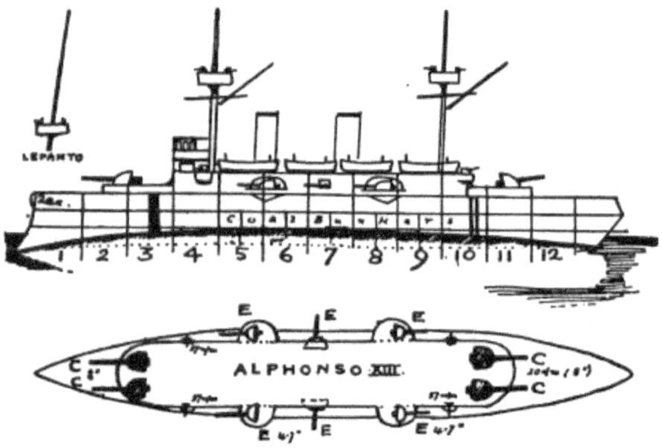

(2.) ALPHONSO XIII. (1891): LEPANTO (1892).

Circa 5000 tons. Complement 400.

L. B. D.: 318 × 50½ × 24 feet = 97 × 15·40 × 7·31 metres.

Guns: 4C (20 c/m) + 6E (12 c/m) + 16*. 5 torpedo tubes.

4¾" Armour-deck = *c*. Gun shields *e*.

Machinery: I.H.P. 11,000 = 20 kts. (2 hours trial). 2 screws. Sea speed, *circa*

Note.—These 2 ships are not very seaworthy.

(2.) *Regina Regente* (building).

5372 tons. Complement .

L. B. D.: 338 × 52½ × 19¾ feet = 103 × 16 × 6 metres.

Guns: 2C (20 c/m) + 8D* (15 c/m) + . 3" Armour-deck = *c* (76 m/m).

Machinery: I.H.P. (?) 6500 = 20 kts. (*est.*).

ESPAGNE.

(IV.) VITORIA (1865 Blackwall) (1897 La Seyne).

7000 tons. Complement, 560.

L.: 295 ft. = 90 m. B.: 57 ft. = 17·4 m. D.: 28 ft. = 8·5 m.

Guns: 5D* (16 c/m) (?) + 8E* (14 c/m) + 6E* (12 c/m) + 12*. 2 torpedo tubes.

Armour (iron): Belt and battery *f*. (120 m/m = 4·7″). Casemate above battery *f*.

Machinery: 1 screw. Coal, ? .

(II.) PELAYO (1887 La Seyne) (1897 La Seyne).

9950 tons. Complement, 600.

L.: 330 ft. = 100·5 m. B.: 66 ft. = 20 m. D.: 28 ft. = 8·5 m.

Guns (1898): 4B (24 c/m Canet) + 9D* (15 c/m Canet) (8 of these in battery) + 20*. Torpedo tubes, ? .

Armour (*acier* Creusot): Belt *aa* - *a* (450–300 m/m). Barbettes *aa*. Shields to big guns ? . Hoists and supports *e*. Battery *c*. Deck below belt 90 m/m (3·9″).

Machinery: Engines compound. Boilers, 16 Niclausse (in 4 groups). I.H.P. *natural draught* 8000 = *mean* 16·1 kts. (10 *hours trial* with 14 boilers). 2 screws Coal, 500 tons., *max.* 800 tons.

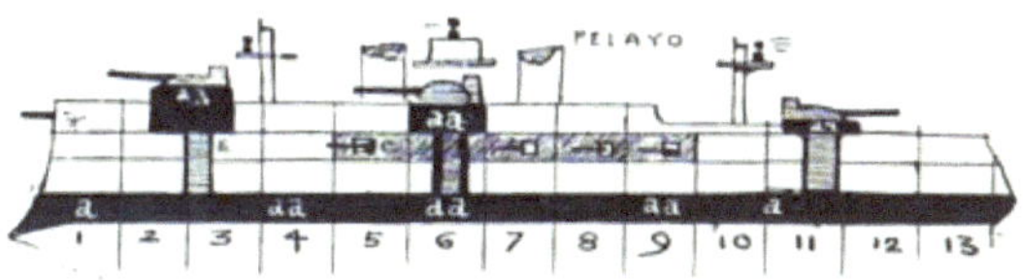

(6.) ALFONSO XII. (1887).

3090 tons.

L.: 280 ft. = 85·5 m.

Guns: 6D + 2Z + 10*. **Armour**: 0. **Sea speed**: 9 kts. I.H.P. 4800 = *trial* 17·5 kts.

Sister to Reina Christina, sunk at Manila and R. Mercedes, sunk at Santiago.

(1.) EMPERADOR CARLOS V. (1895).

9200 tons. Complement 600.

L. B. D.: 403 × 67 × 28½ ft. = 123 × 20·4 × 8·7 m.

Guns: 2A (28 c/m Canet) + 10D* (14 c/m Canet) + 4E* (12 c/m) + 2F* (75 m.m.) + 4 6 prs.* (57 m/m). 6 Torpedo tubes (2 of them submerged).

Armour: *acier special durci*: Deck = *a* (165 m/m = 6½"). Cylinder glacis = *a*. Redoubt (upper and main deck, side and bulkheads) *f*. (50 m/m = 2"). Barbettes *a* (250 m/m = 10"). Shields to A guns *de* (100 m/m = 4"). Hoists to A guns *b* (200 m/m = 8"). C.T. *aa* (300 m/m). Thick coal prot. to water line Shields to E guns *f*.

Machinery: 4 cylinder triple expansion. Maudslay. 2 Screws. Boilers cylindrical. I.H.P. (*forced draught*) 15,000 = 20 kts. (? *hours trial*). Coal: 1200 tons = *circa* 5000 miles at *circa* 12 kts.

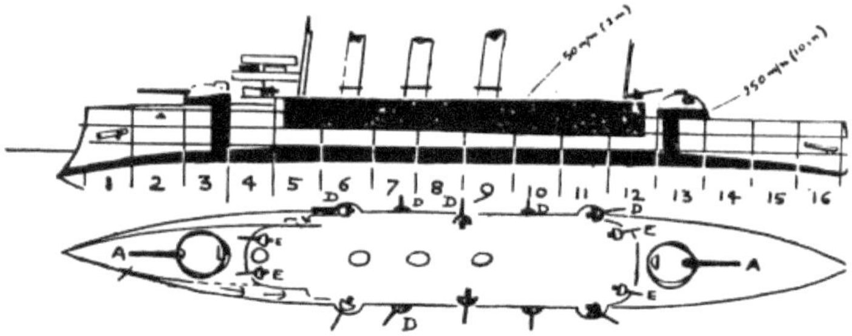

1898.

FRANCE.

ARTILLERIE.

n.	Designation.		Modèle.	Longeur d'arne.	Vitesse Initiale.		Approximate Energie Initiale.		Projectile: Obus de Rupture.	
	in.	m/m		cals.	ft. secs.	met. secs.	ft. tons.	metre tx.	lbs.	kgs.
	13·4	340	1893 & 1887	35 40	2625	800	44000	13600	926	420
	13·4	340	1884 & 1881	30 28·5	2034	620	24900	7710	926	420
	14·5	370	1875–9	28	1968	600	33000	10220	1235	560
	12	305	1893–1	40 45	2625	800	30750	9525	644	292
	10·8	274·4	1893–1 & 1887	45	2625	800	22750	7045	476	216
	13·4	340	1881	21	1820	555	20880	6467	926	420
	9·4	240	1893	40	2625	800	15000	4650	317·4	144
	16·5	420	1875	19	1662	507	17000	5265	1720	780
	16·5	420	1875	22	1738	530	17750	5498	1720	780
	13·4	340	1875	18	1535	486	18000	5575	926	420
	12·6	320	1870–81	25	1804	550	17000	5265	760	345
	7·6	194	1893 1887	40 45	2625	800	7900	2450	165	75
	9·4	240	1884 & 1881	30 28	1968	600	8540	2640	317	141
	10·8	274	1884 & 1880	28·5 25	1968	600	14000	4335	476	216
	10·8	274	1875 & 1870 (*transformé*)	20	1640	500	8865	2745	476	216
	10·8	274	1870	18	1424	434	6700	2075	476	216

Notation.	Designation.		Modèle.	Longeur d'arne.	Vitesse Initiale.		Approximate Energie Initiale.		Projectile: Obus de Rupture.	
	in.	m/m		cals.	ft. secs.	met. secs.	ft. tons.	metre tx.	lbs.	kgs.
D	7·6	194	1870	20	1470	448	2477	767	136·7	62
D	9·45	240	1870	19	1444	440	4590	1400	317·5	144
D*	6·4	164·7	1893–87	45	2625	800	4730	1465	99·2	45
D*	5·4	138·6	...	45	2625	800	3165	980	66	30
D & D*	6·4	164·7	1884	30	2100	640	3060	948	99·2	45
D*	6·4	164·7	1881 (*transformé* *)	28	1968	600	2670	827	92·5	42
DE*	6·4	164·7	,,	28	1821	555	2080	643	70·5	32
DE*	5·4	138·6	...	30	2100	640	2000	619	66	30
E*	5·4	138·6	1884–1 (*transformé* *)	30	1936	590	1700	525	66	30
E*	3·9	100	1892 & 1891	60 50	2625	800	1475	457	30·8	14
E*	3·9	100	modèle 1881–4 (*transforme*)	26	2034	620	885	274	30·8	14
E	5·4	138·6	1875–70	21	1528	466	...	...	...	...
F	2·9	65	...	...	...	...	...	...	...	...

There are several other models differing a very little from the above. All except the oldest models use. or can use smokeless powder. B (320 m/m) is a coast gun mounted in Dévastation only.

In addition to the ships enumerated in the lists following, the French navy also contains: of no fighting value 7 wooden ships *circa* 800 tons, 11 of *circa* 1600 tons, *circa* 26 gunboats 500–200 tons, E guns, and about 30 miscellaneous vessels.

FRANCE.

DESTROYERS.

(2 *funnels.*)

2 Durandal & Hallebarde (1897–8). 309 tons. I.H.P. 4800 = 26 kts.

(4 *funnels.*)

+ *2* Espignole & Fauconneau (1900). 310 tons. I.H.P. 4800 = 26 kts.
4 Espée, Framée, Pique & Yatagan (1899–1900). I.H.P. 5700 = 27 kts.
+ 4 Escopette, Flamberge, Pertuisane & Rapière (bldg.). I.H.P. 5700 = 26 kts.

6 + *6 building + others projected.*

(All the above carry 2 tubes and guns: 1F* (65 m/m) + 6* (47 m/m). All are 2 screws. Except Durandal & Hallebarde (which have 2) all have 4 Normand boilers).

TORPILLEURS.

First Class (*De haute mer*). 2 Screws.

1	(Forban) N.	(1895)	of 130 tons	144 ft. long,	Speed *max.* 30 kts.	2 Torpedo tubes
3 + *5*	(Cyclone *class*)	(1898–9)	of 150 ,,	147 ,,	,, 30 ,,	2 ,, ,,
1	(Mangini)	(1896)	of 129 ,,	147 ,,	,, 27·5 ,,	2 ,, ,,
1	(Chevalier)	(1893)	of 110 ,,	144 ,,	,, 27·2 ,,	2 ,, ,,
1	(Corsaire)	(1892)	of 150 ,,	160 ,,	,, 25·2 ,,	2 ,, ,,
5	(Dragon *class*)	(1892–3)	of 119 ,,	138 ,,	,, 25–21·3 ,,	2 ,, ,,
1	(Mousquetaire)	(1893)	of 125 ,,	154 ,,	,, 24·7 ,,	2 ,, ,,
7 + *1*	(Argonaute *class*)	(1893–9)	of 120 ,,	141 ,,	,, 25–23 ,,	2 ,, ,,
1	(Lansquenet)	(1893)	of 138 ,,	165 ,,	,, 24 ,,	2 ,, ,,
4	(Agile *class*)	(1889–93)	of 104 ,,	139 ,,	,, 20 ,,	2 ,, ,,
2	(Veloce *class*)	(1891–2)	of 114 ,,	147 ,,	,, 24 ,,	2 ,, ,,
1	(Coureur) T.	(1888)	of 120 ,,	147 ,,	,, 23 ,,	2 ,, ,,
3	(Eclair *class*)	(1891)	of 106 ,,	144 ,,	,, 21·5 ,,	2 ,, ,,
5	(Ouragan *class*)	(1887–8)	of 148 ,,	151 ,,	,, 21–20 ,,	4 ,, ,,
+ *2*	(armoured)	(bldg.)	of 180 ,,	,,	,, 30 ,,	

36 + *8*

Second Class. (2 Screws except 9 Balny *class.*)

15 + *29*	(201–234)	(1897–9)	85 tons,	121½ ft. long,	Speed *max.*
6	(195–200)	(1895)	80 ,,	121 ,,	,,
51	{(194–152, 149–126)} N. etc.	(1895–1889)	80 ,,	118 ,,	,,
1	(151)	( )	80 ,,	132 ,,	,,
9	(Balny *class*) N. etc.	(1886–8)	66 ,,	134 ,,	,,

82 + *29* + *11 pro.*

N = Normand boat. T = Thornycroft.

From a photograph.] Torpilleur de haute mer français en marche.

FRANCE.

TORPILLEURS.

Third Class.

14	N. boats	(1890–88)	50 tons,	111 ft. long,	Speed *max.*	21 kts.	2 tubes.
63	Boats	(1892–78)	45 ,,	108 ,,	,,	20–19 ,,	2 ,,
36	Various	(1877–82)	30–35 ,,	85–80 ,,	,,	19–16 ,,	Spar.
7	Various	(1892–76)	15–11 ,,	67–60 ,,	,,	20–16 ,,	1 tube.
120							

Submarine.

Gustave Zédé	(1893)	266 tons,	131 ft. (40 m.) long,	Speed *max.*	14 kts.	1 tube.
Gymnote	(1888)	30 ,,	59 ft. (18 m.) ,,	,,	6·4 ,,	0 tubes.
Narhwal	(1898)	200 ,,	—	,,	12·8 ,,	2 ,,
Morse	(1898)	146 ,,	118 ft. (36 m.) ,,	,,	13 ,,	1 tube.

projected 1898.

Algerien, Francais, Korrigan, Lutin, Gnome, Farfadet (Zédé type building).

SUBVENTIONED MERCHANTMEN.

1 (Navarre). 8600 tons. 17 kts.
1 (Touraine). 8800 tons. 20 kts.
3 (Gascogne, Bretagne and Champagne). 7000 tons. 17·5 kts.
1 (Normandie). 6283 tons. 16 kts.
+ 2 *building* (*Alsace & Lorraine*). 13600 tons. 21 kts.

Le Compagnie Générale Transatlantique.
Guns: 7E (138 m/m).

4 (Armand Béhic, Australien, Polynesien & Ville de la Ciôtat). 6450 tons. 17·5 kts. (Belleville boilers).
1 (Ernest Simons). 4562 tons. 17 kts.
2 (Brésil & La Plata). 5800 tons. 16·5 kts.

Les Messageries Maritimes.
Guns: 5E (138 m/m).

13 + 2

(V.) TONNANT (1880) (1897).

Circa 5500 tons. Complement 171.

L. B. D.: 249 × 58¾ × 18 ft. = 75·8 × 17·8 × 5·5 m.

Guns: 2B (340 m/m) + 6* (47 m/m) + 4 *revolvers.* Torpedo tubes: 2.

Armour (*comp.*): Belt *aa* - *a* (450 - 350 m/m = 18 - 14"). Barbettes *a* (340 m/m). Deck 75 m/m (3").

Machinery: Horizontal compound. 1 screw. Boilers: 4 cylindrical. Coal: 200 tons.

1st trial speed: I.H.P. 1914 = 11·6 kts. (*circa* 1883).

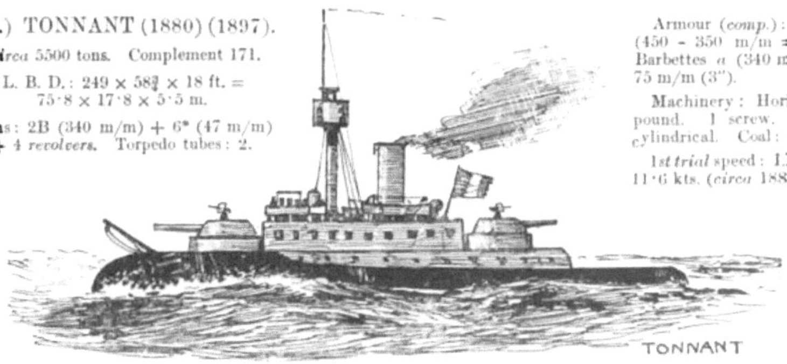

(V.) ONONDAGA (1863) (1897–8).

2600 tons. Complement 102.

L. B. D.: 228 × 51 × 13 ft. = 75·5 × 15·5 × 4 m.

Guns: 2C (194 m/m) + 4*. Armour (*en fer*): Belt = *f*. Turrets = *d*.

Speed: 6 kts. Coal: 200 tons. 2 screws.

(Torpedo boat headquarters S. Malo).

(V.)

ACHERON (1883) & COCYTE (1885).

1640 tons. Complement 112. L. B. D.: 181 × 40 × 11¾ ft. = 55·2 × 12·

& PHLEGETON (1890) & STYX (1891).

1790 tons. Complement 108. L. B D.: 192½ × 40 × 12 ft. = 58·8 × 12·

Guns:
- Acheron: 1C (274 m/m) + 3F* + 2*.
- Cocyte: 1C (274 m/m) + 2F* + 2*.
- Phlegeton: } 1C (274 m/m) + 1DE* (138 m/m) + 9*.
- Styx:

Armour
Belt *b*.
circa 8

Machinery: Horizontal comp. Boilers, 4 cylindrical. 2 screws. Co and Cocyte, 100 tons. Phlegeton and Styx, 72 tons. Trial speeds, Achero 11·6 kts. Cocyte, 1490 = 11·3 kts. Phlegeton and Styx, 1475 = 11·3 k

(V.)

FLAMME (1884) & GRENADE (1891).

1050 tons. Complement 84.

L. B. D.: 164 × 32¾ × 10½ ft. = 50 × 9·9 × 3·2 m.

Guns: 1C (240 m/m) + 1F + 4*. Sea speed: *circa* 8 kts.

Armour (comp.): Belt, *b* (240 m/m, 9·4"). Barbette, *c* (200 Deck, 37 m/m (1·9").

Machinery: Horizontal compound. Boilers, 3 cylindrical. 2 screws. tons.

Trials: Flamme, 1671 H.P. = 12·34 kts. Grenade, 1689 H.P. = 12

(No torpedo tubes to any of these.)

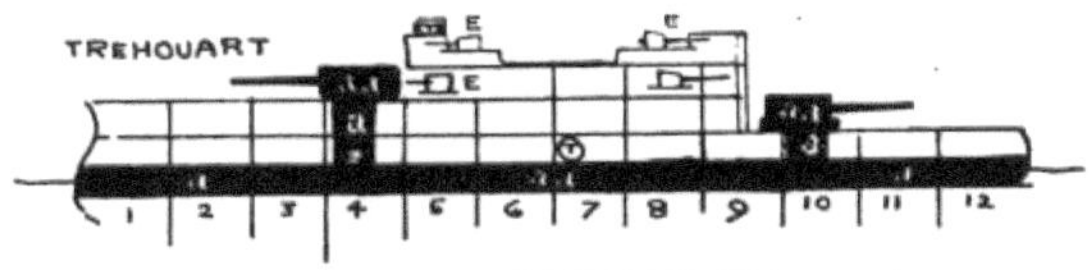

(III.) AMIRAL TRÉHOUART (1893).

6610 tons. Équipage, 329.

L. B. D.: 292 × 59 × 24¼ ft. = 89 × 18 × 7·4 m.

Guns: 2A (305 m/m 45 cals.) + 8E* (100 m/m) + 10* (37 m/m) + 8 *revolvers*. 2 torpedo tubes.

Armour (*acier Creusot*): Belt *aa* - *b* (460 - 250 m/m = 18 - 10"). Deck 10 m/m (:4"). Turrets *aa* (450 m/m). Base of after turret *aa*. Hoist to fore turret *aa*. C. T. *aa*.

Machinery: Inclined triple expansion. 2 screws. Boilers, 16 Belleville. Coal, 300 tons. Speed *trials* 1896, 24 *hours natural draught*, 5220 = 14 kts.; ? *hours*, 6760 = 15 kts. *forced*; 3 hours, 8270 *H.P.* = 15·8 kts. (*mean*).

(V.) TEMPÊTE (1876) & VENGEUR (1878).

4896 tons. 4700 tons.
Équipage, 176.

(V.) TONNERRE (1875) & FULMINANT (1877).

5589 tons. 5621 tons.
Équipage 248.

L.: 249 ft. = 76 m. B.: 58 ft. = 17·6 m. D.: Tempête, 19 ft. (5·8 m.): Vengeur, 18 ft. (5·5 m.); Tonnerre & Fulminant, 22 ft. (6·7 m.)

Guns: Tempête, Tonnerre, & Fulminant, 2C (274 m/m) + 6* (47 m/m) + 4*. Vengeur, 2B (34 c/m) + 6* (47 m/m) + 4 *revolutions*. Torpedo tubes, 2.

Armour (en fer): Belt *c* - *d* (350 - 250 m/m = 12 - 10"). Turret *c*. Redoubt *c*. Deck 50 m/m (2").

Machinery: 1 screw. Boilers cylindrical. Coal, Tempête & Vengeur, 200 tons. Tonnerre & Fulminant, 400 tons. Speed, 1*st trials*, Tempête, 2193 = 10·2 kts. Vengeur, 2030 = 10·8 kts. Tonnerre, 4169 = 12·9 kts. Fulminant, 4533 = 13·7 kts.

(V.) FURIEUX (1883).

Circa 6000 tons. Complement 250.

L.: 249 ft. = 76 m. B.: 58¾ ft. = 17·8 m. D.: *circa* 25 ft. = 7

Guns: 2AB (340 m/m) + 4 F + 10*. 2 torpedo tubes.

Armour (*comp.*): Belt *aa* - *a* (450 - 350 m/m = 18 - 14").]
(350 m/m). Deck 87 m/m (3½").

Machinery: Compound 3-cylinder. 2 screws. Boilers (1898): 8 Be
groups). Speed:

(To be re-armed with 270 m/m guns in *turrets*).

On transforme ce navire.

DRAGONNE.

395 tons. Complement, 63.

Guns: 1D experimental for high angle fire. Torpedo tubes: 2. Armour: 0.

Machinery: 2 screws. Locomotive boilers. Sea speed: 15 kts. (*circa*). I.H.P. 2000.

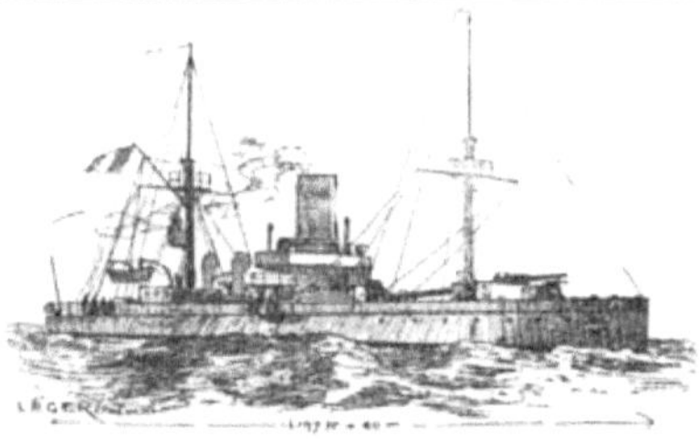

(T.) LÉGER & LEVRIER (1891).

444 tons. Complement 80.

L.: 197 ft. = 60 m. B.: 23 ft. = 7 m. D.: 12 ft. = 3·6 n

Guns: 1* (65 m/m) + 3* (47 m/m) + 2* (37 m/m). 3 torpedo tu

Armour: 0. Machinery: Horizontal triple expansion. Belle
2 screws. Coal: 130 tons.

Speed, 1st *Trials*: {Léger, I.H.P. 2230 = 18·8 kts. / Levrier, chev. 2200 = 18·5 kts.} *forced d.*

(III.) MAGENTA (1890) (laid down 1883).

10,851 tons. Complement, 651.

L. B. D.: 330 × 65½ × 28½ ft. = 100·6 × 20 × 8·7 m.

Guns: 4A (340 m/m 28 cals.) + 16DE$_{*}$ (138·6 m/m) + 6F* (65 m/m) + 18* (47 m/m) + 10*. 3 torpedo tubes.

Armour comp.: = Hoche. (4 barbettes 0 turrets). Some *f* screens in battery.

Machinery: 2 sets 4 cylinder compound Schneider. Boilers, 8 cylindrical. 2 screws. Coal, 800 tons = *circa* 2000 miles at 9 kts. Sea speed: *circa* 14·5 kts.

Trials (1892) *forced draught* = 11,045 H.P. = 16·2 kts.

(Superstructure to be removed.)

(V.) FUSÉE (1891) & MITRAILLE (1886).

1150 tx. Equipage 84.

L.B.D.: 165 × 32⅝ × 11 ft. = 50·3 × 9·9 × 3·4 m.

Guns: 1C (240 m/m) + 2F (100 m/m) + 4*. 1 torpedo tube.

Armour: Belt *b—dc* (240—175 m/m). Barbette *c* (200 m/m). Shield to *c* gun *e* (100 m/m).

Machinery: Fusée, I.H.P. 1640 = 11·8 kts. *1st trial*. Mitraille, I.H.P. 1506 = 10·3 kts. Sea speed: *circa* 8 kts. 2 screws. 3 boilers. Coal: 120 tons.

MARCEAU.

(III.) MARCEAU (1887). 10850 tons. Complement 64
NEPTUNE (1887). 10983 tons. Complement 651.

L. B. D.: 330 × $65\frac{1}{4}$ × $28\frac{1}{2}$ ft. = 100 × 20 × 8·7 m.

Guns: 4A (340 m/m, 30 cals.) + {Mar. 17 / Nep. 16} DE* (138·6 m/m) + 4F* 12* (47 m/m) + (Neptune only) 16* (37 m/m) + 8* revolvers.

Torpedo tubes: Marceau 6, Neptune 5.

Armour (en acier): Belt *aa* - *a* (450–350 m/m). Inclined deck on to Barbettes *aa*. Shields *f*. Hoists *a*. No armour to battery, but *f* scr some guns. Conning tower *a*.

Machinery: 2 set 3 cylinder comp. {M. 8 / N. 12} cylindrical boilers. 2 scr

Coal: 800 tons.

1st. *trials*. {Marceau 4 *hours forced draught* = 11017 H.P. = 16·2 kts / Neptune ,, ,, = 10929 H.P. = 16 kts.}

Sea speed *circa* 13 kts.

Marceau to be reboilered with Niclausse.

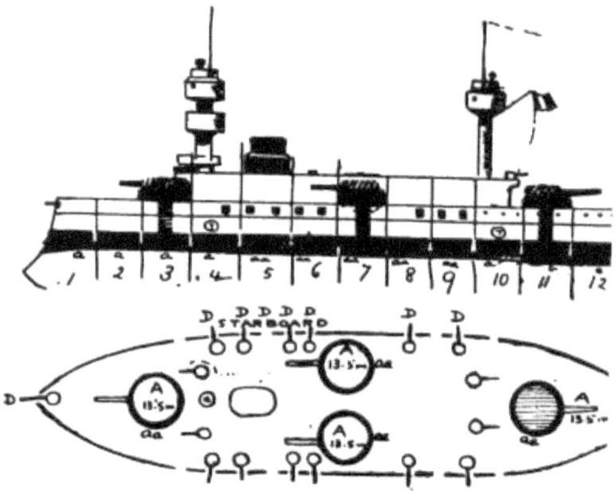

(III.) AMIRAL BAUDIN (1883) (1897): FORMIDABLE (1885) (1897).

Circa 12,000 tons. Complement B. 634, F 655

L. B. D.: 329½ × 70 × 28½ ft. = 100·4 × 21·4 × 8·7 m.

Guns: 2A (370 m/m) + 4D* (164·7 m/m 45 cals.) + 10DE* (138·6 m/m) (Formidable only 8 DE*) + 12* (47 m/m) + 14—18* (Revolver). 6 Torpedo tubes

Armour (en acier): Belt (6½' = 2 m. wide) *aa—a* (545—355 m/m). *Inclined* deck = *d* (90 m/m). Barbettes *aa*. Shields to A guns *f*. Hoists *d*. C.T. *a*. Battery for 16 c/m guns (*acier special*) *d*.

Machinery: 2 sets 3 cylinder compound. 12 cylindrical boilers. 2 Screws. Coal 800 tons. New boilers (1897–8).

Trials: Baudin (1898) 24 *hours natural draught* I.H.P. 6408 = 14·5 kts.
Formidable (1897) 24 *hours natural draught* I.H.P. = 15·2 kts. (oil and coal).
Formidable (1897) 24 *hours natural draught* I.H.P. = 14 kts. (coal only).

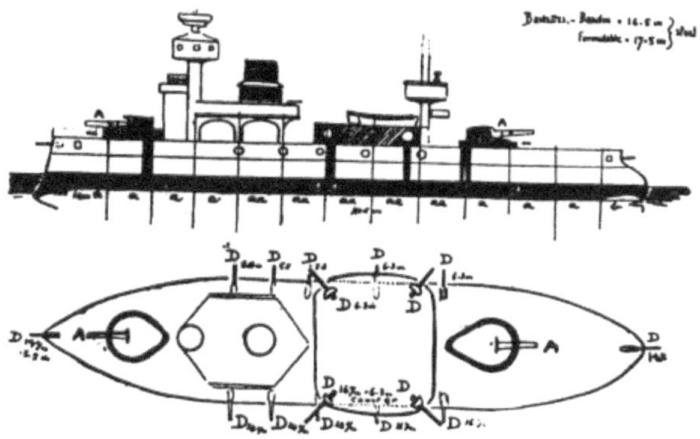

(IV.) FRIEDLAND (1878).

Circa 9000 tons. Complement 680. 1 screw.
L.: 256 ft. = 78 m.
Guns: 8CD + 8E. Armour: *d*. Sea speed: 10 kts.

(IV.) VAUBAN (1882): DUGUESCLIN (1883).

Circa 6000 tons. Complement 440 (*circa*)
L.: 265 ft. = 81 m. B.: 57 ft. = 17·4 m. D.: 25¼ ft. = 7·70 m. (*du plan*).

Guns: 4CD (274 m/m) + 1D (194 m/m '70 model) + 8E* (138 m/m) (Duguesclin 6E*) + 2F (65 m/m) + 6* (47 m/m) + 13*.

Armour (en fer iron). Belt *d* - *e* (250–150 m/m). Barbettes (comp.) *c* (200 m/m). Deck 50 m/m.

Machinery: I.H.P. *circa* 4000. Sea speed: *circa* 10–11 kts. 2 screws.

Dif. Funnel, s.p., etc. Duguesclin has a peculiar stern.

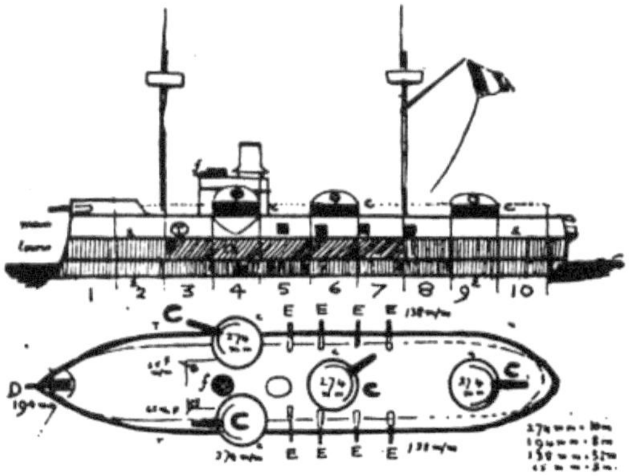

mât

FRANCE.

Photo, M. Bar.

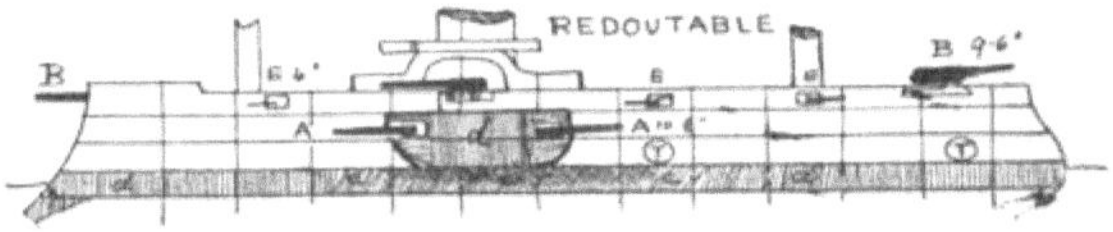

(III.) REDOUTABLE (1876) (1896) (1898).

Circa 9500 tons. Complement 705.

L. B. D.: 318¾ × 64⅝ × 27 ft. = 97 × 19·7 × 8·2 m.

Guns (1898): 4A (274 m/m, model '93) in redoubt + 4B (240 m/m, model '93) + 6E* (100 m/m) + 14* (47 m/m) + 6* (37 m/m) + 15* (*revs.*). Torpedo tubes: 4.

Armour (en fer): Belt 318¾ × 9½ ft. / 97 × 2·7 m. *b c—d* (350—220 m/m) Redoubt *d* (240 m/m). Deck 60 m/m.

Machinery: Horizontal compound. Boilers: 8 cylindrical. 2 screws. Coal. *normal* 600 tons, *max. circa* 1000 tons = 1500 miles at 10 kts. I.H.P. (? *d.*) 6071 = 14·6 kts. (1897). Sea speed: *circa* 13 kts.

Condor has a bigger funnel than the other three.

(5.) VAUTOUR (1889). EPERVIER (1886). & CONDOR (1885). FAUCON (1887).

Circa 1300 tons. Complement 154.

L. B. D.: 223 × 29¼ × 16 ft. = 68 × 8·9 × 4·9 m.

Guns: 5E* (100 m/m 26 cals.) + 6* (47 m/m) + 4*. 5 torpedo tubes.

Armour: deck 1½" (38 m/m) steel.

Machinery: Horizontal compound. 2 screws. 4 cylindrical boilers. Coal: 150 tons.

1st trials *forced draught* (hours): Condor, 3582 H.P. = 17·8 kts. Epervier, 3325 = 17·1 kts. Faucon, 3233 = 17·1 kts. Vautour, 3378 = 17·1 kts.

(5.) WATTIGNIES (1891). FLEURUS (1893).

Circa 1350 tons. Équipage *circa* 160.

L. B. D.: 229¾ × 29¼ × 16 ft. = 70 × 8·9 × 4·9 kts.

Guns: 5E* + 6* (47 m/m) + 4*. 4 torpedo tubes.

Machinery: Horizontal compound. 2 screws. Boilers: Wattignies, 4 cylindrica Fleurus (1897), Niclausse. Coal: 160 tons.

Trials: Wattignies (189–) *forced* 4187 H.P. = 18·1 kts. (*hours*). Fleurus (1898) *natural draught* 2000 = 16·8 kts. (24 *hours*), *forced circa* 3000 = 17·5 kts (*hours*).

(T.) BOMBE & COULEVRINE & DAGUE & FLECHE & SAINT-BARBE (—1885) & LANCE & SALVE (—1886).

" Contre-torpilleurs."

395 tons. Complement 63.

L. B. D.: 197 × 22 × 10½ feet = 60 × 6·7 × 3·2 metres.

Guns: 4* (47 m/m) + 3* (37 m/m revolvers). 2 Torpedo tubes.

Machinery: I.H.P., *natural* 900, *forced* 2000 = 18 kts. (*trial*). Boilers: Locomotive, *except* Lance and Salve, which were fitted with Normand, 1899. 2 screws. Sea speeds, *circa* 15·5 kts. Coal, 100 tons.

(Dragonne, see page 121.)

(IV.) INDOMPTABLE (1883): CAÏMAN (1885): REQUIN (1885): TERRIBLE (1881).

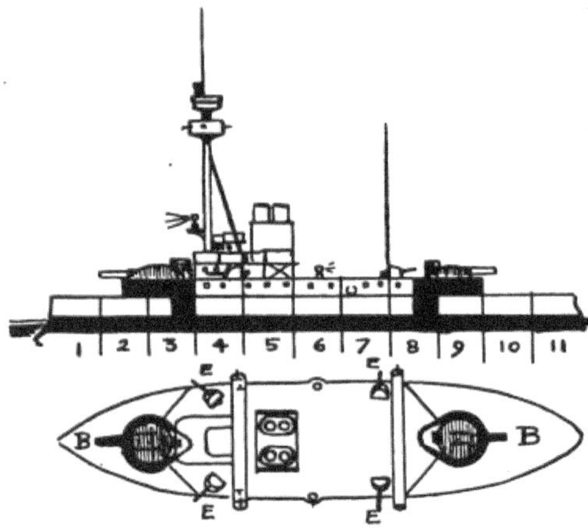

INDOMPTAB

CAÏMAN

TERRIBLE

Circa 7500 tons. Complement 500.

L. B. D.: 279 × 59 × 27 feet = 85 × 18 × 8·2 metres.

Guns: (*new armament*) 2A (274 m/m) + 6E* (100 m/m) + 10* (47 m/m) + 4* (37 m/m).

Torpedo tubes 4.

Armour (compound): Belt *aaa–a* (500–300 m/m). Barbettes *aa* (450 m/m). In *Requin*, and others when reconstructed, barbettes *b* (250 m/m). Hoists *c* (200 m/m). C.T. *a*. Hoods to big guns = *a*.

Machinery: Vertical compound 3 cylinder. 2 screws. Boilers: *Requin*, 10 Niclausse. Others 12 cylindrical. I.H.P., *natural*, 3800.

Trials.

Caïman	*hours forced d.* 6000 = 15 kts.
Indomptable	,, ,, ,, 6605 = 14.8 kts.
Requin	,, ,, ,,
Terrible	,, ,, ,, 6230 = 14·5 kts.

Note.—Indomptable and Caïman are being reconstructed at present. All have recently had rig altered to that noted. Armour has been taken off the barbettes to reduce weight. The extra E* guns in Requin are mounted between the others on the upper deck.

(III.) COURBET (1882) (1899–1900): DÉVASTATION (1879) (1900).

Circa 10,200 tons. Complement, *circa* 690.

L. B. D.: 318¼ × 69 × 28 feet = 97 × 21 × 8·53 metres.

Guns: *Courbet*, 4A (274 m/m, model '93) + 3B (240 m/m, model '93) + 1D* (164 m/m) + 10E* (100 m/m) + 14* (47 m/m) × 17* (37 m/m). 5 torpedo tubes (4 on broadside, 1 in stern). *Dévastation*, 8C (240 m/m, model '75) [4 in battery, 4 upper deck] + 10E* (100 m/m) + 4F* (65 m/m) + 12* (47 m/m) + 20* (37 m/m revolvers). 4 torpedo tubes.

Armour (iron): Belt, *b* - *d*. Bulkhead aft, *d*. Deck aft = *d*. Protection to vitals, *b*. Redoubt and battery, *c*. Barbettes, no armour. Guns in battery and barbettes have shields, *c*, to them. C. T., *f*.

Machinery: 3 cylinder compound. 2 screws. Boilers, Belleville. (1899–1900) I.H.P. *natural* 4500 = 13 kts.; *forced* 8100 = 15·1 kts. Sea speeds, *circa* 12-13 kts. Coal, 900 tons.

(Both reconstructing 1900.)

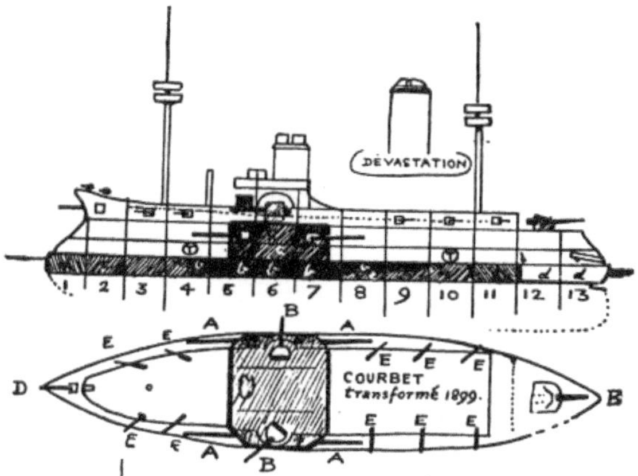

(IV.) AMIRAL DUPERRÉ (1879) (1900).

11,210 tons. Complement 656.

L. B. D.: 329½ × 67 × 27 feet = 97 × 20·42 × 8·23 metres.

Guns (new armament): 3B (340 m/m, 18 cals.) + 5D* (164 m/m) + 14E* (138 m/m) + 18* (37 m/m). 4 torpedo tubes.

Armour (iron): Belt, *a*–*d* (500–250 m/m). Protection to vitals, *a*. Barbettes, *c*. Shields and hoists, *f*. Battery for 4D* guns, *c* (not yet placed).

Machinery: 2 sets compound. Boilers, 12 cylindrical. 2 screws. I.H.P. *natural* 4500; *forced* 5783 = 14·2 kts. (*trial* 1882).

Note.—Present armament is 4B (one in a barbette amidships) + 1D* (in bow) + 14E*.

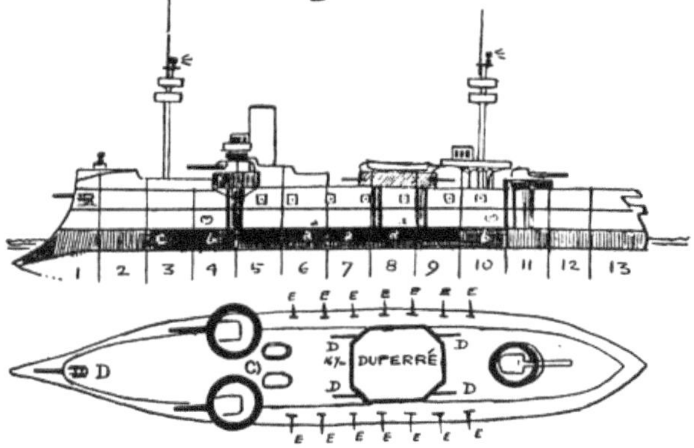

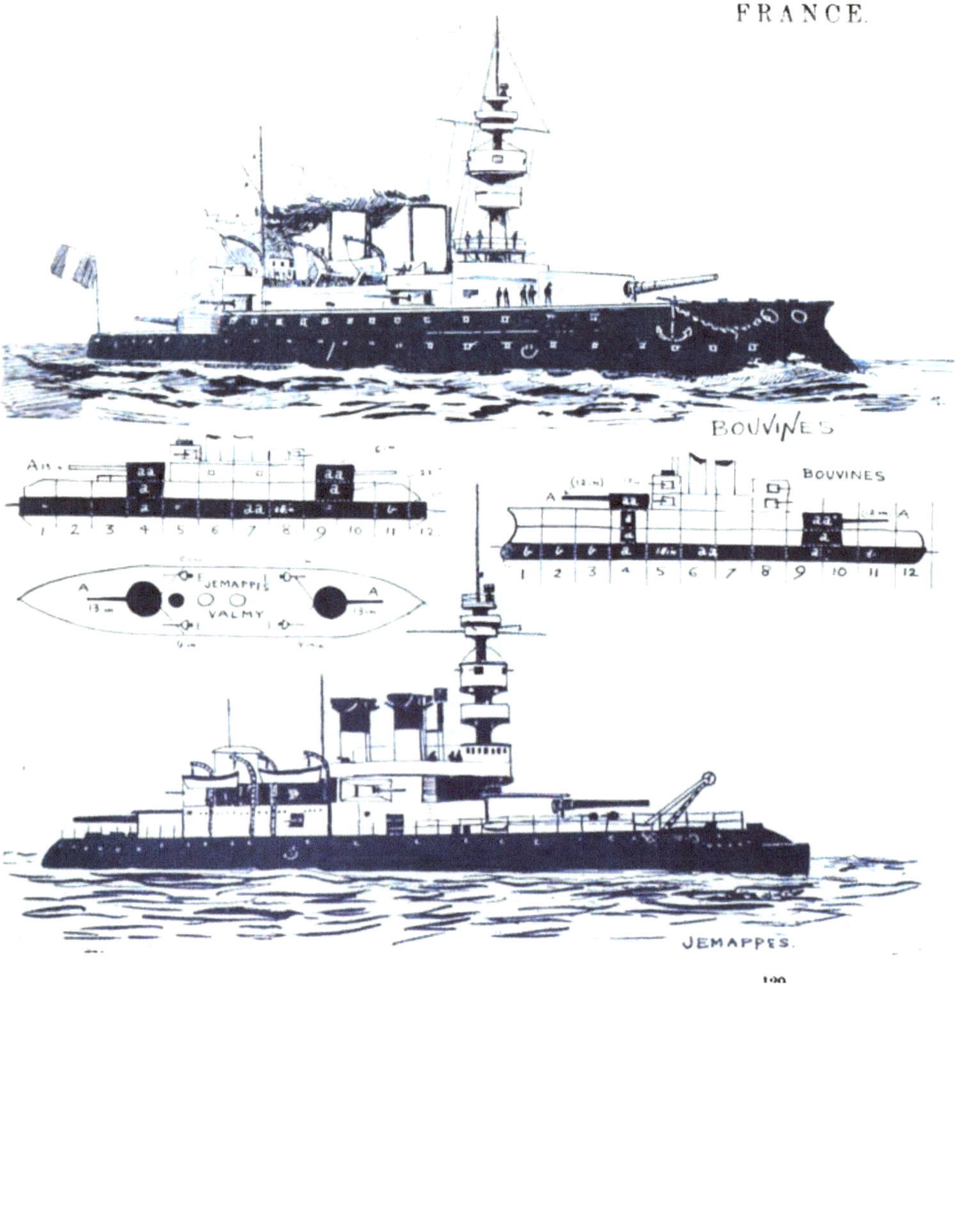

(II.) BOUVINES (1892).

Circa 6800 tons. Complement 315.

L. B. D.: 292 × 59 × 24½ ft. = 89 × 18 × 7·4 m.

Artillerie: 2A (305 m/m, 45 cals.) + 8E* (100 m/m) + 4* (47
10* *revs.* 2 torpedo tubes behind armour.

Cuirasse: ceinture (Acier Creusot) *aa* - *b* (450-250 m/m). Tourell
Passages aux tourelles (bases) *a* (315 m/m). Pont 100 m/m.

Machines: 2 sets *inclined* triple expansion. 16 Chaudières, I
2 helices. Charbon 337 tx.

Essais 1895. *Tirage forcé* 8921 *chev.* = 16·1 no ends. Sea speed

(II.) JÉMAPPES & VALMY (1892).

Circa 6600 tons. Équipage 299.

L. B. D.: 283¾ × 57½ × 23 ft. = 86·5 × 17·5 × 7·1 m

Guns: 2A (340 m/m, 40 cals.) + 4E* (100 m/m) + 6* (47 m/m)
2 torpedo tubes behind armour.

Armour: = Bouvines. Machinery: = Bouvines. Coal: = Bou
= Bouvines.

Trials 1894, *hours forced draught.* Jémappes 9201 = 15·7 kts.
natural draught 7760 = 15 kts.

Valmy, 1895, 2 *hours forced draught* = 8555 H.P. = 15·9 kts.
(with 12 boilers) 7507 = 15·4 kts. *Natural draught* 8 *hours* (16
15·5 kts. Sea speeds: *circa* 12·5 kts.

(3.) SFAX (1884), (1897–8).

(*Transformé*) 4728 tons. Equipage 486.

L. B. D.: 288$\frac{3}{4}$ × 49$\frac{1}{4}$ × 26 ft. = 88 × 15 × 7·9 m.

Guns: (1898) 6D* (164 m/m) + 10D* (138 m/m) + 6* (47 m/m) + 6* (37 m/m) + 6* *revs.* 2 Torpedo tubes.

Armour: Deck = *f* (35 m/m). Cellulose belt. Shields to 6·4 inch guns *e*.

Machinery: Horizontal compound. 12 cylindrical boilers. 2 Screws. Coal: *normal* 700 tons, *max.* 1000 tons. New boilers 1897–8.

Trials (1898): 6 *hours natural draught* = 4200 H.P. = 15·5 kts. *max.*
24 ,, ,, ,, = 13·8 kts. (*mean: moyen*)

(T) DUNOIS (1896) & LA HIRE (1898).

900 tons. Équipage 128.

Longueur, etc.: 256 × 28 × $12\frac{3}{8}$ ft. = 78 × 8·53 × 3·88 m.
Guns: 6F* (65 m/m) + 6* (47 m/m, 3-pdr.).
Machinery: Normand boilers. *Est.* speed 23 kts.

Trials: Dunois (1898-9) *forced 4 hours* I.H.P. 7500 = 21·7 kts.
La Hire (1899) „ 4 „ „ 7100 = 22·11 kts.: *n.d.* 4200 = 20·4 kts.

(T) CASABIANCA (1895) & CASSINI (1894).

L.: 262 ft. = 80 m. 960 tons. Complement 139.

Guns: 1E* + 8*. Armour: $\frac{1}{2}$ *prot.*

L. B. D.: 269 × $27\frac{1}{3}$ × 12 ft. = 82 × 8·3 × 3·7.

Guns: 1E* (100 c/m) + 3F* (65 m/m) + 7* (47 m/m). Torpedo tubes: 2. Armour: 0.

Machinery: 2 sets compound. 3 cylinder. 4 Lagrafel d'Allest boilers. 2 screws. Coal: 117 tons.

Trials: Casabianca (1896). 4 *hours forced d.* = 5612 H.P. = 21·2 kts.
Cassini (1895). 4 „ = 5081 H.P. = 21·2 „
„ (1896). 4 „ = 22 „

(4.) FORBIN (1888).

1920 tons. Complement 200.
L.: 312 ft. = 95 m. B.: $30\frac{1}{2}$ ft. = 9·3 m. D.: 17 ft. = 5·2 m.
Guns: 4D*E* (138 m/m) + 8*. 5 torpedo tubes. Sea speed: 17·5 kts.
f Armour deck = 35 m/m ($1\frac{1}{2}$") over machinery. Cellulose belt.
Coal: 200 tons.

1st ***trials*** (1890-1): Coëtlogon, *forced d.* = 5918 H.P. = 20·3 kts. Forbin, *forced d.* = 6022 H.P. = 20·6 kts. Surcouf, *forced d.* = 6260 H.P. = 20·5 kts.
(Coëtlogon and Surcouf identical to Forbin save that they have 3 masts.)

(T.) D'IBERVILLE (1892)

925 tons. Complement 140.

L. B. D.: $262\frac{1}{2}$ × $26\frac{1}{4}$ × $11\frac{5}{8}$ ft. = 80 × 8 ×

Guns: 1E* (100 m/m) + 3F* (65 m/m) + 6
Torpedo tubes: 6, 1 pair on each broadside. A

Machinery: 2 sets 4-cylinder tandem triple
4 Lagrafel d'Allest boilers. 2 screws. Coal: 1

Trials (1894): 18 *hours natural draught*, I.
19·3 kts. 6 *hours* $\frac{1}{2}$ *forced draught*, I. H. P. 4400
4 *hours forced draught*, I. H.P. 5060 = 21·6 kts

(2.) ALGER (1889).

4200 tons. Complement, 407.

L.: 346 ft. = 105·5 m. B.: 45¼ ft. = 13·8 m. D.: 22½ ft. = 6·9 m.

Guns: 4D* (164 m/m) + 6DE* (138 m/m) + 2F* (65 m/m) + 8* (47 m/m) + 10* (*revs.*) 4 torpedo tubes.

Armour: Deck = *d* (m/m), gunshields *d*.

Machinery: Triple expansion. Boilers, Belleville. 2 screws.

Trials (1891–2): *n. d.* 5089 H.P. = 17 kts.: *forced* = 8127 = 19·6 kts.

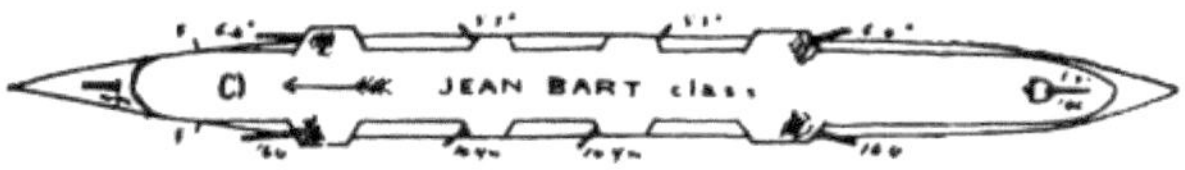

(2.) JEAN BART (1889) (1897): ISLY (1891).

L. B. D.: 353¼ × 43¾ × 22 ft. = 107·7 × 13·3 × 6·7 m.

Guns: Isly = Alger. J. Bart, 138 m/m are D*.

Armour: = Alger.

Machinery: Triple expansion: Isly, vertical: J. Bart, horizontal. 8 cylindrical boilers. 2 screws. Coal, *circa* 880 tons Isly; J. Bart, 750 tons, *max.* 940 tons.

Trials:
- Isly (1891) *forced draught*, I.H.P. 8252 = 18·3 kts.
- J. Bart „ „ „ 7707 = 18·4 „
- J. Bart (1898) *natural draught* „ ? = 17 „ (24 hours).

(2.) CATINAT (1896): PROTET (—1897).

4114 tons. 4065 tons. Complement 384.

DESCARTES (1894): PASCAL (1895).

3988 tons. Complement 378.

L. B. D. 332 × 44½ × 21¼ ft. = 101·2 × 13·6 × 6·5 m.

Guns: 4D* (164 m/m 45 cals.) + 10E* (100 m/m) + 10* (47 m/m) + 4* (37 m/m). Catinat & Protet only 10·47 m/m. 2 Torpedo tubes.

Armour: Deck = *d* (50 m/m). Sponsons *d* (100 m/m).

Machinery: Triple expansion. 16 Belleville boilers in 4 groups of 4. 2 Screws. Coal *normal* 550—600 tons. *Max.* 750 tons.

Trials:

Catinat (1898), 24 *hours natural draught*, I.H.P. 6455 = 17·5 kts. (*moyen*).

Catinat (1898), 4 *hours forced draught*, I.H.P. 9638 = 19·6 kts. (*moyen*).

Catinat (1898), 8 *hours natural draught*, I.H.P. 8044 = 18·3 kts. (*moyen*).

Descartes (1896), 24 *hours natural draught*, I.H.P. 5802 = 17·5 kts. (*moyen*).

Descartes (1896), 24 *hours forced draught*, I.H.P. 8870 = 19·6 kts. (*moyen*).

Pascal (1896-7), 24 *hours natural draught*, I.H.P. 6170 = 17·5 kts. (*moyen*).

Pascal (1896-7), 4 *hours forced draught*, I.H.P. 8720 = 19·7 kts.

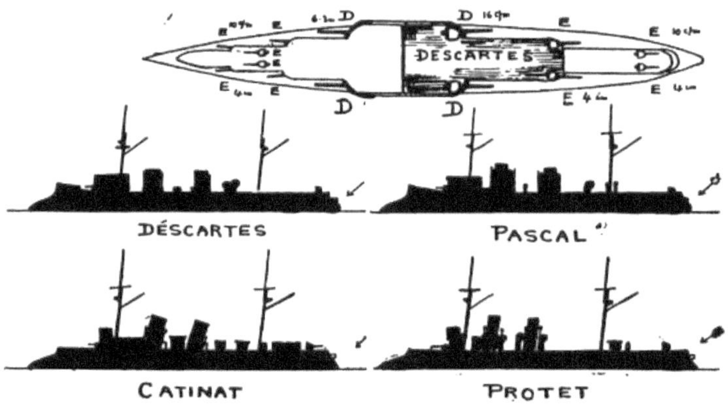

PASCAL. *Photo by M. Bar, Toulon.*

(2.) SUCHET (1893) (1898).

3450 tons. Equipage 346.

L. B. D.: 318¾ × 40 × 20½ ft. = 97 × 12 × 6·3 m.

Guns: 6D* (164 m/m 30 cals.) + 4E* (100 m/m) + 8* (47 m/m) + 8*. Torpedo ...s: 1 in bow *submerged* and 6 above water.

Armour: Deck = *d* (75 m/m).

Machinery: Horizontal triple expansion. 2 Screws. Boilers: Lagrafel d'Allest. ...: 480 tons = *circa* 3500 miles at ? kts. I.H.P. *natural draught* 4000 = 16·7 kts. ...*ed* 9000 = 20 kts., *trial* 1895.

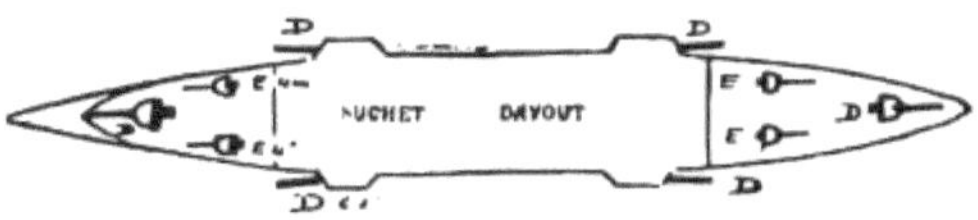

(2.) DAVOUT (1889) (1898).

3027 tons. Complement 336.

...ongueur: 298 ft. = 91 m. Beam: 40 ft. = 12 m. Draught: 20 ft. = 6·09 m.

Guns: 6D* + 4E* + 8*. 4 Torpedo tubes.

Armour: Deck = *ds* (65 m/m). Cellulose belt.

Machinery: Horizontal triple expansion. 2 Screws. 8 Cylindrical boilers. (*To ... Niclausse boilers,* 1898–9). I.H.P. (*trials* 1891–2) *forced draught* 9000 = ... kts. (measured mile). Coal 400 tons, *max. capacity* 600 tons.

SUCHET (1900). *Photo by M. le Marquis de Balincourt.*

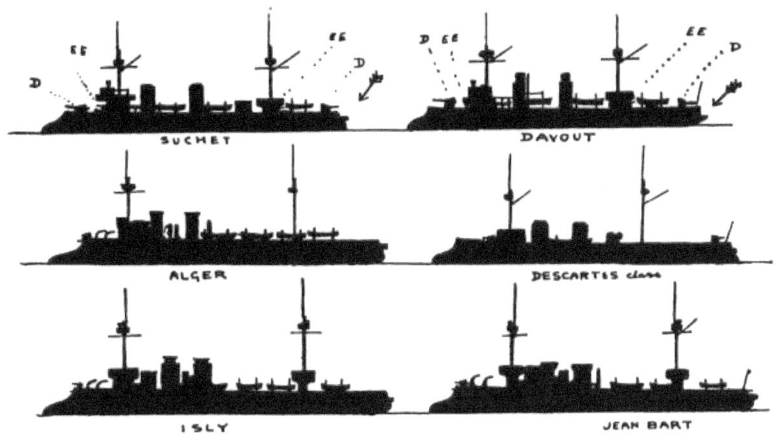

(3.) LINOIS (1894): GALILEE (1896) & LAVOISIER (1897).

Circa 2350 tons. Complement, 248.

L.: 328 ft. = 106 m. B.: 34½ ft. = 10·5 m. D.: 19 ft. = 5·8 m.

Guns: 4D* (138 m/m) + 2E* (100 m/m) + 8* (47 m/m) + 2* (37 m/m) + 4*. 2 torpedo tubes.

Armour: deck (38 m/m *i.e.* 1½") = *e* Gunshields *f* (50 m/m).

Machinery, triple expansion. *Linois*, cylindrical boilers; Galilée and Lavoisier, Belleville boilers. 2 screws.

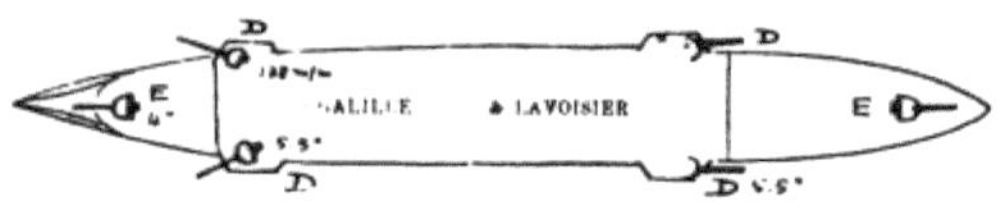

GALILÉE & LAVOISIER.

Trials.	Lavoisier: (1898)	4	*hours forced draught, I.H.P.*		= 21·5	kts. *mean;*	24 *hours natural draught*, H.P. 4560 = 18·6 kts.
	Galilée:	4	,, ,, ,,	6846	= 19·8	,, *moyen;*	,, ,, .
	Linois:	4	,, ,, ,,	6600			.

(FRANCIA) FRANCE (FRANKREICH).

(II.) HENRI IV. (1899).

8948 tons. Complement 400.

L. B. *D.* (*mean*): 354½ × 73 × 23 feet = 108 × 22·20 × 7 metres.

(274 m/m, 45 cal.).
' (138 m/m, 45 cal.).
(47 m/m).
do tubes:
ıbmerged.

Armour (Harvey nickel):
13"–8" Belt *aaa–a* (328-200 m/m).
Deck ?
Prot. to vitals = *aaaaa*.
10" A gun turrets *aa* (250 m/m).
12" A gun supports *aaa* (300 m/m).
5" Redoubt *e* (127 m/m).
5" D gun turret, aft *e* (127 m/m).

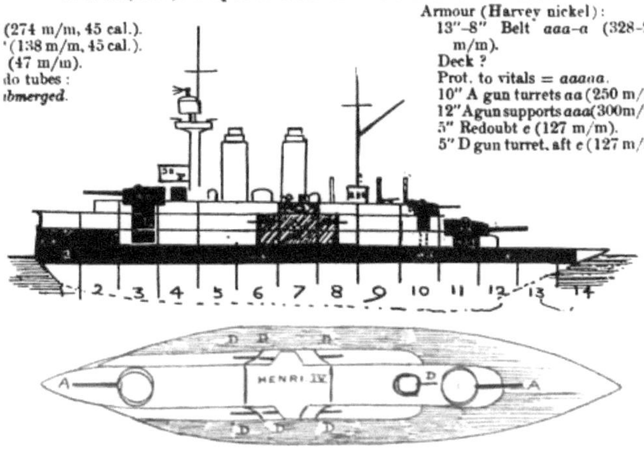

Iachinery: 3 screws. I.H.P. 11,500 = 17 kts. (*est.*). Boilers: Niclausse.

(III.) HOCHE (1886) (1899–1900).

10,800 tons. Complement 631.

L. B. D.: 336 × 65½ × 28¼ ft. = 102·40 × 20 × 8·70 m.

Guns:
2 A B (340 m/m, 21 cals.).
2 C (274 m/m, 28 cals.).
12 E● (138 m/m, 30 cals.).
6 F* (65 m/m).
9* (47 m/m).
11* (37 m/m).
Torpedo tubes: 5.

Armour (compound):
17¾" Belt *aa–a* (440 m/m).
3¼" Deck = *d* (80 m/m).
Prot. to vitals = *aaa*.
16" Turrets and barbettes *aa* (400 m/m).
Bow *f*.
Hoists *d*.
C.T. *aa*.

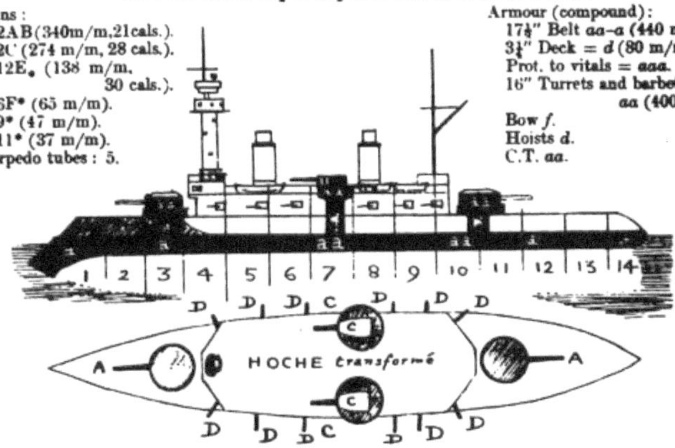

Machinery: 2 screws. I.H.P. 11,300 *f. d.* Boilers: 18 Belleville, with economisers. Sea speed (*est.*): 14 knots.

(II.) BRENNUS (1891).

Photo by M. Bar, Toulon.

11,395 tons. Complement 696. L. B. D.: 374 × 65¼ × 27¾ feet = 114 × 20 × 8·40 metres.

Guns: 3A (340 m/m) + 10D* (164 m/m) + 4F* (65 m/m) + 14* (47 m/m) + 2 Maxims.

Torpedo tubes: 6 *above water* (protected by *d* armour).

Armour (compound): 18–10″ belt *aaa–b* (450-250 m/m). 3″ deck behind belt = *d* (75 m/m). Protection to vitals = *aaaa*. Side above belt *d* (120 m/m). Redoubt *e* (100 m/m): 18″ fore turret, *aaa* (440 m/m): 16″ after turret, *aa* (400 m/m). Supports and hoists *a*. 5″ small turrets, *d* (120 m/m). Cofferdam behind belt. Screens between guns in redoubt.

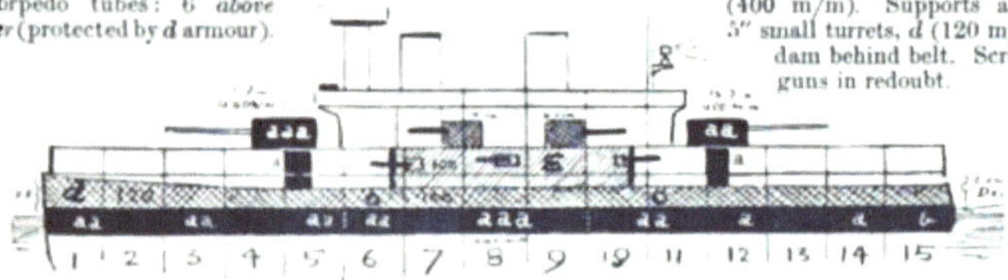

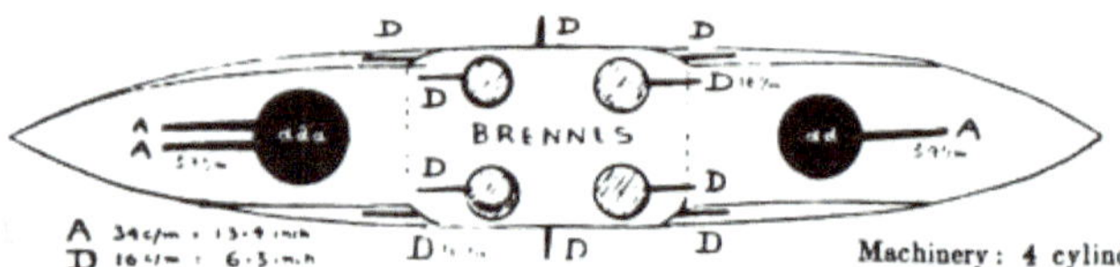

Machinery: 4 cylinder triple expansion. 2 screws. Boilers, 32 Belleville. Coal, 800 tons *normal*. Sea speed, *circa* 16 kts.

Trials (1895–96).
4 hrs. nat. dr. I.H.P. 1142 = 8·2 kts. (8 boilers).
6 „ „ „ 6229 = 14·4 „ (24 „)

Photo by M.

(II.) CARNOT (1894).

Circa 12,000 tons. Complement 625.

L. B. D.: 380½ × 70½ × 29 feet = 116 × 21·50 × 8·80 metres.

Guns: 2A (305 m/m) + 2A (274 m/m) + 8D* (138 m/m) + 4F* (6... (47 m/m) + 6* (37 m/m) + 8 Maxims.

Armour (Creusot): 18-9″ belt. *aaa–b* (450-225 m/m). 2¾″ Armour-de... Protection to vitals = *aaaaa*. Turrets *aa* (370 m/m). Hoists and supports *a*. (100 m/m). Hoists to them *e*.

Machinery: 2 screws. Boilers, 12 d'Allest (welded tubes). Sea speed, 17 kt... 500 tons, *maximum* 700 tons.

Trials (1896–7)—24 *hrs. natural d.* I.H.P. 11,547 = 17·5 kts. *mea...*
4 *hrs. forced d.* I.H.P. = 17·8 kts.

[Plan, see C. Martel, p. 139.]

Photos by M. Bar.

FRANCE.

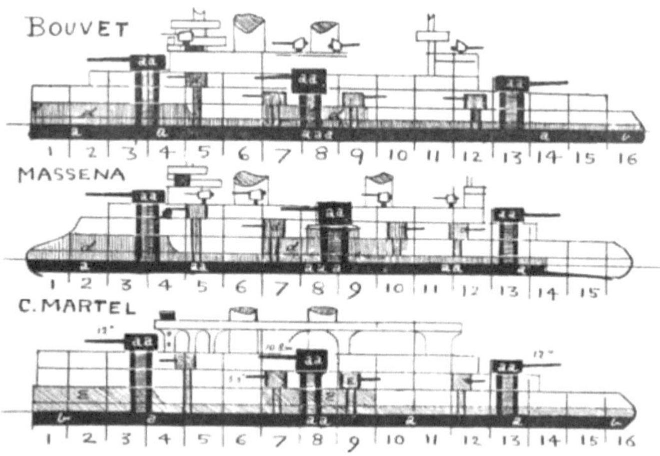

(I.) MASSENA (1895).

Circa 12000 tons. Complement 610.

L. B. D.: 380½ × 66¼ × 27 ft. = 116 × 20·2 × 8·2 m.

Guns: 2A (305 m/m 40 cals. model '93) + 2A (274 m/m 40 cals. model '93) amidship 8D* (138 m/m 45 cals.) + 8E* (100 m/m, model '92) + 2F* + 12* (47 m/m) + 5* (37 m + 7*. Torpedo tubes: 2 *submerged* + 4 *above water* behind *de* armour.

Armour (*acier special durci*) (incomplete): Belt *aaa* - *a* (450 - 250 m/m). Deck be belt curved = *dc* (70 m/m = 2·7 inches). Protection to vitals = *aaaaa*.

Upper belt *de* (100 m/m). Turrets *aa* (350 m/m = 14 inches). Hoists, etc., *aa*. D* turrets *de* (100 m/m = 4 inches). Splinter deck below protection deck, 37 m/m Coffer Bulkhead aft of belt *a*. C.T.'s *aaa*.

Machinery: 3 Triple expansion à pilon. 3 Screws. 24 Lagrafet D'Allest boilers groups. Coal 635 tons = *circa* 3500 miles at 10 kts. Sea speed 17 kts. I.H.P. 13,500 *fo draught, nominal*.

Trials 1898. 24 *hours natural draught*, 9650 H.P. = 15·49 kts. (16 boilers).
4 ,, *forced* ,, ,, = ,, (24 boilers).
24 ,, *natural* ,, 13,460 ,, = 17 ,, (24 boilers).

(I.) BOUVET.

12,250 tons. Complement 630.

L. B. D.: 397 × 70⅛ × 27⅓ ft. = 121 × 21·4 × 8·4 m.

Guns: = Massena. **Armour**: = Massena (*but belt complete*).

Machinery: Triple expansion à pilon. 3 Screws. Belleville boilers. Coal: 620 tons. speed: 17 kts. I.H.P.: *nominal* 14,000.

Trials 1898. 24 *hours natural draught* H.P. = kts.
4 ,, *forced* ,, ,, = 18·8 ,, (dirty weather).

(II.) CHARLES MARTEL (1893).

Circa 12,000 tons. Complement 632.

L. B. D.: 377 × 70 × 26¼ ft. = 115 × 21 × 8 m.

Guns: 2A (300 m/m '87 model) + 2A (274 m/m '87 model). *Amidships* + 8D* (138 m/m 45 cal.) + 4F* + 12* (47 m/m) + 12* (37 m/m). Torpedo tubes: 2 *submerged* + 4 *above water behind e armour*.

Armour (*acier Creusot*): Belt *aa* - *b* (450 - 250 m/m). Deck curved behind belt = *d* (70 m/m) Prot. to vitals = *aaaa*. Turrets: *aa* (370 m/m): D* turrets *e*. Lower deck belt *e*. C.T.: *aa*.

Machinery: Triple expansion. 2 Screws. Boilers: Lagrafel D'Allest. Welded t I.H.P.: *nominal max*. 13,500. Sea speed: 17 kts. Coal: 800 tons.

Essais 1897. 24 *hours natural draught* H.P. = 17 kts. (*moyen*) at sea.
,, ,, ,, 12,000 ,, = 17·8 ,, (consumption trial).
4 ,, *forced* ,, 14,996 ,, = 18·13 ,, (*mean*).

Note.—C. Martel, Bouvet, & Carnot kept station 1898, Algiers to Toulon at 17 kts.

(I.) CHARLEMAGNE (1895): ST. LOUIS (1896): GAULOIS (1896).

11,260 tons. Complement, 631.

L. B. D.: 380¾ × 66¾ × 28 ft. = 116·2 × 20·3 × 8·53 m.

Guns: 4A (305 m/m 40 cal. '93 model) + 10D* (138 m/m 45 cal.) + 8E* (100 m/m) + 20* (47 m/m 3-pounders) + 4* (37 m/m) + 13*. Torpedo tubes: 4 *submerged* + 6 above water.

Armour (*acier special durci*): Belt (2 metres wide) *aaa* – *a* (400 – 250 m/m). Curved deck behind belt = *c* (70 m/m = 2·7"). Protection to vitals = *aaaa*. Upper belt (*above water*) *e* (75 m/m). Battery *ed* (100 m/m) with 50 m/m screen between guns. Turrets *aa* (300 m/m = 12"). Hoists .

Machinery: 4 cylinder triple expansion. 3 screws. I.H.P. *nominal* 14,500. Boilers, 20 Belleville in groups. Economisers. Coal, *normal* 680 tons, *max.* 1100 tons *nominal* 50 days at 8 kts.

Trials. Charlemagne (1898). *hours natural draught.* 9123 = 16·15 kts.: *hours* 14,400 H.P. = 17·7 kts. *natural draught*; hours *forced draught* = kts.

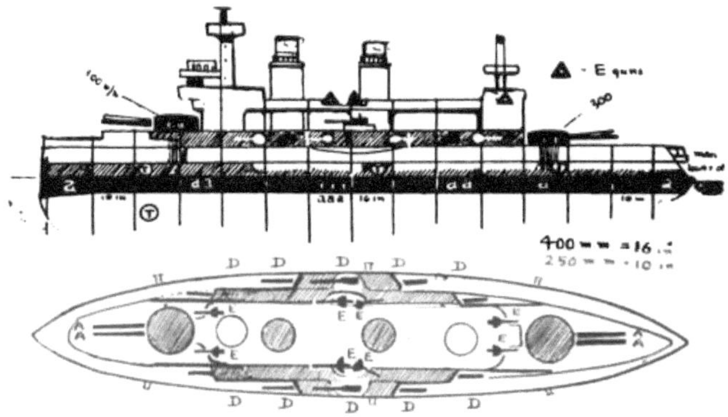

CHARLEMAGNE (*without her guns*).

Photo. "Le Yacht."

(I.) JÉNA (1898).

(*Completing.*)

12,052 tons. Complement 630.

L. B. D.: 400¾ × 68 × 27½ feet = 122 × 20·80 × 8·37 m.

Guns:
- 4A (305 m/m).
- 8D* (164 m/m).
- 8E* (100 m/m).
- 16* (47 m/m).
- 18* (37 m/m).

Torpedo tubes:
- 2 *submerged.*
- 2 *above water.*

Armour (Creusot special).
- 14" Belt amidships *aaa* (350 m/m).
- 6" Belt ends *c* (152 m/m).
- 5"–4" Upper belt *c–d*.
- Protection to vitals = *aaaaa*.
- 4" Battery *d* (100 m/m).
- 12" Turrets *aa* (300 m/m).
- Supports to turrets *aa*.

Machinery: Triple expansion. I.H.P. 15,500 = 18 kts. *forced d.* (*estimated*). 3 screws. Boilers: 20 Belleville, with economisers in 4 groups. Furnaces to burn liquid fuel + coal. Coal carried: *normal* 850 tons, *maximum capacity* 1100 tons.

Notes.—Jéna is an improved Charlemagne. No wood is used in construction. There is a cofferdam behind the belt. All the machinery for the guns, and the hoists are inside the supports.

(I.) *SUFFREN* (building).

12,728 tons. Complement 730.

L. B. D.: 410 × 70 × 27½ ft. = 125 × 21·33 × 8·37 m.

Guns:
- 4A (305 m/m).
- 10D* (164 m/m).
- 8E* (100 m/m).
- 20* (37 m/m, 3 pdr.).
- 2* (37 m/m).

Torpedo tubes:
- 2 *submerged.*
- 2 *above water.*

Armour (Krupp):
- 12" Belt amidsh
- Belt ends.
- 3" Deck (flat on t
- Protection to vi
- 5" Lower deck
- 12" Turrets and
- 6" Redoubt *a* (1
- 5" Small turret

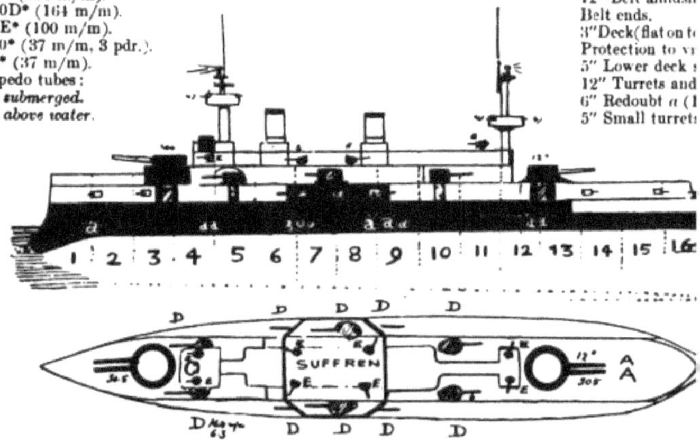

Machinery: Triple expansion. I.H.P. 16,200 = 18 kts. *forced d.* (*estimated*).
Boilers: Niclausse. Liquid fuel + Coal: *normal* 820 tons, *maximum* 1150 tons.

(I.) *New ships* 6 *pro.* 14,865 tons. Complement 822.

L. B. D.: 439⅓ × 79½ × 27½ ft. = 134 × 24·25 × 8·37 m.

Guns: 4A (305 m/m) + 18D* (164 m/m) (6 in the redoubt, 12, in pairs, in turret (47 m/m) + 2*. Torpedo tubes: 2 *submerged* + 3 *above water*. Armour as Suffren, sloping and protection to vitals = *aaaaa*.

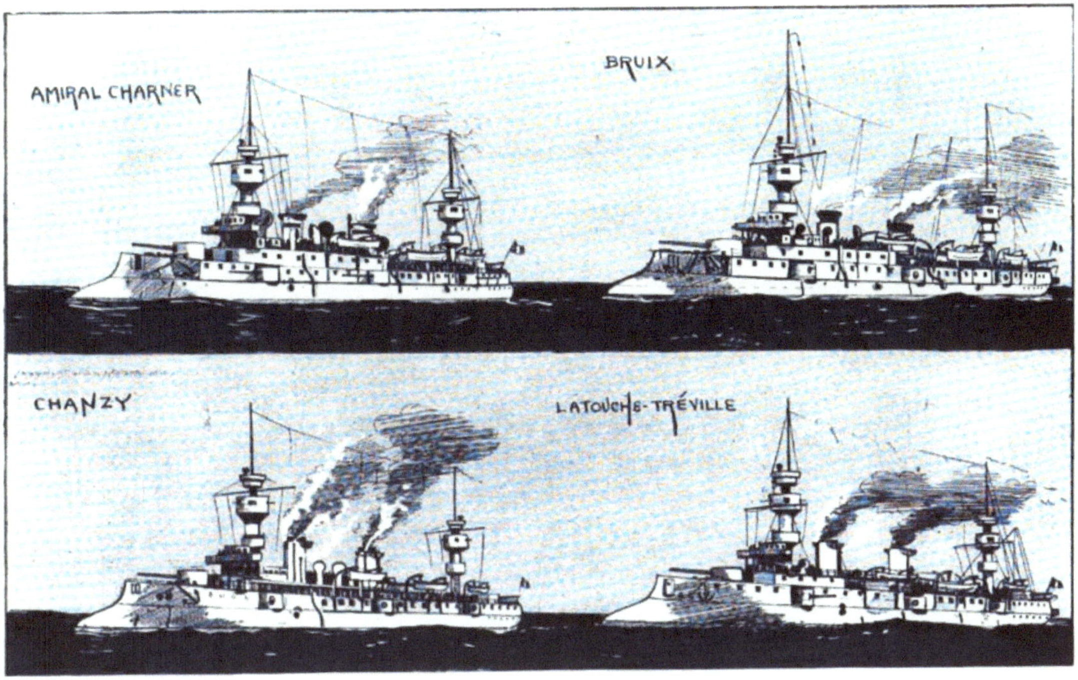

(III.)

AMIRAL CHARNER (1893): BRUIX (1894): CHANZY (1894). LATOUCHE TREVILLE (1892).

4750 tons. Complement 370.

L. B. D. $347\frac{1}{2} \times 46 \times 19\frac{1}{2}$ ft. $= 106 \times 14 \times 6$ m.

Guns: 2C (194 m/m) + 6D* (138 m/m, 45 cals.) + 6F* + 14* (47 m/m) + 6* (37 m/m). 5 torpedo tubes.

Armour: (Acier special) Belt *de* (92 m/m). Turrets *de* (92 m/m). Deck 90 m/m = 3·9".

Machinery: horizontal triple expansion. 2 screws. Boilers: 16 Belleville in 4 groups. Coal *circa* 400 tons.

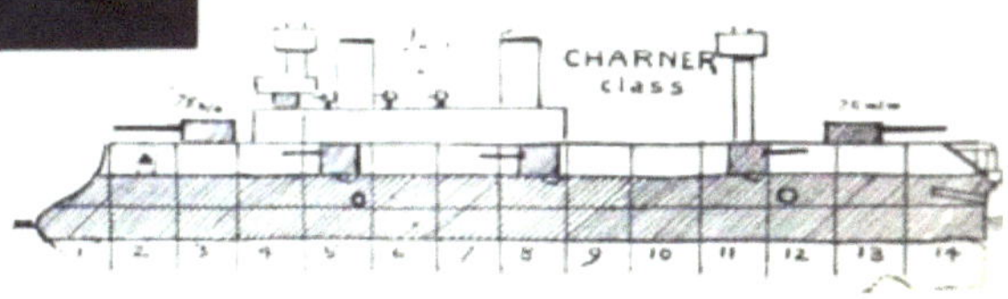

Trials	Bruix . .	24 *hours natural draught*	H.P. =	kts.	*hours forced draught*	H.P. =		kts.		
	Chanzy . .	" "	" =	"	4 " "	8537 "	=	17·8 "	*moyen.*
	Charner . .	" "	" =	"	4 " "	9148 "	=	18·4 "	"
	Latouche-T.	" "	" =	"	4 " "	8310 "	=	18·1 "	*mean.*

DUPUY-DE-LÔME. *Photo by Symonds & Co.*

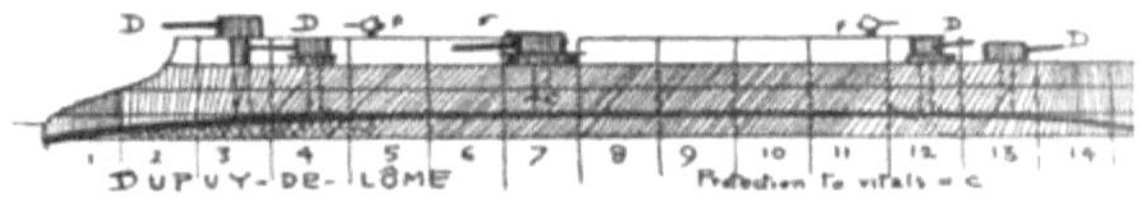

(III.) DUPUY-DE-LÔME (1890) (1893).

Circa 6500 tons. Complement 521.

L.: 374 ft = 114 m. B.: 52 ft. = 15·8 m. D.: 26⅛ ft. = 8 m.

Guns: 2C (194 m/m, model '87) + 6D* (164 m/m 30 cals.) + 6F* + 8* (47 + 8* (37 m/m). 4 torpedo tubes.

Armour (Creusot steel): Belt 374 × 17⅜ ft. / 114 × 5·3 m. (i.e. from 4½ ft. / 1·5 m. below wa upper deck) *d* (120 m/m). C gun turrets *ed* (100 m/m). Cellulos Deck 1·5″ steel = 35 m/m. Splinter deck. C. T. *d*.

Machinery: 2 sets inclined + 1 set vertical triple expansion. 3 13 cylindrical boilers. Coal: 900 tons = *circa* 3000 miles at 13 kts.

Trials 1895: 24 *hours natural draught* = 6250 H.P. = 17 kts.
4 *hours forced draught* = 13000 H.P. = 19·8 kts. *Foul*
and in shallow water.

(I.) JAUREGUIBERRY (1893).

Circa 12000 tons. Complement 624.

L. × B. × D.: 364 × 72½ × 30 ft. = 111 × 22 × 9 metres.

Guns: 2A (305 m/m, model '91) + 2A (274 m/m, model '87) + 8D* (138 m/m) + 4F* (65 m/m) + 12* (47 m/m) + 12* (37 m/m). Torpedo tubes: 2 *submerged* + 4 above water behind *de* armour.

Armour (Creusot steel): Belt *aaa* - *a* (450 - 250 m/m). Upper belt *de* (100 m/m). Large turrets and hoists *aa* (360 m/m = 14⅛″). Turrets for D* gun *ed* (100 m/m). Deck 67 m/m = 2·7″.

Machinery: Vertical triple expansion. 2 screws. 24 Lagrafel d'Allest boilers. Coal: *normal*, 750 tons; *max.*, 1100 tons.

Trials 1896: 24 *hours natural draught* 7000 H.P. = 16·5 kts. (*moyen*).
6 ,, *full power* 12400 ,, = 17·5 ,,
6 ,, *forced draught* 13819 ,, = 17·7 ,,
2 ,, ,, 15931 ,, = 18·1 ,, (*max.*)

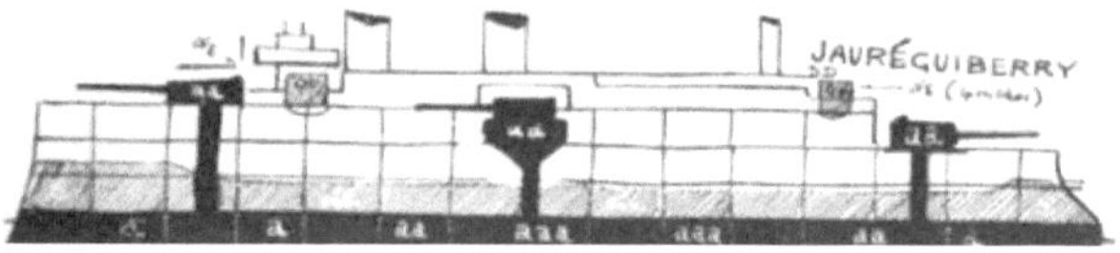

TROUDE (1888) & COSMAO (1889): LALANDE (1889).

Circa 2000 tons. Complement 200.

L.: 312 ft. = 95 m. B.: 31 ft. = 9·5 m. D.: 17 ft. = 5·2 m.

ns: 4E* (138 m/m, 30 cal.) + 7* (47 m/m) + (*except Cosmao*) 3* (37 m/m)
5 torpedo tubes.

mour: deck = *f* (37 m/m = 1½").

chinery: horizontal compound. 5 cylindrical boilers. 2 screws. Coal 200

trials Troude *forced draught 4 hours* 6265 = 20·7 kts.
Cosmao ,, ,, 6272 = 20·6 ,,
Lalande ,, ,, 6532 = 20·9 ,,

(6.) MILAN (1884).

1730 tons. Complement 193.

L. B. D.: 303 × 32⅝ × 16 ft. = 92·4 × 10 × 4·9 m.

Guns: 5E* (100 m/m) + 8*. Armour: 0.

2 torpedo tubes.

Machinery: 2 horizontal compound. 2 screws. Belleville boilers.

Coal: *normal* 300, *max.* 400.

Trials (189–) *4 hours forced draught* = 3960 H.P. = 18·4 kts. (*moyen*).

(6.) DUQUESNE (1876) (1894): TOURVILLE (1876) (1893).
6000 tons. Complement 500. 5580 tons. Complement 500.
Guns: 7D* (164 m/m. 30 cals.) + 14DE* (138 m/m) + 4* (47 m/m) + 14*. Torpedo tubes: 0. Armour: 0. 1 screw. Re-boilered 1893-94. Coal: Duq., 900. Tourville, 800 tons. *Trials* '94-'95.—Duquesne. I.H.P. 7313 = 16·5, *forced d. 4 hours.* Tourville, I.H.P. 7466 = 16·9, *forced 4 hours.*

Duguay Trouin, etc., like Tourville, but smaller.

(6.) DUGUAY TROUIN (1877). 3661 tons. Guns: 5D (164 m/m) + 5E (138 m/m) + 9*. Torpedo tubes: 2. *Trial speed*, I.H.P. 4399 = 15·9 kts. 1 screw. KERSAINT (1897). 1243 tons. Guns: 1D* (138 m/m) + 5E* (100 m/m) + 7*. Torpedo tubes: 0. I.H.P. 2200 *f. d.* = 15 kts. Coal: 200 tons. NIELLY (1880) & PRIMAUGUET (1882). 3300 tons (wood). Guns: 14E (138 m/m). Sea speed, 9 kts.

(4.) COETLOGON (1888) & SURCOUF (1889). 1932 tons. Guns: 4D* (138 m/m) + 7*. (See *Forbin*, p. 131.)

(4.) DESTRÉES (1898) & INFERNET (1899). 2452 tons. (Sheathed and coppered.) Complements 234. L. B. D. (*mean*): 312 × 39⅓ × *16* feet = 95 × 12 × *4·87* m. Guns: 2D* (138 m/m) + 4E* (100 m/m) + 8* (47 m/m). Armour-deck: 1½ inches = *c* (33 m/m). Machinery: I.H.P. 8500 = 20·5 kts. 2 screws. Coal: 345 tons, ***normal***; 480 tons, ***maximum***. (Like Coetlogon, but a D gun bow and stern.)

COETLOGON.

(T.) FOUDRE (1895).

(Carries 6 boats.)

6086 tons. Complement 410.

L. B. D.: 374½ × 51¾ × 25 ft. = 114·20 × 15·70 × 7·62 m.

Guns: 8E* (100 m/m) + 4F* (65 m/m) + 5* (47 m/m). Torpedo tubes: 4. Armour-deck = *c* (85 m/m). Machinery: I.H.P. 11,930 = 19·57 kts. (***trial***, *4 hours forced d.*). Boilers: 24 Lagrafel d'Allest. 2 screws. Coal: 840 tons.

(*Note.—Steam pipes have been altered since this photo was taken.*)

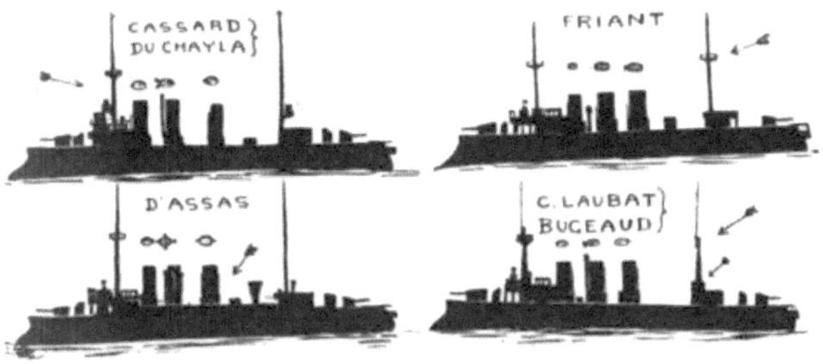

(2.) BUGEAUD : FRIANT & CHASSELOUP-LAUBAT (—— 1893).

3772 tons. Complement, 358.

L.: 308 ft. = 94 m. B.: 43 ft. = 13 m. D.: $20\frac{2}{3}$ = 6·30 m.

Guns: 6D* (164 m/m 45 cal.) + 4E* + 4* (47) + 11* (37). 2 torpedo tubes.

Armour: Deck = *c* (75 m/m = 3"). Machinery: Triple expansion boilers B., Belleville; Friant, 20 Niclausse; C.-Laubat, Lagrafel d'Allest. 2 screws. Coal, 587 tons. I.H.P. *nominal forced draught*, 9000.

&

(2). CASSARD: D'ASSAS: DU CHAYLA (—1895–96).

4000 tons. Complement, 393.

L. B. D.: $325\frac{1}{2}$ × 45 × 23 ft. = 99·25 × 13·7 × 7 m.

Guns and armour = Friant.

Machinery: Vertical triple expansion. 2 screws. Boilers 20 Lagrafel d'Allest. Coal, 624 tons. I.H.P. *nominal forced draught*, 9500.

Trials.	Bugeaud	24	*hours natural draught,*		*H.P.*	=		kts., 4	*hours forced H.P.*	9965	= 19	kts., *natural draught* 6	*hours* 7471	*H.P.* = 17·7 kts. (very economical steamer).
	Friant	24	,, ,,		,,	=		,, 4	,, ,,		= 18·1	,, ,, 6	,,	,, = ,,
	C.-Laubat	24	,, ,,		,,	=		,, 4	,, ,,		= 19·2	,, ,, 6	,,	,, = ,,
	Cassard	24	,, ,,	6100	,,	=	17·5	,, 4	,, ,,	10,143	= 19·8	,, ,, 6	,, 3790	,, = ,,
	D'Assas	24	,, ,,		,,	=		,, 4	,, ,,	9,500	= 19·8	,, ,, 6	,,	,, = ,,
	Du Chayla	24	,, ,,	6200	,,	=	17·8	,, 4	,, ,,	10,009	= 20·2	,, ,, 6	,, 3500	,, =15·15 ,,

Photo by W. Nye, Esq.

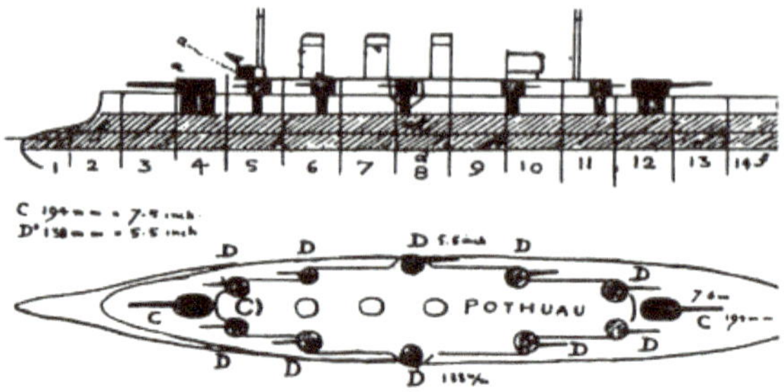

(III.) POTHUAU (1895).

5360 tons. Complement 463.

L. B. D.: 370¾ × 50 × 22½ ft. = 113 × 15·2 × 6·9 m.

Guns: 2C (194 m/m) + 10D* (138 m/m = 5·5 in.) + 16*. Armour (durci): Belt *d* - *f* (100 - 50 m/m). Deck, behind belt = *b* (85 m/m = 3⅓"). to vitals = *aa*. Turrets *b* (180 m.m.). D gun turret *c* (140 m/m). Lower deck *d* - *f*.

Machinery: Horizontal triple expansion. 16 Belleville boilers. 2 screws. 540 tons.

Trials (1897): 24 *hours natural d.* = H.P. = 17·8 kts.
4 ,, *forced d.* = 10378 ,, = 19·2 ,,

(1.) TAGE (1886) (1898).

Circa 7600 tons. Complement 530.

L.: 390 ft. = 119 m. B.: 54 ft. = 16·5 m. D.: 26 ft. = 7·9 m.

Guns: 8D* (164 m/m 28 cals.) + 10E* (138 m/m) + 12* (47 m/m) + 10* *revs.* 7 torpedo tubes *above water.*

Armour (*acier*): Deck = *ef* (45 m/m). Cellulose belt. Bulkheads to battery *ef* (100 m/m).

Machinery: Horizontal 3-cylinder compound. 2 screws. 12 cylindrical boilers. Coal: 1000 tons.

Trials (1896): 24 *hours natural d.* = H.P. = 17 kts. *moyen.*
4 ,, *forced d.* = 12114 ,, = 19·1 kts. *moyen.*

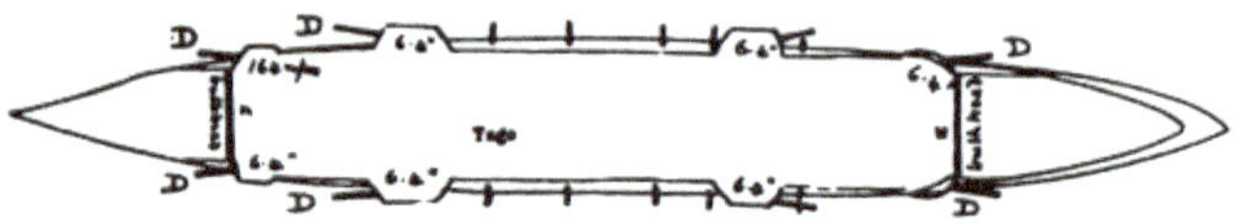

(Reconstructing again, 1900.)

(1.) D'ENTRECASTEAUX (1896.)

8014 tons. Complement, 521.

L. B. D. = 393½ × 58½ × 26 ft. = 120 × 17·85 × 7·9 m.

Guns: 2B (240 m/m, 40 cals.) + 12D* (138 m/m, 45 cals.) + 12*.

Torpedo Tubes: 2 *submerged* + 4 *above water*.

Armour (acier special): Deck = b (100 m/m), *i.e.* 4"). Also splinter deck 20 m/m. Turrets = *aa* (240 m/m). Casemates D* guns in battery *e* (70 m/m). Cofferdam. Prot. to vitals = *a*.

Machinery: 2 sets triple expansion with 4 double and 1-ended boilers. 2 screws.

Coal, *normal* 650 tons; *max.* 950 tons = *nominal* 5500 and 17,000 miles at 10 kts., or *nominal* 1250 and 8000 miles at 19 kts

Trials (1898) 24 *hours, natural d.*, 8800 *H.P.* = 17·8 kts. (moyen). 4 *hours. forced d.*, 14,630 *H.P.* = 19·2 kts.

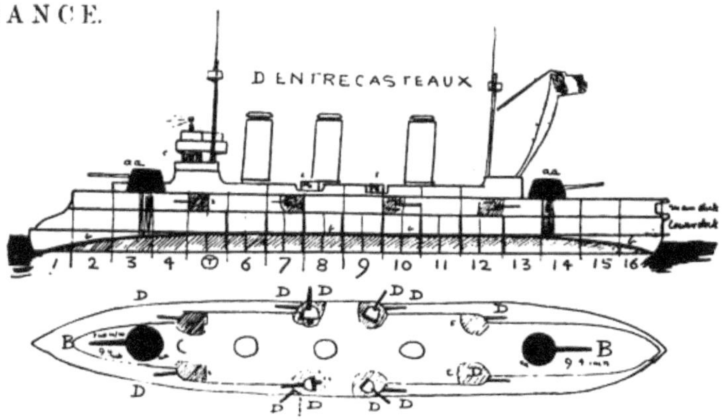

From Le Yacht.

Photo by Marius Bar, Toulon.

(1.) CÉCILLE (1888.)

Circa 6000 tons. Complement. 526.

L. B. D. = 379 × $49\frac{1}{4}$ × 24 ft. = 115·5 × 15 × 15 × 7·3 m.

Guns: 8D* (164 m/m) + 10E* (138 m/m) + 10* (47 m/m) + 10* (37 m/m) + 10* *revs.* 4 Torpedo tubes.

Armour: Deck = *ef.* Bulkheads to battery, *ef.* (100 m/m).

Machinery: Comp., 3 cylinder: 12 cylindrical boilers. 2 screws. Sea speed. *circa* 17 kts.

Coal. 940 tons.

Trials (1896) 2 *hours.* *Forced Draught.* I. H. P. 11,000 = 19·5 kts.

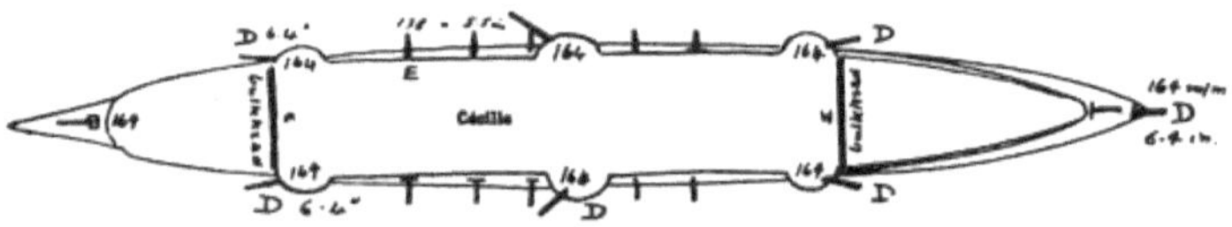

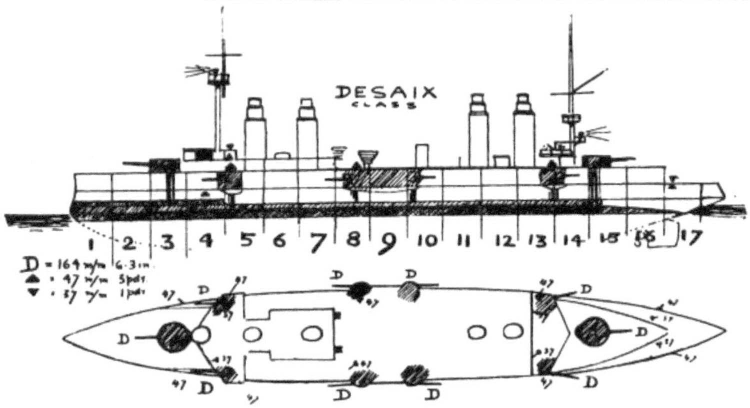

(III.) *Desaix & Kléber & Dupleix* (building).

(Sheathed and coppered.) 7700 tons. Complement .

L. B. D.: $426\frac{1}{2}$ × $58\frac{1}{3}$ × $24\frac{1}{4}$ ft. = 130 × 17·75 × 7·4 m.

Guns: 10D* (164 m/m, 45 cal.) + 10* (47 m/m) + 6* (37 m/m). tubes *above water*.

Armour (Krupp): Belt *c* – *e*. Deck = *c*. Prot. to vitals = *aaa*. T casemates *c*. Bulkhead aft *c*. Hoists *c*.

Machinery: vertical triple expansion. 3 screws. Boilers, Kléber, 20 in 4 groups. Desaix, 20 Belleville. Dupleix, Normand. Coal *normal* 880 1200 tons + petroleum (for use mixed with coal).

CHATEAU-RENAULT. *Photo by M. Bar.*

(2.) CHATEAU-RENAULT (1897).

(Commerce destroyer.)

8018 tons. Complement 600.

L. B. *D.* (*mean*): 443 × 56 × 22¼ ft. = 135 × 17 × 6·85 metres.

Guns: 2D* (164 m/m, 45 cals.)+6D* (138 m/m, 45 cals.)+10* (47 m/m). Torpedo tubes: 0

Armour-deck 2⅝ inches = *d* (67 m/m). Gun shields *f* (50 m/m).

Machinery: Vertical triple expansion. 3 screws. Boilers: 28 Niclausse. I.H.P. 13,800 *natural*; and 23,000 *forced* = 23 knots (*estimated*). Coal: *normal* 1400 tons + liquid fuel. *Maximum* 2100 tons.

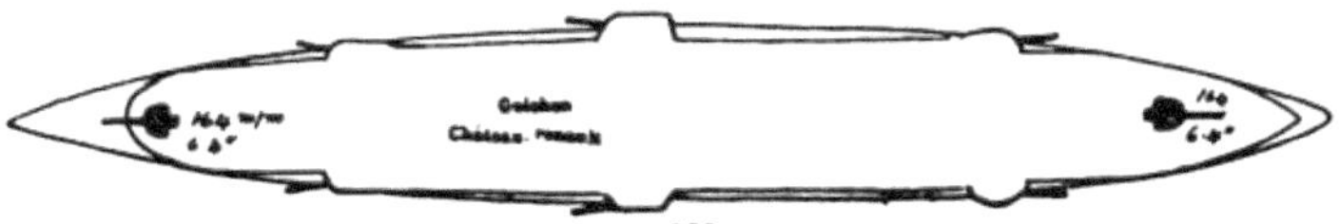

GUICHEN, 1900. *Photo by M. Bar, Toulon.*

(2.) *Jurien de la Granière* (building).

5500 tons. Complement 511.

L. B. *D.* (*mean*): 440 × 48⅔ × *22* feet = 134 × 14·80 × 6·70

Guns: 8D* (164 m/m, 45 cals.) + 12* (57 m/m).
Torpedo tubes: 2 *above water*.
Armour-deck 3" = *c* (76 m/m).
Machinery: I.H.P 17,000 = 23 kts. (*estimated*). 3 screws
Boilers: Guyot. Coal, *normal* 600 tons; *maximum* 900 tons.

(Commerce destroyer.) (2.) GUICHEN (1897).

8277 tons. Complement 625. L. B. *D.* (*mean*): 436⅓ × 55 × 24½ feet = 133 × 16·80 × 7·50 metres.

Guns: 2D* (164 m/m) + 6D* (138 m/m) + 10* (47 m/m).
Armour-deck 2½" = *d* (63 m/m). Gun shields *e* (50 m/m).
Machinery: I.H.P. 24,000 *forced d.* = 24 kts. (*est.*). Boilers, Lagrafel D'Allest. 3 screws. Coal, *normal* 1460 tons; *maximum* 2000 + liquid fuel.
Trials.—*24 hours natural draught* 18,500 I.H.P. = 20 kts. 4 *hours forced draught* 25,455 = 23·55 kts.

(Plan , see Chateau-Renault.)

(III.) *DUPETIT-THOUARS* (1900) & *GUEYDON* (1899) & *Montcalm* (bldg.). 9517 tons. Complement 540.

L. B. *D.* (*mean*): $452\frac{3}{4} \times 63\frac{5}{8} \times 24\frac{1}{2}$ = 138 × 19·40 × 7·45 m.

Guns: 2C* (194 m/m) + 8D* (164 m/m) + 4E* (100 m/m) + 16* (47 m/m) + 6* (37 m/m). Torpedo tubes: 2 *submerged*.

Armour (= Harvey nickel): Belt *b* (152 m/m). Deck = *d*. Prot. to vitals = *aa*. C gun turrets *b*. Supports *d*. Casemates *d*. Upper belt *d*. Bow *d* & *e*.

Machinery: Vertical triple expansion. 3 screws. Boilers: *D.-Thouars*, 28 Belleville; *Gueydon*, 28 Niclausse; *Montcalm*, 20 Normand. I.H.P. 19,600 = 21 kts. (*est.*). Coal: *normal* 1000, *max.* 1600 tons.

(III.) *Condé* & *Gloire* & *Sully* (bldg.) 10,000 tons; & *Aube* & *Marseilleise* (bldg.) 10,014 tons. Complements 600.

L. B. *D.* (*mean*): $452\frac{3}{4} \times 65\frac{1}{8} \times 24\frac{1}{2}$ feet = 138 × 19·85 × 7·45 m. D. of *Aube* & *Marseilleise* is 3 inches (7·6 c/m more).

Guns: 2C* (194) + 8D* (164) + 6E* (100) + 18* (47) + 8* (37). Torpedo tubes: 2 *submerged* + 3 *above water*.

Armour (Krupp): Belt amidships *a* (152 m/m). Deck = *c* (in *Aube* & *M.* it is = *b*). Prot. to vitals = *aaa* (in *Aube* & *M.* = *aaaa*). C gun turrets *a*. Supports *c*. Casemates and D gun turrets *c* Lower deck side (upper belt) amidships *b* (in *Aube* & *M.* it is *c*). Bow *c* & *e*.

Machinery: I.H.P. 20,500 = 21 kts. (*est.*). 3 screws. Boilers: *Condé* & *Gloire*, Niclausse; *Sully* & *Aube*, Normand.

(III.) JEANNE D'ARC (1899). 11,270 tons. Complement 626.

L. B. *D.* (*mean*): $477 \times 63\frac{5}{8} \times 26\frac{1}{2}$ feet = 145·40 × 19·40 × 8·12 m.

Guns: 2C* (194 m/m) + 14D* (138 m/m) + 16* (47) + 8* (37). Torpedo tubes: 2 *submerged*.

Armour (= Harvey nickel): Belt *b* (152 m/m). Deck = *d* (50). Prot. to vitals = *aa*. Lower deck side *e*. Bow *e*. C gun turrets *b*. D gun turrets *c*. Hoists and supports *d*.

Machinery: I.H.P. 28,500 = 23 kts. (*est.*). 3 screws. Boilers: 48 du Temple. Coal: *normal* 1400 tons + liquid fuel, *maximum* 2000 tons.

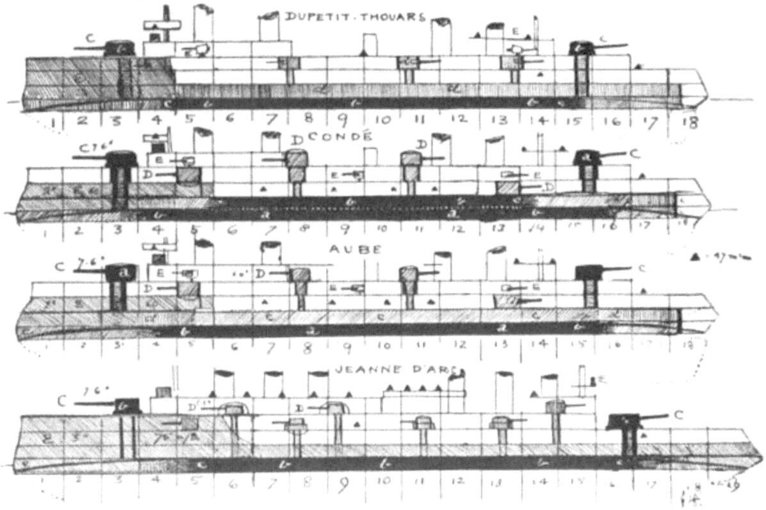

(II.) "*C 11*" & 5 *others* (projected). 12,416 tons. Complement 710.

L. B. *D.* (*mean*): $476 \times 71 \times 26\frac{1}{4}$ feet = 145 × 21·65 × 8·04 m.

Guns: 4C* (194 m/m) + 16D* (164 m/m) + 20* (47) + 4* (37). Torpedo tubes: 2 *submerged* + 3 *above water*.

Armour (Krupp): *Aube's* Type *Aube* + *Jeanne D'Arc* improved.

Machinery: I.H.P. 24,000 = 21 kts. (*est.*). Coal: *normal* 1350 tons.

The B 27 c/m (10·6″) Canet gun of 36 cal. has m. velocity 2230 f.s. (680 m.s.): m. energy *circa* 17500 ft. (5420 m.t.); velocity of the 30 cal. gun is *circa* 600 m.s. (1968 f.s.); m.e. *circa* 4650 m.tx. (15000 ft. tons).

TORPEDO BOATS.

Second Class.

6 Stettin boats (1885), 85 tons, 128 ft. long. *Trial* speed: 19 kts. 3 tubes.

6
—

Third Class.

6 Yarrow boats	(1881), 45 tons,	100 ft. long.	*Trial* speed:	19	kts.	2 tubes.
4 La Seyne boats	(1880), 50 ,,	72 ,,	,,	19	,,	1 tube.
6 ,, ,,	(1881), 35 ,,	89 ,,	,,	17·5	,,	1 ,,
2 Yarrow ,,	(1878), 18 ,,	75 ,,	,,	16	,,	? ,,

18
=

Also about 12 spar torpedo boats of 20 tons.

The Greek navy also contains the old ironclad Basilissa Olga, 4 E guns, 2000 tons; 3 unprotected ships of 1000 tons; and 9 gun-vessels of *circa* 400 tons carrying 2 or 1 F guns.

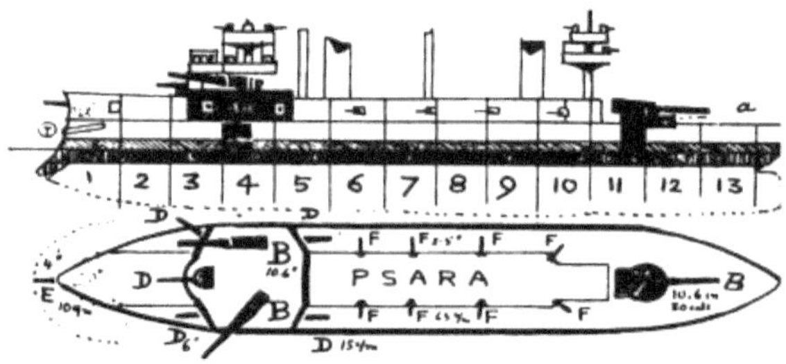

Psara.
(Data *see next page*.)

NAVARCHOS MIAULIS

(6.) NAUARCHOS MIAULIS

(La Seyne, 1879).

1800 tons.

Guns: 3D (long 17 c/m Krupp) + 3E (short 17 c/m Krupp).

(IV.) BASILEOS GEORGIOS (Blackwall).

L.: 200 ft. = 61 m

Guns: 2D + 1D. Armour: *s*. Sea speed:

The positions of fighting-tops as here given accurate.

Nous ne saurions pas garantir la mâture exac

Die hier gegebene Stellung der Marsen kann weise nicht genau sein.

E possibile che la posizione delle coffe n indicata, non sia precisa.

1774 tons.

GRÈCE.

(III.) HYDRA (1889 St. Nazaire) & SPETSAI (1889 Graville): PSARA (1890 Graville) (1897 La Seyne).

4890 and 4950 tons. Complement **400.**

L. B. D.: 331 × 52 × 23½ ft. = 101 × **15·7 × 7·2 m.**

Guns: (Hydra & Spetsai reconstructing as Psara): **2B + 1B + 5D* (15 c/m 45 cal.) + 1E*** (10 c/m 50 cal.) + 8·65 m/m* + 4·**47 m/m* + 12* (300 m/m - 120 m/m).**

Torpedo tubes: 3 (1 of them in bow.

Armour: Belt (under water 4 ft. = 1·2 m. wide) ***a.—d.*** **Above water line (4 ft. = 1·2 m. wide)** *e.* (75 m/m). Redoubt and Battery ***aa.*** **(350 m/m). After barbette** *a.* (300 m/m) shield *d.* Flat deck 75 m/m. In Hydra **62 m/m.**

Machinery: I.H.P. 6900—6700 = 17 kts. (*trial*). **Sea speed: 14—15 kts. 2** screws.

Psara.

Photo by Bar.

HOLLANDE.

ARTILLERIE.

	Calibre.		Long.	Initial Velocity.		Initial Energy.		Obus Ordinaire.	
	c/m	n.	cals.	ft. secs.	metre secs.	ft. tons.	tx. metre.	lbs.	kgs.
B	28	11	35	2133	650	17740	6494	297	135
C	28	11	12	1332	406	6563	2033	536	243
C	28	11	19	1558	475	9423	2918	476	216
C	21	8	35	1740	530	6470	2004	308½	140
D	17	6·8	22	1558	475	2226	690	112	51
D*	15	6	40	2500	762	4334	1341	100	45·3
D	15	6	35	2000	609	3100	960	112	51
E	15	6	23	1558	475	1450	449	86	39
E*	12	4·7	40	2150	655	1442	447	45	20·4
E	12	4·7	35	1755	535	1224	379	57	26
E	12	4·7	35	1804	550	1260	390	57	26
F	7·5	3	40	2379	725	500	154	12·7	28
Z	12	4·7	16	970	296	800	248	30·5	14
Z	7·5	2·9	15	...	...	...	...	...	...

TORPEDO BOATS.

First Class.

13 *to|be built or building.* 160 ft. long.

Second Class.

				On completion max. trial.	
5 boats	(1888–9),	90 tons,	128 ft. long,	Speed 21–19 kts.,	3 torp
1 Yarrow	(1888)	91 ,,	128 ,,	,, 24 ,,	3
5 Yarrow, etc.	(1887–6),	83 ,,	125 ,,	,, 21–20 ,,	2
3 ,,	(1893–4),	83 ,,	125 ,,	,, 21 ,,	2
1 ,,	(1891),	85 ,,	127 ,,	,, 21 ,,	2

15 + 5 other 83 ton boats belonging to the Indian Marine.

Third Class.

2 Yarrow	(1882),	45 tons,	100 ft. long,	Speed 21 kts.,	2 torpe
19 Thornycroft	(1886–78),	29 ,,	78 ,,	,, 18 ,,	2 spar
3 boats	(1890),	37 ,,	84 ,,	,, 18 ,,	1 ,,
1 White	(1883),	45 ,,	15 ,,	,, ? ,,	1 ,,
+ 4 *pro. to be built.*			100 ,,		

25 + 4

In addition to the ships enumerated, Holland possesses 30 gunboats carr 1C (28 c/m) gun, and 27 other gunboats for police duties, chiefly in the Indi These are from 850 to 350 tons, and generally carry 3E (12 c/m) guns. S 6 kts.

HOLLANDE.

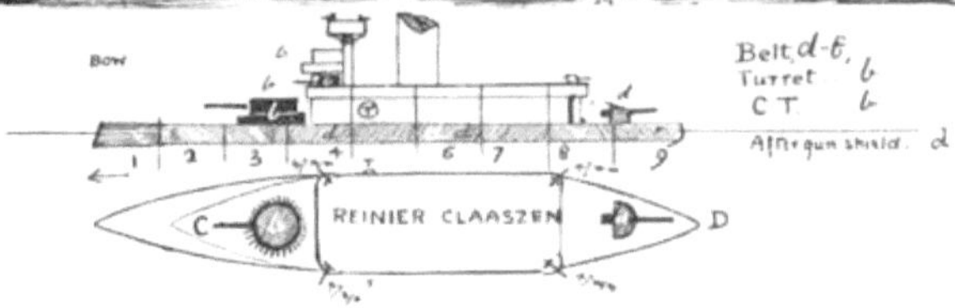

(V.) REINIER CLAESZEN (1891).

2490 tons. Complement .

L.: 229 ft. = 69·5 m. B.: $42\frac{1}{2}$ ft. = 13 m. D.: 15 ft. = 4·6 m.

Guns: 1C (21 c/m) + 1D (17 c/m) + 4 *. Speed: 10 kts.

Armour: Belt *de - e* (120–50 m/m). Turret *b* (280 m/m compound). C. T. *b*. Shield to D gun *d*. Deck 50–76 m/m.

I.H.P. 2035 = 12·5 kts. (*trial*). 2 screws.

Coal: 90 tons, *max.* 150 tons.

(V) 3 *Monitors to be built type A*, 3 *of type B*.

Type A 1500 tons. *Type B* 1400 tons.

L.B.D.: 187 × 42 × $10\frac{1}{3}$ (A type) $9\frac{2}{3}$ (B type) ft. = 57 × 13 × 3·1 (A) (3 B) m.

Guns: (*A*) 2C (21 c/m 40 cal.) + 4* (6 pdr. 7 m/m) + 4*. (*B*) 1C (21 c/m 40 cal.) + 1E* (12 c/m) + 4* (57 m/m) + 4*.

Armour: Belt *b - d*. Barbettes *b*. Curved deck behind belt = *d*.

Machinery: I.H.P. 700 = 9·5 kts. 2 screws. Coal 60 tons.

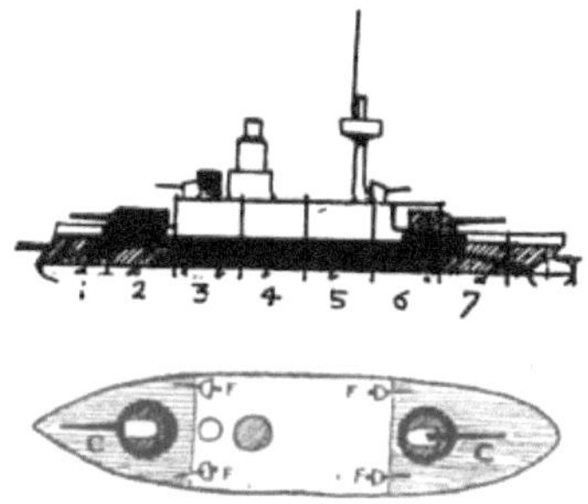

(III.) *KONINGIN REGENTES* (1900) & *No.* 2 (building).

4950 tons.

L. B. D.: 312 × 48 × 18 feet = 95 × 14·63 × 5·48 metres.

Guns:
2B (24 c/m).
4D* (15 c/m).
8F* (7·5 c/m).
2* (37 m/m).
Torpedo tubes:
2 *submerged.*
1 *above water* (in bow).

Armour (Krupp):
Belt *a–c.*
Deck = *c.*
Protection to vitals = *aaa.*
Barbettes *aaa.*
Hoods to B gun = *c.*

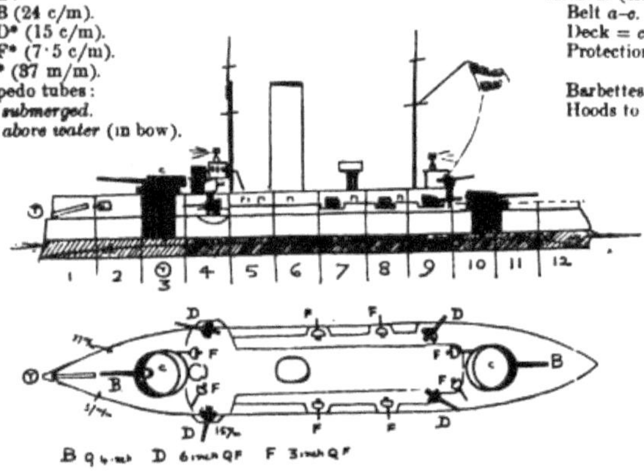

Machinery: Triple expansion. 2 screws. Boilers: Yarrow. I.H.P. 5300 = 16 kts. (*est.*).
Coal: 680 tons.

Notes.—Armour thicknesses are: Belt 6–4 in. (152–100 m/m). Deck 2 in. (50 m/m). Barbettes 10 in. (250 m/m).

(V.) LUIPARD (1876) & HAII & WESP (— 1871) & HIJENA & PANTER (— 1870) BLOEDHUND & CERBERUS (— 1869) & KROKODIL & HEILIGERLEE (— 1868 Birkenhead).

1600–1520 tons. Guns: 1C (28 c/m) + 1Z (7·5 c/m). Armour: Belt *d–e.* Turret *d.* Speeds: *circa* 7 kts.

(V.) STIER (1868 Birkenhead): SCHORPIOEN (1868 La Seyne) & MATADOR (1878) & DRAAK (1877).

2000–2100 tons. Guns: (Stier & Schorpionen) 1C (28 c/m) + 2Z (7·5 c/m). (Draak & Matador) 2C (28 c/m) + 7Z (7·5 c/m). Armour: Belt *d–e.* Turret *d.*

(IV.) EVERTSEN : PIET HEIN : KORTENAAR (1894).

L. B. D.: 284 × 47 × 19 ft. = 86·5 × 14·3 × 5·8 m.

3520 tons. Complement 260.

Guns: 3C (21 c/m) + 2D (15 c/m 35 cal.) + 6F* (75 m/m) + 6 * + 2 Torpedo tubes.

Armour: Belt (complete) *dc.* (152 m/m). Barbette *a* (240 m/m). Shield to C gun aft *c*. D gun shields *e*. Deck 60 m/m (2·3"). Prot. to vitals = *c*.

Machinery: I.H.P. 4800 = 16 kts. *trial*. Sea speed 13·5 kts. Coal 250 tons. 2 screws.

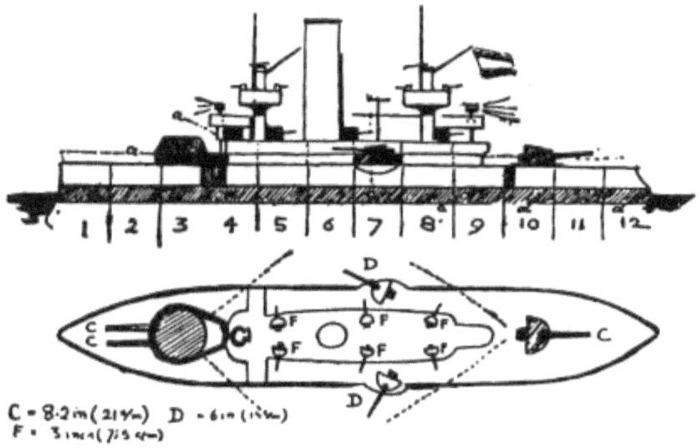

EVERTSEN.

Photo by le Marquis de Balincourt.

Photo, Geiser.

(1.) KONINGIN WILHELMINA DER NEDERLANDEN (18

L.B.D.: 328 × 49¼ × 21¼ ft. = 100 × 15 × 6·5 m.
4600 tons. Complement .

Guns: 1B (28* c/m 11 inch) + 1C (21 c/m) + 2D (17 c/m) + 4F* + 11* (37 m
3 Torpedo tubes (1 of them in the bow).

Armour: Deck = *d* (75 m/m). Engine hatches = *b* (100—150 m/m). T
(for B gun) *b* (280 m/m). Hoists *e* (100 m/m). C.T. *b*. Cellulose belt.

Machinery: *re-boilered and re-engined* (1899). Coal 450 tons. 2 screws.
I.H.P. *forced* 5900 = 17 kts.

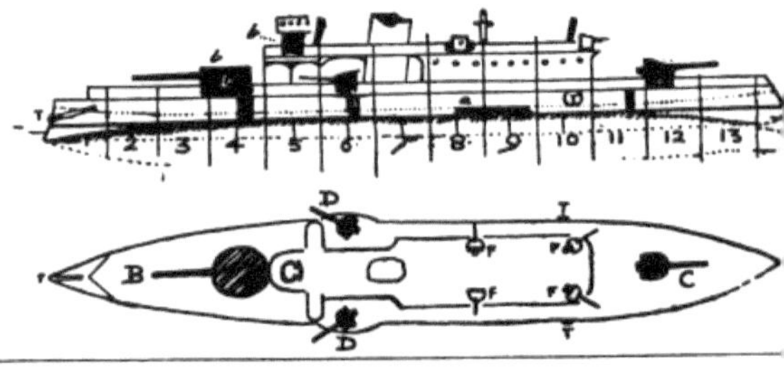

(V.) PRINS HENDRIK DER NEDERLANDEN (1866, Birkenhe

Guns: 4D + 4E + 2Z + 2*. **Armour**: *e* & *d* (guns). **Sea speed**: 6 kts. (?
3375 tons. Complement 243. 1 screw.

(6.) DE RUYTER (1880) & VAN SPEIJK (1882).

L.: 263 ft. = 80 m.

Guns: 6D (17 c/m) + 8E + 10*. **Armour**: 0. **Sea speed** *circa* 9 kts.

3400 tons. Complement 300. 1 screw.

HOLLANDE

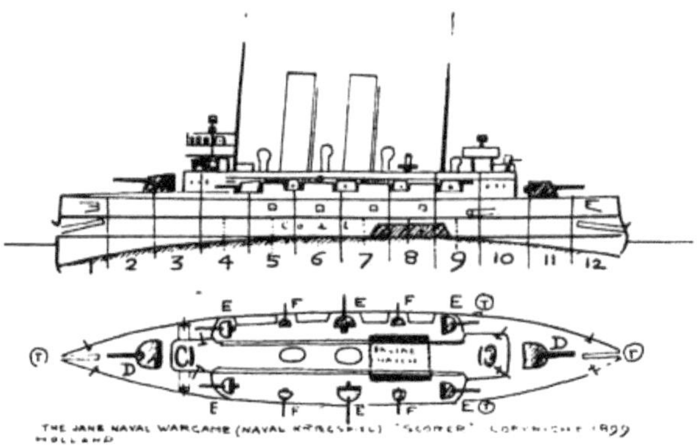

(2) HOLLAND: ZEELAND & FRIESLAND (—1896).

3900 tons. Complement .

L.B.D.: 294 × 48 × 17¾ ft. = 89·6 × 14·6 × 5·4 m.

Guns: 2D* (6 in. = 15 c/m) + 6E* (4·7 in. = 12 c/m) + 4F* (3 in. = 75 m/m) + 12*. 4 torpedo tubes.

Armour: Deck = *e* (50 m/m). Gun shields to D guns *d*, to E guns *e*, to F guns *f*.

Machinery: 2 sets 3-cylinder triple expansion (8 Yarrow, 2 cylindrical) boilers. 2 screws. Sea speed: kts.

Coal: 800 tons.

1897-8. 4 ***hours' trials with all boilers.***—Holland, 6236 H.P. = 16·9 kts., 10548 = 19·6 and 11712 = 20 kts. Zeeland, 9818 H.P. = 19·4 kts., 5203 = 16·46 kts., 10589 = 20 kts. Friesland, 5982 H.P. = 17·27 kts., 10416 = 19·87 kts., 10850 = 20 kts. Speed with 4 Yarrow boilers = *circa* 12 kts.

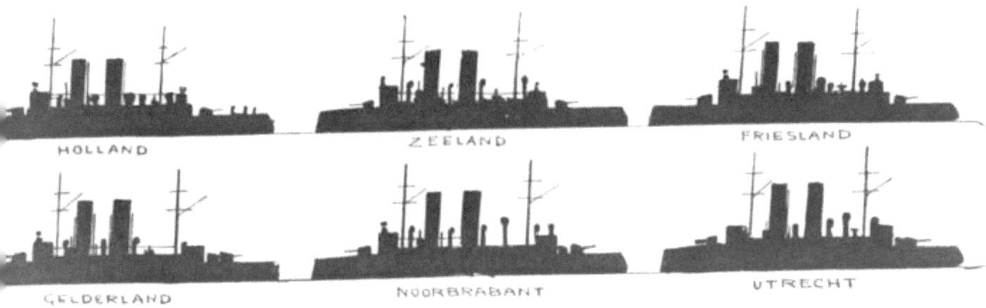

(2) GELDERLAND: NOORDBRABANT: UTRECHT (—1898).

3950 tons. Complement .

L.B.D.: 318½ × 48½ × 17¾ ft. = 97 × 15 × 5·4 m.

Guns, etc. = Holland.

Machinery: Boilers, Yarrow. I.H.P. 9000 – 10,000*.

Trials: Gelderland (1899) 4 ***hrs. natural*** I.H.P. 10,067 = 20·1 kts.

(Others of this class building or projected.)

(3.) SUMATRA (1890).

L.: 229 ft. = 70 m. B.: 36½ ft. = 11 m. D. 15½ ft = 4·7 m.

Guns: 1C (21 c/m) + 1D + 2E + 6 *. Armour: *prot.* Sea speed: 15 kts. 1700 tons. 2 torpedo tubes.

Armour: Deck = *f.* Cellulose belt. Gun shields *f.* C.T. *d.*

Machinery: I.H.P. *nat.* 2000 = 15 kts.: *forced* 3320 = 17·1 kts. (*trial*). screws. Coal 170 tons. Deck 1½" (375 m/m).

(6.) TROMP (1877) & ATJEH (1876) & KONINGIN EMMA DER NEDERLANDEN (1879).

L.: 262 ft. = 80 m.

Guns: 6D + 8E + 2Z + 6 *. Armour: 0. Sea speed: 8 kts. 3400 tons. Complement 300. I.H.P. 3000. 1 Screw. 2 Torpedo tubes.

(IV.) KONING DER NEDERLANDEN (1874).

L.: 268 ft. = 81·5 m.

Guns: 4C (11 in *m.l.*) + 4E + 2Z + 8 *. Sea speed: *circa* 8 kts. 5500 tons. Complement 256. I.H.P. 4500. 2 Screws.

Armour: belt *d—f.* Turrets *c.* Redoubts *d.* Flat deck 3". Torpedo tubes: 0.

ARTIGLIERIE.

Cal.			Lunghezza in Calibres.	Velocità Iniziale.		Energie Initiale, *circa*.		Proiett.	
in.	c/m			ft. secs.	met. secs.	ft. tons.	Dinamodi.	lbs.	kgs.
17·7	45	Duilio 100 ton Armstrong. *Bouch. m.l.*	20	1700	518	40000	12388	2000	908
17	43	105 ton Lauria *class*	29·5	1988	606	55000	17030	2000	908
17	43	100 ton Italia . .	26·5	1685	514	45000	13940	2000	908
17	43	102 ton Lepanto .	27·5	1929	588	50000	15485	2000	908
13·5	34	69 ton Sardegna. .	...	2014	614	35200	10902	1250	567
12	30·5	to be adopted.							
10	25·4	Elswick	40	2250	686	16930	5243	500	227
10	25·4	...	30	2030	619	12300	3717	448	226
	m/m								
6	152	Piemonte . . .	40	2220	676	3417	1038	100	45
6	152	6 ton	40	2000	610	3000	930	99	45
6	152	...	32	1956	596	2554	791	100	45
6	152	converted . . .	32	...	...	...	...	100	45
6	152	. . .	26	1946	593	2100	650	100	45
5·87	149	Flavio Gioa . . .	...	? 1673	510	?	?	?	?
4·7	120	Piemonte. . . .	40	1786	536	995	308	44	20
4·7	120	Duilio	35	1679	520	940	291	36	16
4·7	120	...	23	1591	485	916	283	36	16
4·7	120	...	20	1345	410	650	201	36	16

TORPEDO BOATS.

First Class. *Max.*

1	(Pellicano)	(1898),	147 tons,	155 ft. / 47 metres	long, speed 27 kts.,	2 tubes.
1	(Condor)	(1898),	136 ,,			
5	Schichau	(1888),	130 ,,	152 ft. / 46 metres	,, ,, 26·6 ,,	3 ,,
2	Boats (79, 87)	(1887),	110 ,,	135 ft. / 41 metres	,, ,, 24 ,,	5 ,,
2	Yarrow (77, 76)	(1887)	100 ,,	141 ft. / 43 metres	,, ,, 25 ,,	5 ,,
11						

Second Class. *Max.*

50	Schichau	(1887–92)	85 tons,	39 metres long,	speed 22 knots,	2 tubes.
81	Boats	(1893–9)	85 ,,	40 ,, ,,	,, 22 ,,	2 ,,
20	Schichau	(1887)	65 ,,	39 ,, ,,	,, 22·5 ,,	2 ,,
101						

Third Class. *Max.*

2	Yarrow	(1882)	40 tons,	30·5 metres long,	speed 22 kts.,	2 tubes.
30	Thornycroft	(1882–6)	34 ,,	30·5 ,, ,,	,, 22 ,,	2 ,,
4	boats	(1888)	34 ,,	31 ,, ,,	,, 21 ,,	2 ,,
2	Thornycroft	(1881)	33 ,,	28 ,, ,,	,, 22 ,,	2 ,,
2	,,	(1878)	30 ,,	23 ,, ,,	,, ? ,,	? ,,
31	boats	(1883)	16–8 ,,	... ...	,, 15–12 ,,	? ,,
71						

Submarine. *Max.*

1	Delfino	(1895)	? tons,	15 metres long,	speed ?
1					

In addition to the ships illustrated, the Italian Navy contains 8 despatch vessels 1000–2000 tons, 4E guns, sea speeds *circa* 12 kts.: 6 gunboats, *circa*; 12 wooden gunboats; and 16 subventioned merchantmen, 7000–2000 tons, speed 16–12 kts.

DESTROYERS.

2 funnels.

+4 Dardo, Freccia, Lampo & Strale, 320 tons, 6000 I.H.P., 2 tubes, 1F* + 5* (57 m/m) guns. Speed: 30 knots. *Schichau boilers. Schichau.*

4 funnels.

1 Fulmine, 260 tons, 4800 I.H.P., 3 tubes. Guns: 5* (57 m/m). Speed: 28 kts. *Sestri. Bleychenden boilers.*

+ 4 Meteora, Nembo, Turbine, Tuono, 350 tons, 6000 I.H.P., 2 tubes. Guns: 1F* + 3* (57 m/m). Speed: 30 knots. *Naples, Pattison. Bleychenden boilers.*

(T.) SAETTA & FOLGORE (—1886).

L.: 187 ft. = 57 m.

Guns: 6*. Armour: 0. Sea speed: *circa* 17·5 kts.

370 tons. Complement 55. 4 Torpedo tubes. I.H.P. 2200 = 20 kts. (*trial*). 2 locomotive boilers. Coal 60 tons.

(T.) CONFIENZA (1887).

L.: 230 ft. = 70 m.

Guns: 1E* (12 c/m) + 6* (57) + 2* (37). 4 torpedo tubes.

765 tons. Armour deck = *f*. Complement 105. I.H.P. 2000 = 17·5 kts. (*trial*). Coal 120–180 tons.

(IV.) SAN MARTINO (1863 La Seyne) & CASTELFIDARDO (1863 St. Nazaire): MARIA PIA (1863 La Seyne), & ANCONA (1864 Bordeaux).

L.: 256 ft. = 78 m.

Guns: {San Martino / Maria Pia} 8D (15 c/m 32 cal.) + 6E* (S. Martino 5E*) + 10*. Armour: *e*. Sea speed: 11 kts.

Guns: {Ancona / Castelfidardo} 6D* + 6E* + 8*. Armour: *e*.

Sea speed: {Ancona 11 kts. / Castelfidardo 11·5 kts.}

Torpedo tubes: 3. 1 Screw. Boilers: 6 Cylindrical. Coal 485 tons (fitted for liquid fuel).

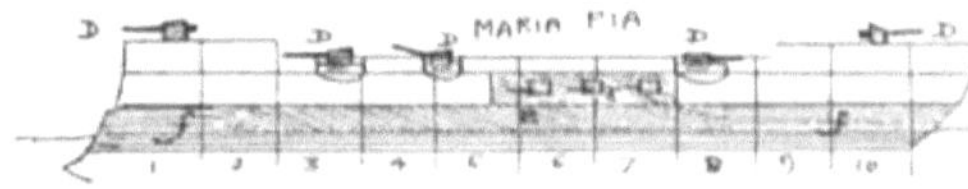

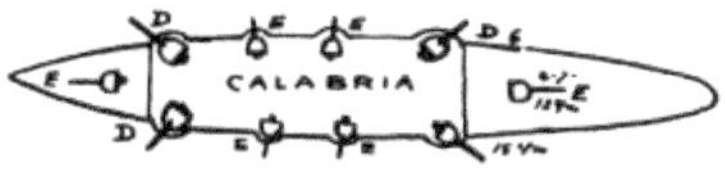

(3.) CALABRIA (1894).

L.: **249 ft. = 76 m.** B.: 42⅔ ft. = 13 m. D.: 17 ft. = 5·2 m.
Guns: 4D* + 6E* + 8* Armour: deck = *f*.
2470 tons. Complement: 240. 2 *submerged* torpedo tubes.
I.H.P.: trial 1897, *natural* 2439 = 14·5 kts., *forced* 4094 = 16·4 kts.
2 screws. Coal 500 tons.

& *PUGLIA* (1898).

2550 tons.

(6.) VOLTURNO (1887) & CURTATONE (1888).

L.: 177 ft. = 54 m.
Guns: {Volt. 4E + 6* / Curt. 6E + 6*}. Armour: 0. Sea speed: *circa* 8 kts.
1155 tons. I.H.P.: 1100 = 11 kts. (*trial*). 1 screw.

&

GOVERNOLO (1894).

L.: 185 ft. = 56·5 m.
Guns: 4E* + 6*. Armour: 0. Sea speed: 13 kts.
1255 tons. I.H.P.: 1100. 1 screw.

(4.) FLAVIO-GIOIA (1881) & AMERIGO-VESPUCCI (1892).

L.: 256 ft. = 78 m.
Guns: {F. Gioia: 6E + 4* / A. Vespucci: 6D + 4*.} Armour: deck = *f*. Sea speed: {F. Gioia: 12·5 kts. / A. Vespucci: 11 kts.}
3050 tons. I.H.P.: *circa* 4000. Coal: 500 tons (liquid fuel). 1 screw.

(II.) AMMIRAGLIO DI ST. BON (1897) & EMANUELE FILIBERTO (1897).

9800 tons. Complement 600.

L. × B. × *D.* (*mean*): 344½ × 69 × 25 feet = 105 × 21 × 7·60 metres.

Guns: 4B (25·4 c/m) + 8D* (15 c/m) + 8E* (12 c/m) + 6* (57 m/m) + 20* + 5 torpedo tubes, *submerged*.

Armour (acier nickel): Belt *aa–d*. Prot. deck = *b*. Prot. to vitals = *aaaa*. Lower deck redoubt *b*. Battery *b*. Barbettes *aa*. Shields to big guns *b*.

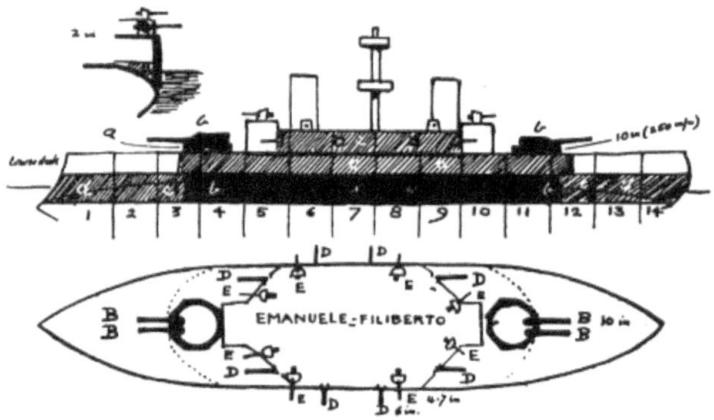

Machinery: 2 sets 4 cylinder triple expansion. 12 cylindrical boilers. 2 screws. I.H.P. *natural draught* 9000 = 16 kts.: I.H.P. *forced* 13,500 = 18 kts. Coal: *normal* 600 tons, *maximum* 1000 tons = *nominal* 7500 miles at 10 kts. Liquid fuel is also carried.

Note.—Armour belt is 10–4" (250–100 m/m). Armour deck 3·1" (80 m/m). On the main deck is a 6" (152 m/m) redoubt, the deck above which is 2" (50 m/m) thick. Battery 6" (152 m/m). Barbettes 10" (250 m/m).

(III.) GIUSEPPE GARIBALDI & VARESE (–1900) *F. Ferruccio* (building).

7400 tons. Complement 500.

L. B. *D.* (*mean*): 344 × 59 × 23 feet = 105 × 18 × 7 m.

Guns: 1B (25 c/m 45 cal.) + 2C* (20·3 c/m) + 14D* (15 c/m 45 cal.) + 2F + 10* (57 m/m) + 8* (37 m/m). 4 Torpedo tubes.

Armour (Ternii): Belt *b–c*. lower deck and battery *b*, barbettes *b*. Deck = ε. to vitals = *aaaa*.

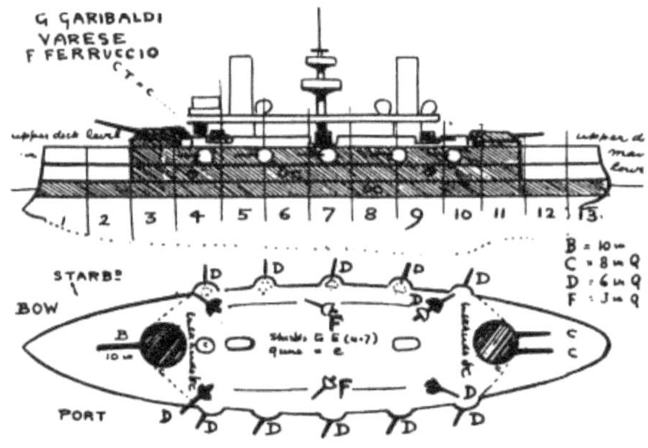

Machinery: Boilers, *G. Garibaldi* & *F. Ferruccio*, Niclausse, *Varese*, Bellev 13,500 = 20 kts. Coal: *normal* 650 tons; *maximum* 1200 tons.

Note.—All armour is 6" (152 m/m) thick, except ends of belt, 4½" (113 m/m).

(III.) DUILIO (1876).

11,138 tons. Complement *circa* 450.

L. B. D.: 341 × 65½ × 28 ft. = 104 × 20 × 8·50 metres.

c/m, 100 ton *M.L.*).
0 m/m, 35 cals.).
m/m).
n/m, 6 pounders).
m/m, 1 pounder).
ubes: 4 *above water*.

Armour (steel):
21½" Belt amidships *aa* (550 m/m).
Protection to vitals = *aa*.
3" Deck = *de* (75 m/m).
17" Redoubt below turrets *a* (425 m/m).
16" Bulkheads *a* (400 m/m).
18" Turrets *aa* (450 m/m).

inery: I.H.P. 7710 = 15 kts. (*trial*). Present sea speed, *circa* 8 kts. 2 screws.
0 tons, *normal*.

(Dandolo, p. 172.)

Photo, *Conti Vecchi*.

II.) RUGGIERO DI LAURIA (1884): ANDREA DORIA: FRANCESCO MOROSINI (— 1885).

11,000 tons. Complement 500.

L. (*p.p.*) B. D.: 328 × 65⅓ × 30 feet = 100 × 20 × 9·14 metres.

. (105 ton).
c/m, 26 cals.).
c/m).
m/m).
m/m).
m/m).
ubes: 2 *submerged*.
water.

Armour (compound):
17¾" Belt amidships *aa* (450 m/m).
3" Deck = *de* (75 m/m).
Protection to vitals = *aa*.
17¾" Redoubt amidships *aa* (450 m/m).
14" Bulkheads *a* (350 m/m).
1" Hoods (25 m/m). C. T. *aa*.

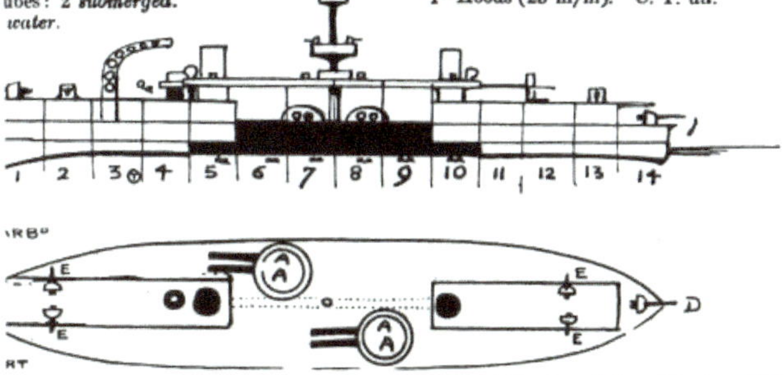

inery: I.H.P. *natural*, *circa* 7500 = 15–15·5 kts. (*trial*); *forced draught*, *circa* 10,200
7 kts. (*trial*). 2 screws. Coal, *normal* 850 tons + petroleum.

ANDREA DORIA.

Photo by *Fleming, Southsea*.

ITALIA.

(T.) TRINACRIA (1884 Clydebank).

L.: 433 ft. = 132 m.

Guns: 4*. Armour: 0. Torpedo tubes: 0.

9207 tons. 1 screw. Boilers: 6 2-ended, 11 1-ended.

Coal: 1550 tons. Speed *trial natural d.* 6500 H.P. = 16 kts.; *f*
9000 H.P. = 18·5 kts.

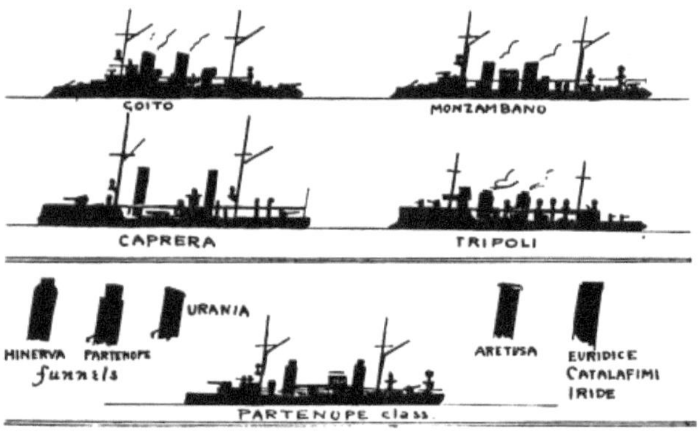

(T.) TRIPOLI (1886), GOITO (1887) } 820 tons.

Guns: 4* Armour: *prot.* Sea speed: 15 kts.

&

MONZAMBANO (1887), MONTEBELLO (1887) } 850 tons.

L.: 230 ft. = 70 m.

Guns: 6*. Armour: *prot.* Sea speed: *circa* 13 kts.

Draught 16 - 17 ft. = 4·9 - 5·2 m.

Goito 2 screws. Others 3 screws.

(T.) PARTENOPE (1889
CALATAFIMI (189
MINERVA (1892)
ARETUSA (1891)
EURIDICE (1891)
IRIDE (1891)
URANIA (1891)

Guns: 1E* + 6*. Armou
speed: *circa* 16 k

&

CAPRERA (1894).

L.: 230 ft. = 70

Guns: 2E* + 4*. Armou
speed: 18 kts. D.: 13
5 torpedo tubes

(2.) LOMBARDIA (1890) & ETRURIA (1891): UMBRIA (1891): LIGURIA (1893).

2280 tons. Complement 246.

L. B. D.: $262\frac{1}{2} \times 39\frac{1}{3} \times 17$ ft. = $80 \times 12 \times 5{\cdot}2$ m.

Guns: 4D* (15 c/m 32 cal.) + 6E* (12 c/m) + 8* (6-pdrs.) + 6* (1-pdr.) Torpedo tubes, 3.

Armour: deck = *f* (acier 25 m/m).

Machinery: 2 sets horizontal triple expansion. 4 2-ended boilers. 2 screws.

Coal: 400 tons.

Trials.	*Natural draught.*	*Forced draught.*
Etruria . .	I.H.P. 4602 = 16·9 kts. . .	I.H.P. 7019 = 18·8 kts. for 6 hours.
Liguria . .	.. 3444 = 15·9	.. 5536 = 18·1 .. .. 10 ..
Lombardia .	.. 2610 = 15	.. 6010 = 18·4 .. .. 6 ..
Umbria . .	.. 4910 = 17·7	.. 7400 = 18·9 .. .. 6 ..

LOMBARDIA (1899). *Photo by Conti Vecchi.*

(2.) ELBA (1892).

2730 tons. Complement, 247.

Guns: 4D* (32 cal. 15 c/m) + 6E* + 8*. **Armour**: deck = *f* (25 m/m 1").

L. B. D.: $272\frac{1}{3} \times 42\frac{2}{3} \times 18$ ft. = $83 \times 13 \times 5{\cdot}5$ m.

Machinery: 2 sets horizontal triple expansion. 2 screws. 4 cylindrical boilers. I.H.P. *trial* 1895, 4928 = 15·9 *natural*, *forced* 7471 = 17·9. Coal, 500 tons. Torpedo tubes, 2 above water.

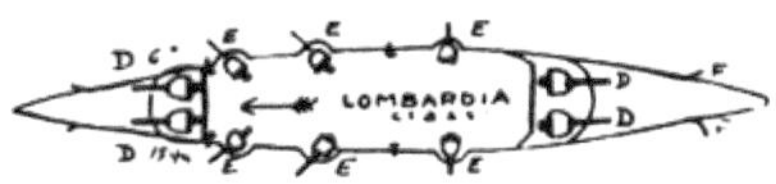

(1.) MARCO POLO (1892).

4583 tons. Complement, 308.

Guns: 6D* + 10E* + 9*. 4 torpedo tubes behind armour 1 *submerged* in bow.

L. B. D.: 328 × 49¼ × 19¼ ft. = 100 × 15 × 5·9 m.

Armour: Deck = *f* (25 m/m). Belt amidships *e* + Cofferdam. Gunshields *d*.

Machinery: Ansaldo after Maudslay. 2 sets vertical triple expansion. Boilers, 4 cylindrical. 2 screws. Coal, 600 tons. Speed: 10 hours ***natural draught, trial*** 1895 7047 H.P. = 16·86 kts. (*mean*). 3 ***hours forced*** = 10,664 H.P. = 17·8 kts. (***mean***).

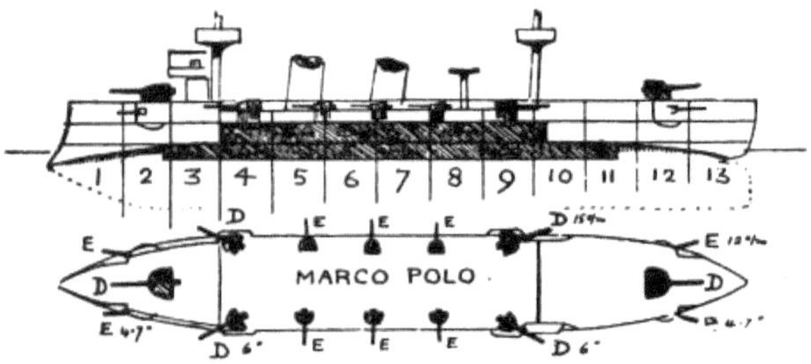

Photo by Conti

(3.) FIERAMOSCA (1888) ETNA (1885) STROMBOLI (1886) VESUVIO (1886)

FIERAMOSCA (1888) { 3600 tons. L. B. D.: 290 × 46 × 20 ft. = 88·5 × 14 × 6·1 m.

ETNA (1885) 3530 tons
STROMBOLI (1886) 3475 tons
VESUVIO (1886) 3427 tons
} L. B. D.: 283½ × 42½ × 20 = 86·5 × 13 × 6·1 m. Complement 308.

Guns: 2C (35 c/m) + 6D (15 c/m 26 cal.) - (D_* in Fieramosca) + 5 6-pr. 57 m/m* + 8* (37 m/m). 4 torpedo tubes, 1 in bow ***submerged***. Fieramosca, 2 ***submerged***. Armour: deck = *fe* (50 m/m).

Machinery: ***forced draught*** I.H.P. Fieramosca 7700 = 17·5 kts. Etna 7480 = 17·8 kts. Stromboli 6252 = 17 kts. Vesuvio 6820 = 17 kts. 2 screws. Coal *circa* 600 tons.

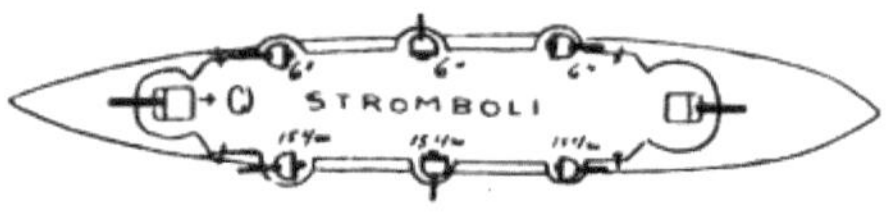

STROMBOLI. (All alike.) *Photo. West*

Photo, Mark Ferrez, Rio de Janeiro.

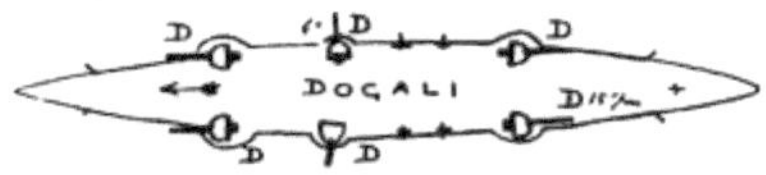

(3.) DOGALI (1886) (Elswick).

Guns: 6D (15 c/m 28 cals.) + 9*. Sea speed: 17·5 kts.

Dislocamento 2220 tons. Equipaggio 246.

L. B. D.: 250 × 36½ × 16 ft. = 76·2 × 11·1 × 4·9 m.

Armour: Deck = *d*t (62 m/m). Torpedo tubes 4.

Macchine: I.H.P. 7600 = 19·7 kts. (*trial*). Coal: 480 tons. 2 Screws.

Photo, Conti Vecchi

(2.) GIOVANNI BAUSAN (1883) (Elswick).

L.: 280 ft. = 85·5 m.

Guns: 2C (25·4 c/m) + 6D (28 cal. 15 c/m) + 4*. Sea speed: 15 kts.

3068 tons. Complement 274. D.: 20 ft. = 6·1 m.

Torpedo tubes: 1 submerged, 2 above water. Armour: Deck = *f* (37 m/m). Cork on water-line.

Machinery: I.H.P. 6550 = 17·5 kts. *trial* 1885. Coal 480 tons, *max.* 600 tons. 2 Screws.

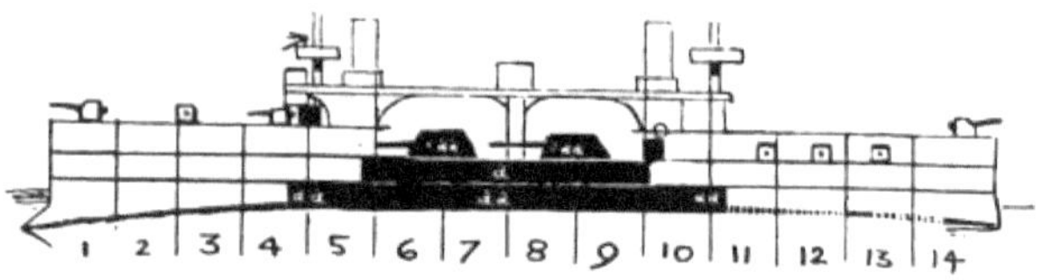

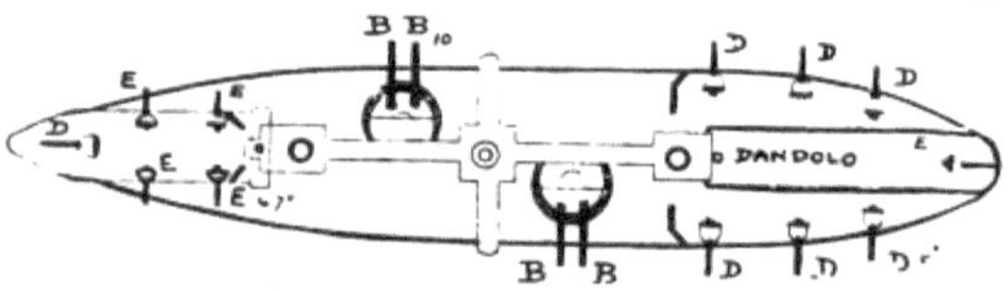

(II.) DANDOLO (1878) (1897).

L. B. D. (?): 341 × 65½ × 28ft. = 104 × 20 × 8·5 m.

12,265 tons. Complement, *circa* 450.

Guns: 4B (25·4 c/m) + 7D* (15 c/m 32 cals.) + 5E* + 10* (57 m/m 6 pdrs.) + 14* (37 m/m). Torpedo tubes 4.

Armour: Deck = *d* (75 m/m). Redoubt amidships side *aa* (550 m/m). Bulkhead *a* (400 m/m). Lower deck redoubt *a* (430). B gun turrets = *aa* (254 m/m).

Machinery: Express 2 sets vertical compound. Maudslay. Boilers (1898)? I.H.P. (8 *hours' trial*) 1898. Natural draught 7500 = 15 kts.

Coal: *max.* 100 tons. Normal ——?.

DANDOLO *Photo by Conti*

(2.) PIEMONTE (1888).

Guns: 6D* (15 c/m 40 cal.) + 6E* (40 cal.) + 10*. 3 torpedo tubes. Sea speed: 20 kts.

L. B. D.: 305 × 38 × 16 ft. = 93 × 11·6 × 4·9 m.

Displacement: 2650 tons. Complement: 263.

Armour deck = *de* (76 m/m). C.T. *f*.

Machinery: I.H.P. *trial natural draught* 7040 = 20·4 kts.; *forced draught* 12980 = 22·3 kts. 2 screws. Coal 200 tons.

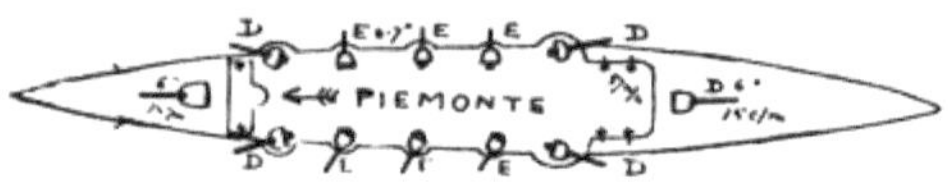

C. ALBERTO. *Photo, Conti Vecchi.*

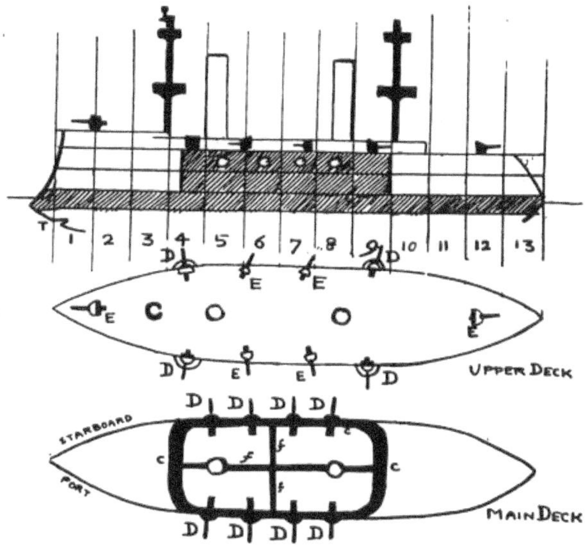

(III.) CARLO ALBERTO (1895): VETTOR PISANI (1896).

6500 tons. Complement 451.

L. B. D.: 324⅔ × 59 × 24 ft. = 99 × 18 × 7·3 m.

Guns: 12D* (15 c/m 40 cals.) + 6E* (12 c/m 40 cals.) + 10* (6-pdrs.) + 10* (1-pdrs.). Torpedo tubes: 5 *submerged.*

Armour: Harvey belt *c - d* (150-110 m/m). Battery *c* (150 m/m, 6"). Lower deck *c*. Bulkheads *c*. Gun shields *d*. Screens across battery *f*.

Machinery: {C.A. by Ansaldo after Maudslay / V.P. by Hawthorn Guppy} 2 sets 4 cylinder triple-expansion, 8 cylindrical boilers. 2 screws. Coal: *normal* 600 tons, *max.* 1000 tons and liquid fuel.

Trial, 1898, Carlo Alberto, *natural draught* I.H.P. 7860 = 17·3 kts.
1892, Vettor Pisani. .. .. .. 8290 = 17·5 .. (8 hours).

ITALIA.

(X.) (V.) AFFONDATORE (1865 Blackwall) (1887) (1

4062 tons. Complement 253.

L. B. D.: 292 × 39⅓ × 21⅓ ft. = 89 × 12 × 6·5 m.

Guns: 2C (25 c/m) + 6E* + 12*. **Sea speed**: 8 kts. 4 torpedo tu
Armour: Belt and turret *f*.
Machinery: I.H.P. 2620. 1 screw. **Coal**: 460 tons.

(4.) CRISTOFORO COLOMBO (1875).

L.: 246 ft. = 75 m.

Guns: 8E + 8*. **Armour**: 0 **Sea speed**: *circa* 14 kts.

2316 tons. Complement . D.: 19 ft. = 5·8 m.

I.H.P. 3782. 1 screw. Coal 500 tons.

(III.) *Amalfi* & *Genova* & Pisa & Venezia (to be built).

Displacement *circa* 8000 tons. Length. 400 feet = 122 metres.

Guns: 12C* (20·3 c/m) + 12F* (75 m/m). I.H.P. 19,000 = 22 kts.

Type: Benedetto Brin (see p. 175), with 2 C guns in each turret, and no guns in battery. All armour 6" Krupp *a* (152 m/m). 3 funnels, fore and aft; 2 masts.

(III.) RE UMBERTO (1888). 13,251 tons. SICILIA (1891). 13,375 tons. Complement 794.

L. B. D. = 400¾ × 75½ × 29 ft. = 122 × 23 × 8·8 m.

Guns: 4A (34 c/m) + 8D* (15 c/m 40 cals.) + 16E* (12 c/m cals.) + 10* (57 m/m) + * (37 m/m) + 5 Torpedo tubes.

Armour: Deck = *c* (75 m/m). Belt partial, to 1 metre below water line *e* (100). Bulkheads Barbettes (inclined at angle) = *aaaa* (450 m/m) Hoists *b*. C.T. *b*. Shields to D* guns *e*.

Machinery: {Re Umberto: Maudslay / Sicilia: Ansaldo} 4 sets comp. 2 on each shaft. 18 cylindrical boilers. Screws.

Coal: *normal* 1200 tons, *max.* circa 3000 tons.

Re Umberto *trials*: 1893 hours *forced draught*. I.H.P. 17,000 = 18·2 kts.

Sicilia *trials*: 1895—6 hours *natural draught*. I.H.P. 15,974 = 19·2 kts. (Ship light.)

(III.) SARDEGNA (1890).

13,860 tons. Complement 794.

L. B. D. = 410 × 77 × 29 ft. = 125 × 23·5 × 8·8 m.

Guns: 4A (34 c/m) + 8D* (15 c/m 40 cal.) + 16E* (12 c/m cal.) + 10* (6 prs.) + 20* (1 prs.) + 5 Torpedo tubes, 2 of them submerged.

Armour: Deck = *c* (75 m/m). Belt (partial to 1 metre below water line) *e* (4"). Lower deck *e* (4"). Bulkheads *f*. Barbettes (slanting) = *aaaa* (350 m/m). Hoists *b*. C.T. *b*. Shields to D* guns *e*. 3 Bottoms.

Machinery by Hawthorn Guppy, 4 sets triple expansion, 2 on each shaft. 18 cylindrical boilers. 2 screws. Coal: *normal* 1200 tons; *max. circa* 3000 tons.

Trials 1894 at displacement 13,950 tons.
3 hours *natural draught*. I.H.P. 13,169 = 19 kts.
3 hours *natural draught*. I.H.P. 14,190 = 19·2 kts.
2 hours *forced draught*. I.H.P. 16,220 = 19·6 kts.

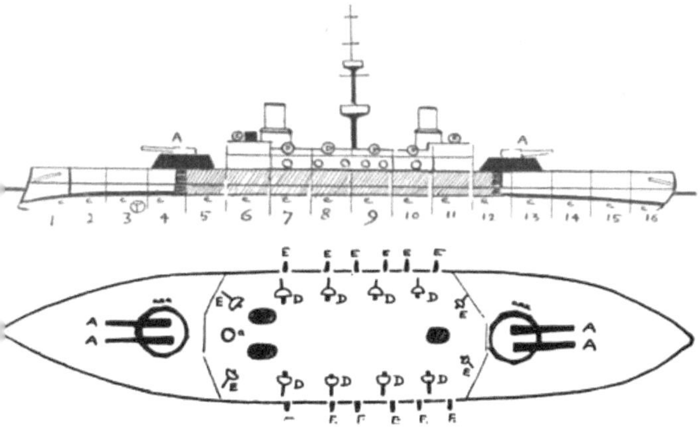

Re Umberto. *Photo by Fleming. Southsea.*

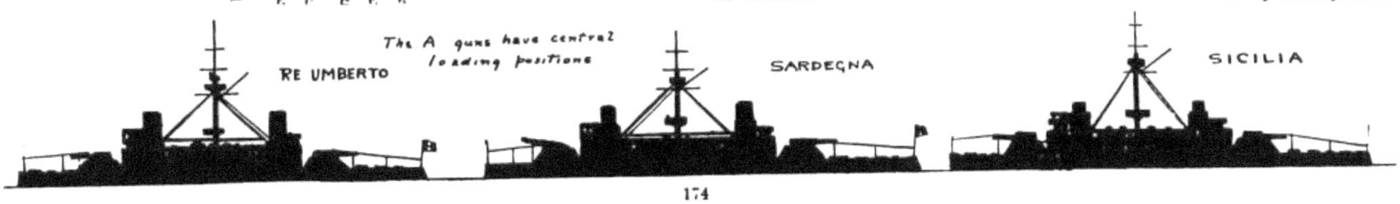

(ITALY) ITALIA (ITALIEN).

(I.) *BENEDETTO BRIN & REGINA MARGHERITA* (building) & *Principessa Elena* (projected).

12,765 tons.

L. (*p.p.*) B. D. (*mean*): 413 × 78 × $27\frac{1}{4}$ feet = 134·50 × 23·84 × 8·25 metres.

Length over all, 430 feet (131 metres).

Guns:
- 4A* (30 c/m).
- 4C* (20 c/m).
- 12D* (152 m/m.)
- 10F* (75 m/m).
- 6* (47 m/m).
- 2 Maxims.

Torpedo tubes:
- 4 submerged.

Armour (Krupp):
- 6" Belt, lower deck and battery, amidships *a* (152 m/m).
- 2" Belt, at ends *e* (50 m/m).
- 3" Armour deck = *a* (75 m/m).
- Protection to vitals = *aaaa*.
- 10" Barbettes and bulkheads, *aaa* (250 m/m).
- Hoods to big guns = *aaaa*.
- 6" C gun turrets, *a* (152 m/m).

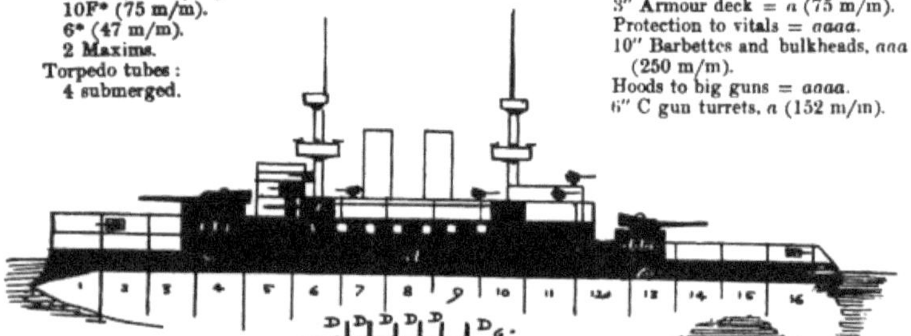

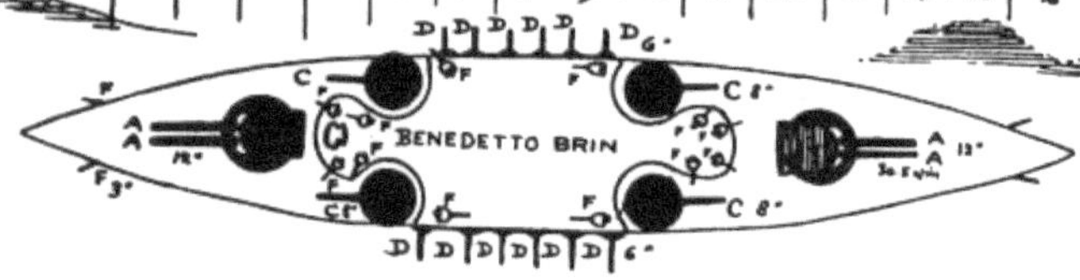

Machinery: 2 sets 4 cylinder triple expansion. 2 screws. Boilers: *B. Brin*, Belleville: *R. Margherita*, Niclausse. I.H.P. *natural draught* 14,000 = 18 kts. (*est.*); *forced d.* 18,000 = 21 kts. (*est.*). Coal, *normal* 1000 tons + petroleum; *maximum* 2000 tons = *nominal* 10,000 miles at 10 kts.

Note.—The belt is $10\frac{1}{2}$ feet (3 metres) deep. Guns, rapid loaders, central pivot mounting, load at any elevation.

(1.) ITALIA (1880). 15,654 tons. Complement 675.
LEPANTO (1883). 15,900 tons. Complement 700.

L. B. D.: $400\frac{1}{4}$ × $75\frac{1}{2}$ × $32\frac{3}{4}$ (*circa*) feet = 122 × 23 × 10 (*circa*) m.

Guns:
- 4A (43 c/m).
- (In *Italia* 100 tons, in *Lepanto* 102 tons).
- 8D* (152 m/m, 26 cals.).
- 4E* (120 m/m, 28 cals.).
- 12* (57 m/m).
- + in *Italia* 26*, in *Lepanto* 34* (37 m/m).

Armour (compound):
- 3" Deck = *de* (75 m/m).
- Very thick coal bunkers.
- Protection to vitals = *c*.
- 16" Funnel bases, slanting = *aaa*
- 19" Redoubt for A guns (inclined)
- 18" Big hoist amidships *aa* (450 m

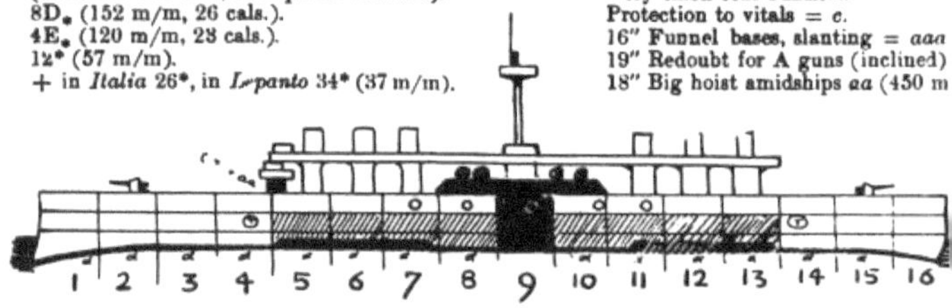

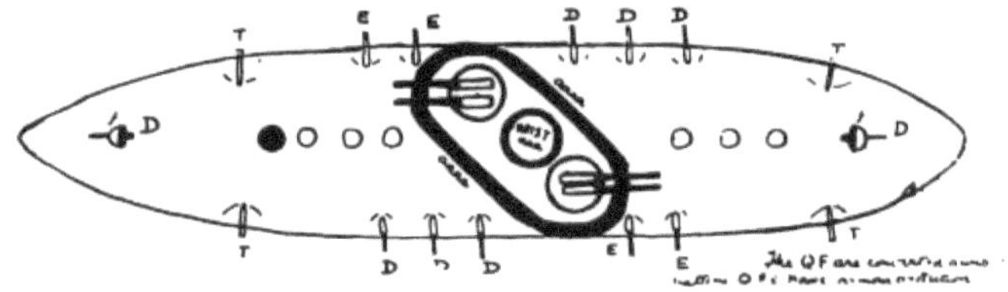

Machinery: 4 sets vertical 3 cylinder compound. 2 screws. Boilers: *Italia*, 26 oval: *I* + 16 locomotive. I.H.P. *Italia* 11,986 = 17·8 kts. (*trial* 1887): *Lepanto* 16,150 = 18·4 *trial* 1888). Coal: *normal* 1650 tons; *maximum circa* 3000 tons.

Both to be re-constructed.

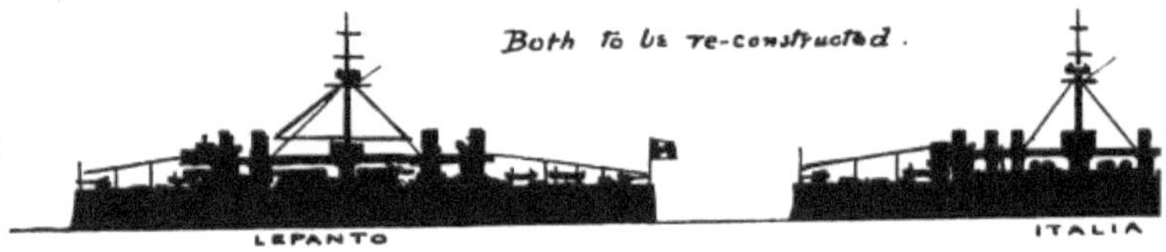

JAPON 本日 NIPPON.

Revised by Lieut. Com. Takarabe, Imperial Japanese Navy.

Nominal calibre.		—	Long.	Initial velocity.		Initial energy.		Shell.		Gun.
n.	c/m		cals.	foot secs.	metre secs.	ft. tons.	metre tx.	lbs.	kgs.	tons.
2·5	32	Canet([1]) . . .	40	2306	611	35220	10900	990	449	66
2	30·5	Elswick([3]). . .	40	2423	738	34600	10716	850	385	49
2	30·5	Krupp([4]) . . .	20	1755	535	14750	4359	725	328	35·4
)	25·4	Elswick([4]). . .	32	2030	619	13000	4026	448	203	26
)·2	26	Krupp([6]) . . .	25	1640	500	8400	2600	450	204	..
l	28	Armstrong, *m.l.*([7])	13	1332	405	6563	2032	536	242	25
3·2	21	Krupp	30	1935	590	6167	1910	..	..	13
3	20·3	Elswick . . .	45	2650 / 2480	808 / 755	10226 / 10662	3167 / 3302	210 / 250	95 / 113·4	20
3	20·3	Elswick . .	40	2242 / 2068	683 / 630	7319 / 7413	2267 / 2296	210 / 250	95 / 113·4	15½
;	15	Elswick . . .	45	2570	783	4580	1396	100	45·3	7
;	15	Elswick . . .	40	2500	762	4334	1311	100	45·3	6½
;	15	Elswick . . .	40	2220	676	3417	1058	100	45·3	6
;	15	..	35	1958	597	2554	791	100	45·3	5
;·6	17	Krupp	..	1657	505	2600	805	132	60	..
4·7	12	Elswick . . .	45	2570	783	2060	638	45	20·4	2⅔
4·7	12	Elswick . . .	40	2150	655	1442	447	45	20·4	2
4·7	12	Elswick([1]). . .	32	1938	590	900	279	36	16·3	1⅔
3	7·5	Elswick . . .	40	2200	670	420	130	12	5·4	2

latsushima class: ([3]) Shikishima, Fuji, and Yashima; ([4]) Chin Yen; ([4]) Cruisers; Yuen-Go.

—There are no "converted Q.-F." guns in the Japanese Navy. The velocity given for the 32 c/m ie Service velocity. The maximum velocity is 720 metre sec.

DESTROYERS.

2 funnels.

6 Kagero, Murakumo, Ousougoumo, Shinonome, Shiranöi, Yugiri (1897-99), 275 tons. I.H.P. 5400. *Thornycroft* boilers, *Thornycroft*. Speed 30 kts. (L. 210 feet = 64 metres.)

4 funnels.

6 Akebono, Ikadsuchi, Inazuma, Niji, Oboro, Sazanami (1897-99), 306 tons. I.H.P. 6000. *Yarrow* boilers, *Yarrow*. Speed 31 kts. (L. 220 ft. = 67 metres.)

12

Note.—All the above carry 2 torpedo tubes. Guns: 1F* (75 m/m) aft, + 5* (57 m/m). Bullet-proof conning towers.

TORPEDO BOATS.

First Class.

			max. on trial.	
1 ***Kotaka*** (Yarrow, 1886),	190 tons.	Speed	19 kts.	6 tubes.
5 (Yarrow, 1898)	135 ..	..	24 ..	3 ..
1 (Germania)	128 ..	..	19 ..	6 ..
4 (Normand, 1898-99)	150 ..	..	24 ..	3 ..
+ 4 (Schichau, 1899-'00)	? ..	..	..	..
+ 8 *pro.*	..	..	..	..

11 + ***12 building or projected.***

Second Class.

				max. trial.	
2 Schichau (1891),	90 tons,	128 ft. long.	Speed	23 kts.	3 torpedo tubes.
1 Normand (1891),	75 ..	118 ..	..	23 ,,	2 ,,
21 various (1889),	56 ..	114 ..	..	20 ,,	

24

Third Class.

4 Yarrow (1879), 40 tons, 100 ft. long. Speed 20 kts. 3 torpedo tubes.

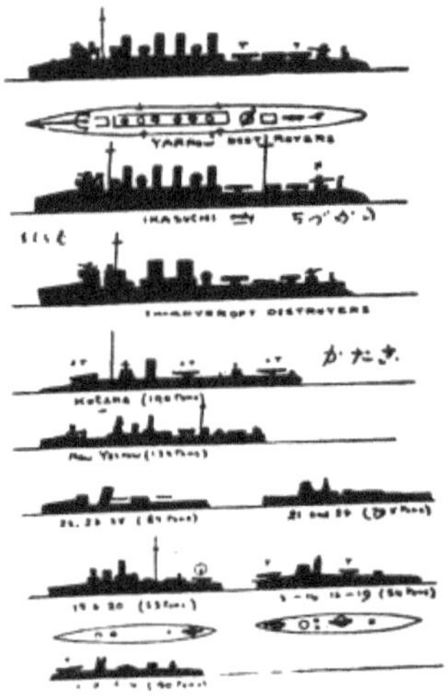

(T.) *Toyohaschi* (projected or building).

Torpedo depot ship 6750 tons

L. B. *D.*: (*mean*) 347½ × 59 × 23 feet = 106 × 18 × 7 metres.

Guns: 8* (47 m/m) + 24 Maxims, etc.

Machinery: I.H.P. 12,000 = 20 kts. (*est.*). 2 screws Belleville boilers.

SUBSIDISED MERCHANTMEN.

(Gross tonnage in brackets after the names.)

14 kts.

Futami (3496), Kasuga (3797), Kamakuru (6123), Kobé Saikio (2900).

13·5 kts.

Sado (6220).

13 kts.

Bingo (6240), Hakata (6150), Hitachi (6170), Inaha Kanagava (6150), Kawachi (6100), Kinschiu (4000), Matsuyama Nagato (1854), Omi (2473), Rio-jo (4800), Sanuki (6100), § (1866), Tamba (6100), Tosa (5794), Wakasa (6266), Yamaguchi Yamaschiro (2530), Yokohama (2300).

12·5 kts.

Ikiai (3076), Kokura (2567), Sakura (2953), Ouakanoura (2

12 kts.

Kagoshima (4370), Taïren (2897), Tokio (2194).

(V.) HEI-YEN (*ex* Ping Yuen) (1890 Foochow).

2600 tons. Complement 250.

L. B. D.: 200 × 40 × 19 ft. = 61 × 12 × 5·89 metres.

Guns: 1C (26 c/m Krupp) + 2D* (15 c/m 45 cals., Elswick) + 9*.

Torpedo tubes: 4.

Armour (compound): Short belt amidships *c*. Deck = *e*. Prot. to vitals *c*. Barbette *e* C T. *e*.

Machinery: I.H.P. 2400 = 8 kts. 2 screws.

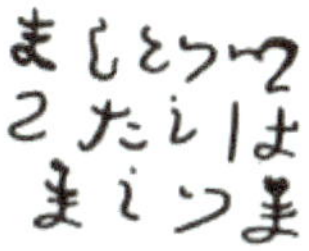

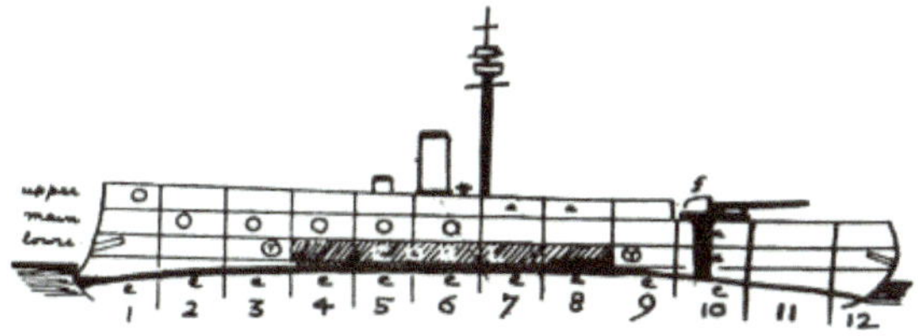

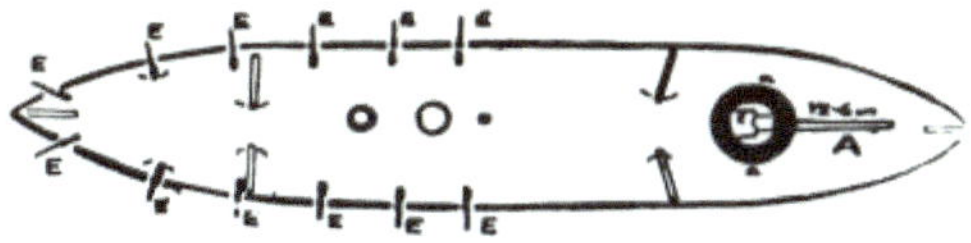

(2.) ITSUKUSHIMA (1889 La Seyne): HASHIDATE (1891): MATSUSHIMA. (1890 La Seyne).

4277 tons. Complement, 360.

L. B. D.: 295 × 50½ × 21¾ ft. = 90 × 15·3 × 6·5 m.'

Guns: 1A (32 c/m Canet) + 11E* (12 c/m ·32 calibres) + 5 6-pdr.* (57 m/m) + 11 3-pdr.* (47 m/m) + 6*. 6 torpedo tubes. (Matsushima: 1A + 12E* + 5·57 m/m + 16·47 m/m + 6*).

Armour: Deck = *f* (50 m/m) cellulose belt. Engine hatches *a*. Barbette *a* (300 m/m.) Shield to big gun *e*.

Machinery: I.H.P. ***natural draught*** 3400 = 15·7 kts. (trial), ***forced draught*** 5400 = 16·5 kts. trial. 2 screws. Sea speed: 14·5 kts. Coal, 400 tons. 6 boilers.

(All are now *rigged* as Hashidate).

HASHIDATE, red band. ITSUKUSHIMA, black band. The two are identical, save that Hashidate has E* guns slightly sponsoned.

(NIPPON) JAPON (JAPAN).

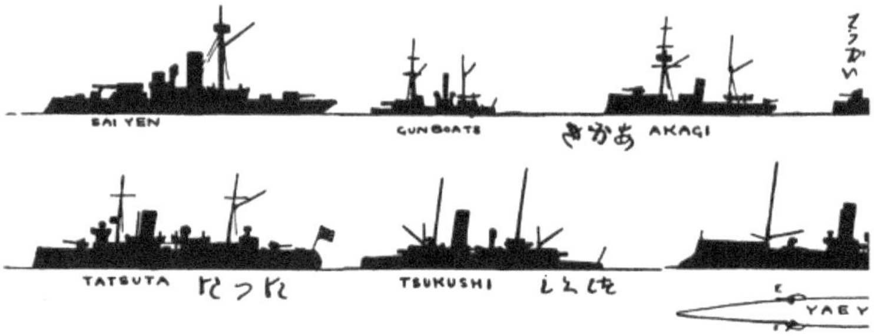

(3.) **SAI-YEN** (*ex* TCHE-YUEN) (1886, Stettin). 2300 tons. Guns: 2D (21 c/m Krupp) + 1D (17 c/m) + 8*. Torpedo tubes, 4. Armour-deck = *d.* Turret (C guns) *b.* I.H.P. 2800 = 15 kts. (*trial*). Sea speed, 12 kts. Coal, 230 tons = *circa* 1000 kts. at 9 kts

(6.) **CHIN-PEN** (κ), **CHIN-CHU** (ι), **CHIN-NAN** (η), **CHIN-TO** (ε), **CHIN-SEI** (ζ), **CHIN-HOKU** (θ), (—1877–81, Elswick). *Circa* 500 tons. Guns: 1C (27 c/m 11-inch Armstrong M.L.) + 2*. Speeds, *circa* 4 kts.

(6.) **AKAGI** (1887). 614 tons. Guns: 2E* (12 c/m special pattern French guns) + 6*. Speed, 8 kts.

(6.) **ATAGO** & **CHIOKAI** & **MAYA** (1887). 614 tons. Guns: 1D (21 c/m Krupp) + 1E* (12 c/m) + 2*, *except* Maya, which carries 2D (17 c/m Krupp) + 4*. Speeds, *circa* 8 kts.

(T.) **TATSUTA** (1894, Elswick) & *CHIHAYA* (building) & **AKASAKI** (1895). 875 tons. Complement 90. Length 240 feet = 73 metres. Guns: 2E* (12 c/m Elswick) + 4* (47 m/m). Torpedo tubes: 5 (one of them in bow). I.H.P. *forced* 5500 = 21 kts. (*trial*). 2 screws. Coal, 188 tons.

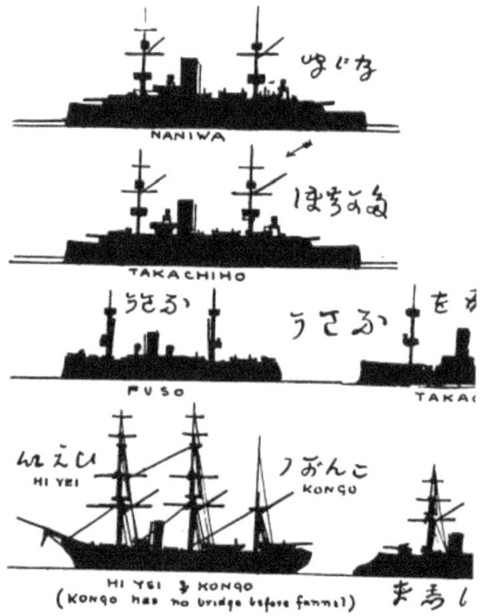

(5.) **TSUKUSHI** (1881. Elswick). 1350 tons. Complement 190. Length 210 feet = 64 metres. Guns: 2C (25 c/m Krupp) + 4E* (12 c/m) + 4*. Torpedo tubes, 2. Armour, 0. I.H.P. 2887 = 16·4 kts. (*trial*). 2 screws. Coal, 250 tons.

(5.) **YAYEYAMA** (1889). 1600 tons. Complement 200. Guns: 3E* (12 c/m Krupp). Torpedo tubes, 2. Armour: Deck, over engines only, ½ inch steel (13 m/m). I.H.P. *forced* 5630 = 20·7 kts. (*trial*). 2 screws.

(6.) **TAKAO** (1888). 1765 tons. Complement 220. Guns: 4D (15 c/m, 35 cals., Krupp) + 1E* (12 c/m). Torpedo tubes, 2. Armour, 0. Machinery: I.H.P. 2300 = 15 kts. (*trial*). Sea speed, *circa* 12 kts. 2 screws. Coal, 300 tons.

(3.) **NANIWA** & **TAKACHIHO** (Elswick, 1885). 3700 tons.

L. B. D.: 300 × 46 × 20 feet = 91·50 × 14 × 6 metres.

Guns: 2C (21 c/m Krupp) + 6* (15 c/m, 35 cals., Elswick) + 2* (57 m/m, 6 pdrs.) + 14*. Torpedo tubes, 4.

Armour-deck = *d* (75 m/m). C. T. *f.* Gun shields *f.* Glacis to C guns = *f.*

Machinery: I.H.P. 7120 = 18·7 kts. (*trial*). Sea speed, *circa* 15 kts. 2 screws. Coal, *normal* 350 tons; *maximum* 800 tons = 5000 miles at 12 kts.

(V.) **FU-SO** (1877, London) (1897 wrecked and refloated) (1898–99 reconstructed). 3718 tons. Complement 386. Guns: 8D* (15 c/m Elswick). Torpedo tubes, 2. Armour (iron): Belt *e.* Protection to vitals *e.* Battery *d.* Gun shields *f.* Machinery: I.H.P. 3500. Sea speed, *circa* 9 kts. 1 screw. Coal, 350 tons.

(6.) **OSHIMA** (1890). 640 tons. Guns: 4E* (12 c/m, 4·7 inch) + 8* (47 m/m). Armour, 0. Nominal speed, 13 kts.

(4.) **HI-YEI** (1878, Milford Haven) & **KON-GO** (1877, Hull). 2250 tons. Complement 300. Guns: 3D (17 c/m Krupp) + 6E (old 15 c/m Krupp). Torpedo tubes, 2. Armour belt *f.* I.H.P. 2450. Sea speed, *circa* 10 kts. Coal, 250 tons. (*Note.*—Hi-Yei has a red band round the hull, Kon-go a black band Kon-go has no bridge before the funnel.)

Photo, copyright.

Photo by C. O. Travers, Esq.

Chitose is the same, but has a red band round the hull instead of a black one.

(3) CHIYODA (1889 Clydebank).

2450 tons. Complement 350.

L. B. D.: 308 × 43 × 17 feet = 94 × 13 = 5 metres.

Guns: 10E* (12 c/m Elswick) + 14*. Torpedo tubes, 3.

Armour (Chrome steel): Deck = *f*. Belt amidships *d* (126 m/m). Prot. to vitals = *c*. Cofferdam.

Machinery: I.H.P. 5600 = 19 kts. (trial). Boilers: Belleville (1898). 2 screws. Coal: 400 tons.

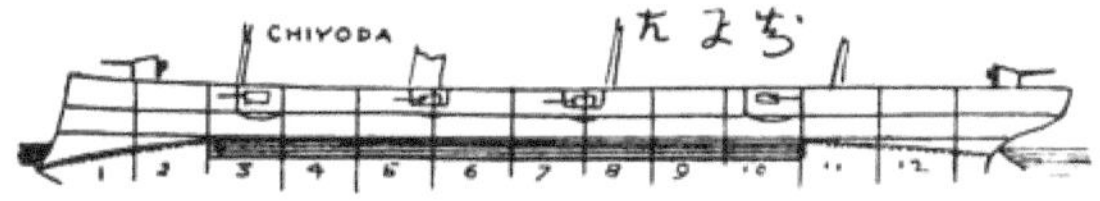

(2) ちとせ (CHITOSE) (S. Francisco 1898) & KASAGI (Cramp's, 1897).

4760 tons. Complement 405.

L. B. D.: 374 × 48 × 24 feet = 114 × 14·60 × 7·31 metres.

Guns (Elswick):
- 2C* (20 c/m).
- 10E* (12 c/m).
- 12F* (75 m/m).
- 6* (2½ pdrs.)

Torpedo tubes:
- 4 above water

Armour (Harvey nickel):
- 4½" Deck = *a* (113).
- Gun Shields, *b*.
- Supports to C guns.
- Hoists (Electric), *e*.
- Cofferdam and bunkers amidships.
- Prot. to vitals = *a*.

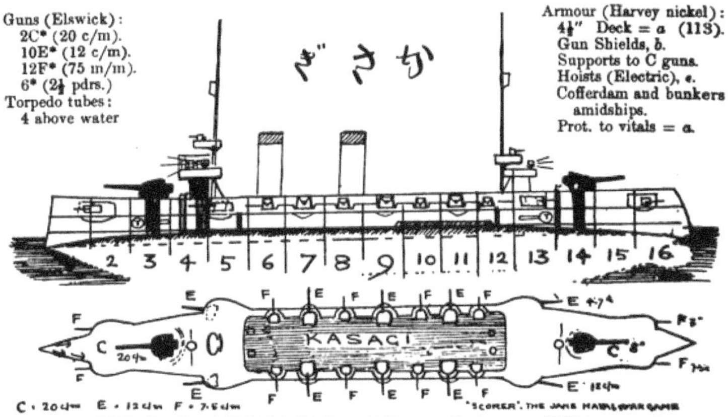

Machinery: I.H.P. 15,000. 8 Cylindrical boilers. 2 Screws. Coal: *normal* 350 tons, *maximum capacity* 1000 tons.

Trials—Kasagi, *natural d.* = 21·68 kts. *Forced* = 22·76 kts.
Chitose „ „ „ = 23·76 kts.
(Average *f.d.* of Chitose was 22·76 for 3 hours.)

Black band round hull. *Photo by Sir W. G. Armstrong, Mitchell & Co., Ltd., Elswick.*

(2.) YOSHINO (1892, Elswick).

4150 tons. Complement 360. L. (*p.p.*) B. D.: 350 × 46½ × 20 feet = 106·56 × 14·15 × …

Guns (Elswick):
4D* (15 c/m).
8E* (12 c/m).
22* (47 m/m).
Torpedo tubes: 5.

Armour:
4½" Steel …
= *a* (1…
Prot. to v…

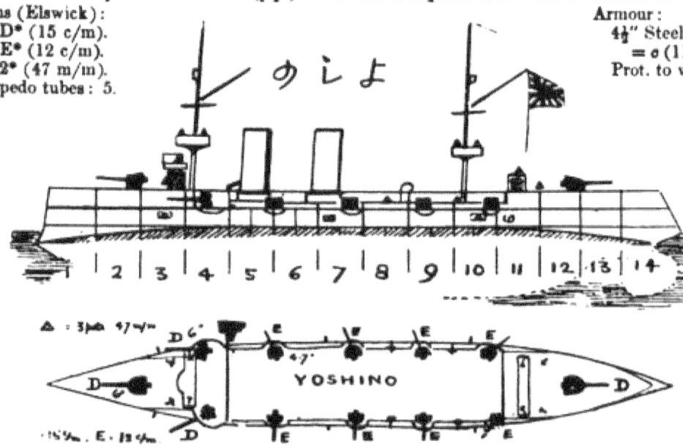

Machinery: I.H.P. 15,000 = 23·08 kts., *forced d. trial*. Boilers: Cylindrical. …
Coal: *normal* 300 tons, *maximum* 1000 tons.

(2.) TAKASAGO (1897, Elswick).

4300 tons. Complement 400. L. (*p.p.*) B. D.: 360 × 46½ × 20 feet = 109·72 × 14·15 × …

Guns (Elswick):
2C* (20 c/m).
10E* (12 c/m).
12F* (75 m/m).
6* (2½ pdrs.).
Torpedo tubes: 5.

Armour (Harvey nickel):
4½" Deck = *a* (113 m/…
4½" Fronts C gun shiel…
= *a* (1…
2½" Sides C gun shields
= *a* (6…
2½" Shields, E guns = …
C.T. *a*. Hoists *a*.
Prot. to vitals = *a*.

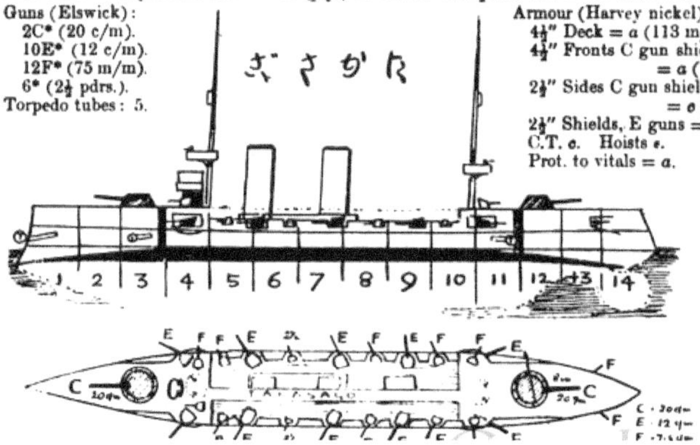

(2) SUMA (1895) & AKASKI (1897).

2700 tons. Complement .

L.: 305 ft. = 93 m. B.: 41 ft. = 12·5 m. D.: $16\frac{1}{3}$ ft. = 5 m.

Guns: 2D* + 6E* + 12*. Armour: *prot.* Sea speed: 16 kts.
Machinery: I.H.P. f. d. 8500 = 20 kts. (*trial*). 2 Screws. 8 cylindrical boilers. Coal 200 tons, *max.* 600 tons.
Armour: Deck = *e*. D gunshields *f*. 2 torpedo tubes.

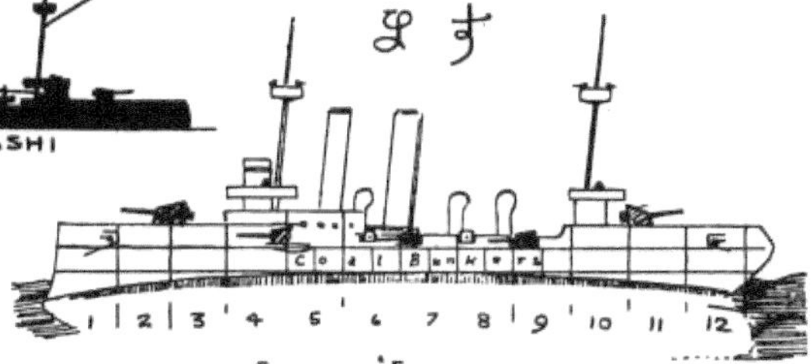

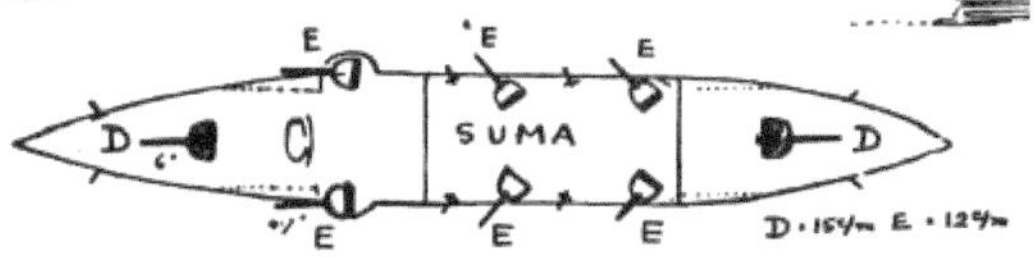

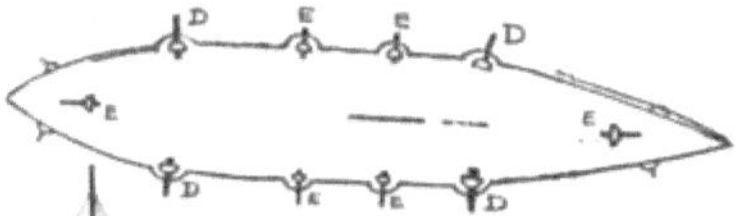

(2.) AKITSUSHIMA (1892).

3150 tons. Complement 330.

L.: 302 ft. = 92 m. B.: 43 ft. = 13 m. D.: $18\frac{1}{2}$ ft. = 5·6 m.

Guns: 4D* + 6E* + 10* (47 m/m). Sea speed: 17·5 kts.
Machinery: I.H.P. Forced draught 8400 = 19 kts. (*trial*). 2 Screws. Coal 500 tons.
Armour: Deck = *d* (75 m/m). 4 Torpedo tubes.

(5.) *MIYAKO* (building).

1800 tons. 6130 I.H.P. = 20 kts. Guns: 2E* + 10*

(Reduced copy of Suma.)

(3.) IDZUMI (1884 Elswick).

3000 tons. Complement 300.

L.: 270 ft. = 82·5 m. B.: 42 ft. = 12·8 m. D.: 19 ft. = 5·9 m.

Guns: 2C + 6E* + 2*. Armour: *prot.* Sea speed: 15 kts.

Machinery: I.H.P., *natural* 6500 = 18·3 kts (*trial* in 1885). 2 Screws. (400 tons. *Max.* = 600 tons = *circa* 15 days at easy speed.

Deck = 1 inch (25 m/m) steel. 3 torpedo tubes (1 of them in bow). Guns, 10-inch Elswick.

Photo, copyright.

(III.) CHIN YEN (CHEN-YUEN-GO) (1882 Stettin) (189?

7350 tons. Complement 280.

L. B. D.: 308 × 59 × 23 ft. = 94 × 18 × 7 m.

Guns: 4B (12-inch Krupp 30 c/m) + 4D* + 2 6-prs. + 8*.

Armour: Deck = *e* (75 m/m) Belt (150 ft., i.e., 46 m. long). Amidships —*b* (355—254 m/m). Bulkheads *b* (355 m/m). Barbettes *a* (305 m/m). C.T. tween big guns) *d*. Shields to B guns *e*; bow turret and gunshields *f*. 3 Tor tubes (1 in stern).

Machinery: I.H.P. 6200 = 14·5 kts. (6 hours *trial* 1882). Sea speed 1894, c 10 kts. 2 Screws. Coal 1000 tons = 2000 miles at 8 kts.

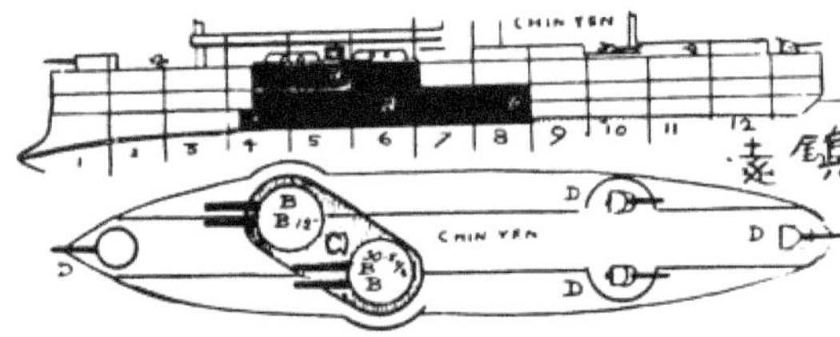

FUJI. Photo by W. Nye, Esq.

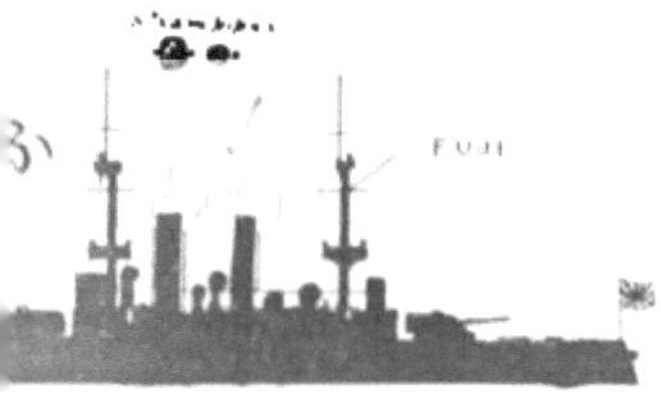

(I.) FUJI (1896 Blackwall) & YASHIMA (1896 Elswick).

12.500 tons. Complement 600.
L.: 385 ft. = 117·5 m. B.: 73¼ ft. = 22·4 m. D.: 30 ft. = 9·1 m.
Guns: 4A (12 in.) + 10D* + 24* + 4T. Sea speed: 15 kts.

Armour (Harvey): Deck = *d* (67 m/m). Belt *aaa–aa* (457–400 m/m). Lower deck *d*. Bulkheads *aa* (370 m/m). Casemates on main deck *c* (150 m/m). Barbettes *aa* (350 m/m). Hoods to A guns slanting = *aa* (150 m/m). Shields to D guns *d*. C.T. *aa*. 5 torpedo tubes (4 Elswick submerged and 1 above water in bow).

Machinery: 2 screws. I.H.P. Fuji *nat.* 10,200 = 16·8 kts. *trial, f.d.* 14,100 = 18·5 kts. Yashima *nat.* 9750 = 17·7 kts., *f.d.* 14,075 = 19·2 kts. (*trial*). Coal 700 tons, *max.* 1300 = *circa* 5000 miles at 12 kts.

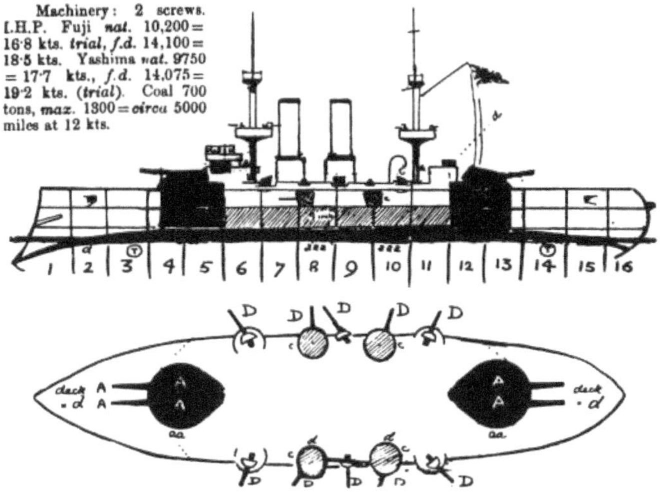

(I.) ASAHI (Clydebank, 1899).

15,200 tons. Complement, 740.

L. (*p.p.*) B. D.: 400 × 76 × 27 feet = 122 × 23·16 × 8·22 metres.

(Completing for sea, April, 1900).

Guns (Elswick):
- 4A* (12 in. 40 cals. R.L.).
- 14D* (6 in. 45 cals. Q.F.).
- 20F* 3 in.).
- 8* (3 pr., 47 m/m) in tops.
- 6* (2½ pr.)

Torpedo tubes:
- 4 *submerged* (18 in.).

Armour (Harvey nickel):
- 9" Belt amidships *aa* (228 m/m).
- 4" „ at ends *d* (100 m/m).
- 4" Deck on slopes = *a* (100 m/m).
- Prot. to vitals = *aaaaa*.
- 14" Barbettes *aaa* (355 m/m).
- 10" Hoods, in front = *aaaa* (250 m/m).
- 8" „ sides *a* (200 m/m).
- 14" Bulkheads *aaaa* (355 m/m).
- 6" Lower deck, side *b* (152 m/m).
- 6" Casemates *b* (152 m/m).
- Fore C.T. *aaa*. After C.T. *d*(100 m/m).

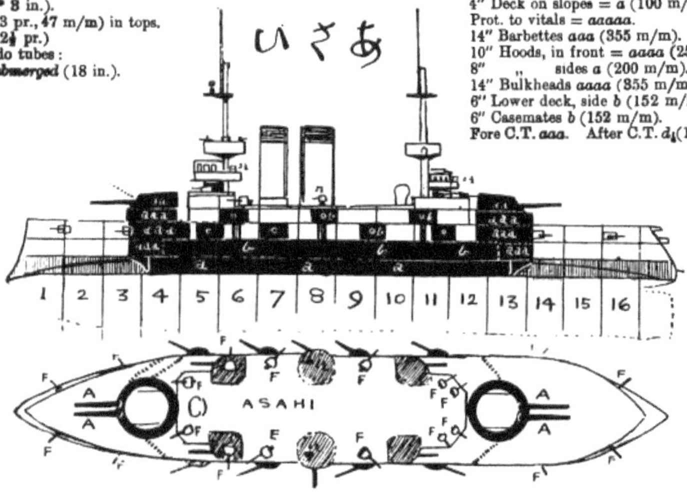

Machinery: 2 sets triple expansion. 2 screws (4-bladed). Boilers: 25 Belleville, with economisers. I.H.P. *est.* 15,000 *forced* = 18 kts.

Coal: 1000 tons.

Trials: *hrs. nat. draught* I.H.P. = kts. *hrs. forced* = kts.

Note.—This ship, except for her funnels, is practically identical with the Shikishima. The big guns have, however, improved hoists. They are on central pivot mountings and load at any elevation. The 2½ pounders are placed differently to those in Shikishima. The ship has an even el, and the deadwood is not cut away aft. The barbettes rise about 7 feet above the deck.

Asahi. *Photo, Symon*

ちきと

(III.) ASAMA & TOKIWA (Elswick, 1898).

"Armoured Cruisers."

9750 tons. Complement

L. (*over all*) × B. × D. (*mean*):
408 × 67 × 24¼ feet = 124·5 × 20·5 × 7·38 metres.

Guns (Elswick):
4C* (8 in. 45 cals.).
14D* (6 in. 45 cals.).
12F* (3 in.).
7* (2½-pdrs.).
Torpedo tubes (Elswick):
1 in bow *above water*.
4 *submerged*.

Armour (Harvey nickel):
7" Belt amidships *b* (175 m/m).
3½" Belt at ends *d* (87 m/m).
2½" Deck, on slopes = *d* (75 m/m).
Prot. to vitals = *aa*.
6" C gun turrets and bases *b* (152 m/m).
6" Casemates (10) *b* (152 m/m).
5" Lower deck side *c* (125 m/m).

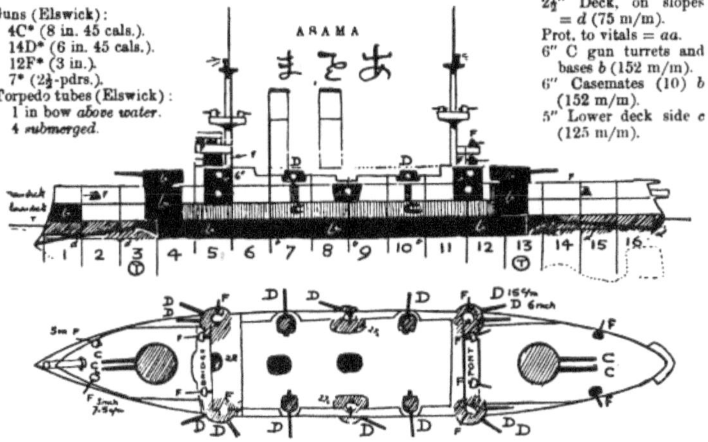

Machinery: Two sets triple expansion. 2 screws. Boilers, cylindrical.
Coal: *normal* 550 tons; maximum 1300 tons = *nominal* 10,000 miles.

Trials.

Asama, 6 *hours n. d.* 13,000 I.H.P. = 20·37. 6 *hours forced d.* 19,000 I.H.P. = 22 kts.
Tokiwa, 6 „ „ ? „ = 20·85. 4 „ „ ? „ = 22·7 kts.

Note.—The 8 in. guns load in any position or elevation. The deadwood is cut away aft. There is a patch of 6" (152 m/m) armour over the bow torpedo tube.

(Asama and Tokiwa are exactly alike.)

ASAMA. *Photo by Sir W. G. Armstrong, Mitchell & Co., Ltd.*

(III.) *ADZUMA* (St. Nazaire, 1900).

まらあ

"Armoured Cruiser."

9436 tons. Complement

L. (*over all*) × B. × D. (*mean*): 446 × 59 × 24¼ feet = 136 × 18 × 7·38 metres.

Guns (Elswick):
4C* (8 in. 45 cals.).
12D* (6 in. 45 cals.).
12F* (3 in.).
12* (37 m/m and machine).
Torpedo tubes (Elswick):
4 *submerged*.

Armour (Krupp mostly):
7" Belt amidships *a* (175 m/m).
3½" „ at ends *c* (87 m/m).
2½" Deck, on slopes = *d* (75 m/m).
Prot. to vitals = *aaa*.
6" C gun turrets and bases *a* (152 m/m).
6" Casemates *a* (152).
5" Side above belt *b* (125).

[Plan, *see* Yakumo (page 187) for arrangement of guns.]
(The Adzuma has, however, two funnels, like Asama.)

Machinery: Two sets vertical triple expansion. 2 screws. Boilers, Belleville. I.H.P. 17,000 *forced* = 21 kts. (*estimated*).

(III.) IDZUMO & IWATE (1899, Elswick 1900).

"Armoured cruisers." Completing for sea.

9800 tons. Complement

L. (*p.p.*) B. D.: 400 × 68½ × 24¼ feet =
Length over all, 408 feet (124·5 m.).

Guns (Elswick):
4C* (8 in. Q.F.).
14D* (6 in. Q.F. 45 cals.).
12F* (3 in. Q.F.).
12*.

Torpedo tubes:
4 *submerged*.

Armour (Krupp):
7" Belt amidships *a* (175 m/m).
3½" Belt ends *c* (87 m/m).
2½" Armour deck on slopes = *e* (65 m/m).
Protection to vitals = *aaa* (belt & deck).
6" C gun turrets, and bases *a* (152 m/m).
6" Casemates *a* (152 m/m).
5" Lower deck, side *b* (126 m/m).

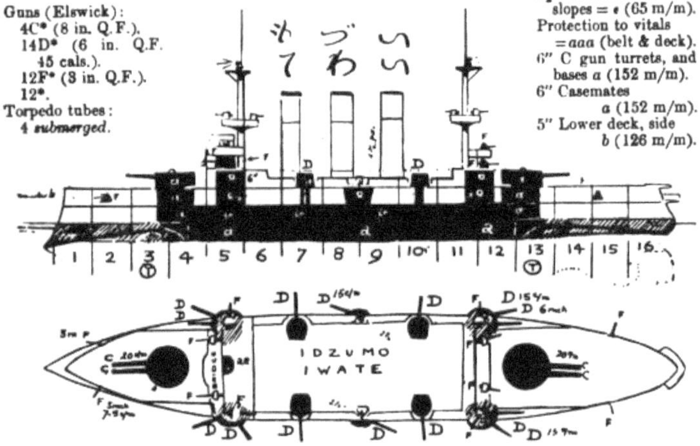

Machinery: 2 sets triple expansion. 2 screws. Boilers, Belleville (with economisers). I.H.P. *est.* 15,000 *forced draught* = 21 kts.

Coal: *normal* 550 tons; *maximum* 1300 tons = *nominal* radius of 10,000 miles.

Trials.

Idzumo	hrs. I.H.P.	=	kts.	hrs. *forced d.* I.H.P.	=	kts.
Iwate	" "	=	"	" " "	=	"

Note.—These ships belong to the Asama class, but have a knot less speed, and the thick belt is a little shorter. The bow torpedo tube is suppressed.

(III.) *YAKUMO* (1899, Stettin).

"Armoured cruiser." Building.

9850 tons. Complement

L. (*over all*) B. *D.* (*mean*): 407 × 65½ × 23⅗ feet = 124 × 20 × 7·3 metres.

Guns (Elswick):
4C* (8 in. Q.F.).
12D* (6 in. Q.F. 45 cals.).
12F* (3 in. Q.F.).
12*.

Torpedo tubes:
1 in bow, *above water*.
4 *submerged*.
(Elswick patent.)

Armour (Harvey nickel &
7" Belt amidships *b* (17
3½" Belt at ends *d* (87
2½" Armour deck or *e* (65 m/m).
Protection to vitals deck) = *aaa*.
6" C gun turrets an *a* (152 m/m).
6" Casemates *b* (152 m
5" Lower deck *c* (126

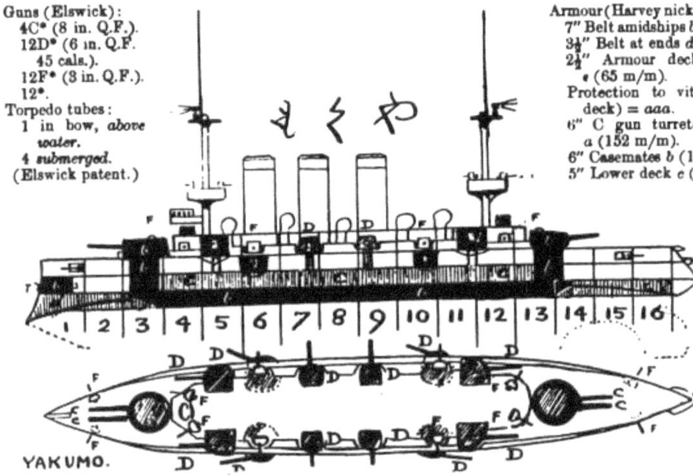

Machinery; 2 screws. Boilers, Belleville (with economisers). I.H.P. 16,000 = 20 (*estimated*).

Coal: *normal* 550 tons; *maximum* 1300 tons.

(Adzuma is like Yakumo in arrangement of guns, but has only 2 funnels.)

(I.) *Mikasa* (Vickers; *building*).

15,200 tons. Complement, 730.

L. (*p.p.*) B. D.: 400 × 76 × 27¼ feet = 122 × 23·16 × 8·30 metres.

Length over all 436 feet (133 metres).

lswick):
12 in. 40 cal. R.L.).
(6 in. 45 cal. Q.F.)
(3 in., 12 prs.)
pr. 47 m/m).
½ pr.)
tubes:
merged (18 in.).

Armour (Krupp):
9″ Belt amidships *aaa* (228 m/m).
4″ „ ends *c* (100 m/m).
4″ Deck on slopes = *a* (100 m/m).
Prot. to vitals = *aaaaa*.
14″ Barbettes *aaaa* (355 m/m).
10″ Hoods, in front = *aaaa* (250 m/m).
8″ „ sides *aa* (200 m/m).
14″ Bulkheads *aaaa* (355 m/m).
6″ Lower deck *a* (152 m/m).
6″ Casemates and battery *a* (152 m/m).
Fore C.T. *aaaa*. After C.T. *b* (100 m/m).

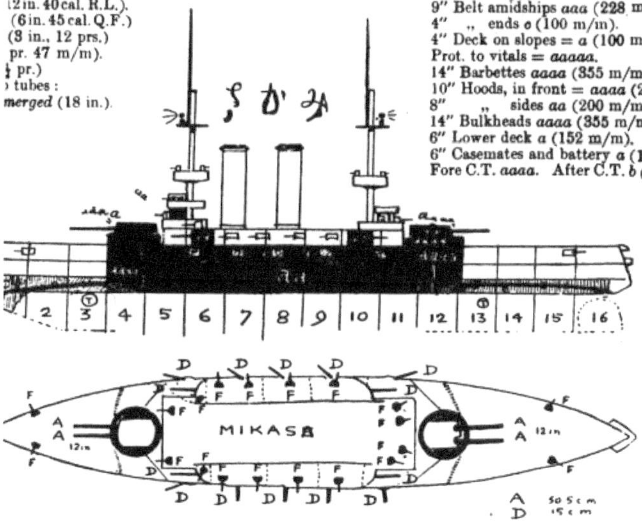

Machinery, Coal, Boilers, etc. = as Asahi.

Note.—The armour (Krupp instead of H. Nickel) is generally of the same thickness as ahi's. There is 6 inches over the main deck battery. Behind each side of belt is a 2-inch) m/m) armoured wall (*e*), and screens isolate each battery gun.

(I.) SHIKISHIMA (Thames Ironworks 1898) & HATSUSE (Elswick 1899).

Displacement {Shikishima 14,850 tons / Hatsuse 15,000 tons} Complement 741.

L. (*p.p.*) × B. × *D.* (mean) {Shikishima 400 × 75½ × 27¼ ft. = 122 × 23 × 8·30 metres. / Hatsuse 400 × 76¾ × 27 ft. = 122 × 23·31 × 8·22 metres.

Guns:
- 4A* (12 in. 40 cal. R.L.).
- 14D* (6 in. 45 cal. Q.F.).
- 20F* (3 in. Q.F.).
- 8* (3 pounder 47 m/m).
- 6* (2½ pounder).

Torpedo tubes:
4 *submerged*.
Shikishima has also 1 *above* water in the bow.

Armour (Harvey nickel):
- 9" Belt amidships *aa* (227 m/m).
- 4" Belt at ends *d* (100 m/m).
- 4" Armour deck = *a* (100 m/m).
- Prot. to vitals = *aaaaa*.
- 14" Barbettes *aaa*.
- 10" Hoods in front = *aaaa* (250 m/m).
- 10" Hoods at sides *aa* (250 m/m).
- 14" Bulkheads *aaa* (350).
- 6" Lower deck *b* (152).
- 6" Casemates *b* (152).
- 14" Fore C.T. *aaa*.
- After C.T. *c*.

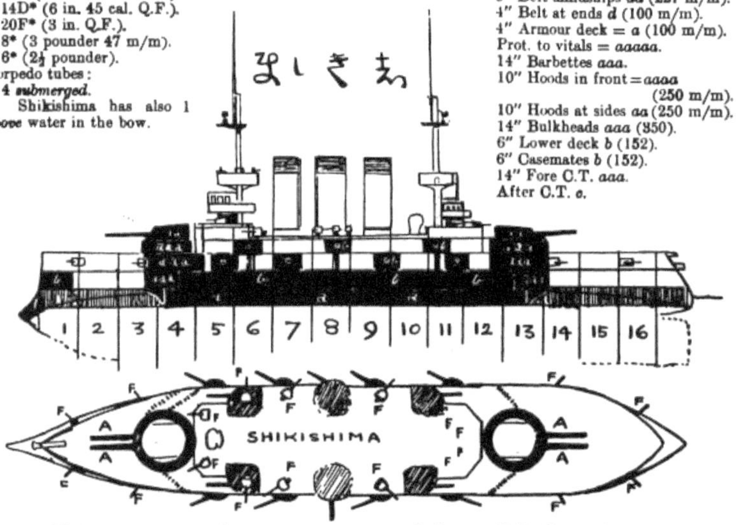

Machinery: 2 sets triple expansion. 2 screws. Boilers, 25 Bellevilles, with economisers. I.H.P. 14,500 *forced d.* = 18 kts.

Coal: normal 700 tons, maximum 1400 tons for Shikishima. The Hatsuse carries a little more.

SHIKISHIMA (1900). *Photo by Sym*

Trials.

Shikishima	hours *n. d.* I.H.P.	= 17 kts.	hours *forced* I.H.P.	= 18·78 k
Hatsuse	" " "	=	" " " "	=

Notes.—The Shikishima only, carries a patch of 6 in. (150 m/m) armour (*b*) in bow a bow tube. The big gun shields slope in front, but are straight on the sides. Th are carried at least 6 feet (2 metres) above the deck. The big guns are on central pivot and load at any elevation as well as in any direction. All hoists are electric, steam The Hatsuse has the deadwood cut away aft, but the Shikishima has not.

OESTERREICH-UNGARN.

GESCHÜTZE.

..libre.	—	Long in Calibres.	Muzzle Velocity.		Approximate Muzzle Energy.		A.P. Projectile.	
c/m			ft. secs.	met. secs.	ft. tons.	met. tnx.	lbs.	kgs.
30·5	...	35	1755	535	21420	6687	1000	454
24	Wien, 25-ton . .	40	2360	719	13100	4057	350	159
24	(Cruisers). 26-ton.	35	?	?	11000	3406	474	215
24	19-ton	30	1657	505	9020	2849	474	215
26	22-ton	20	1575	480	6800	2106	369	167
21	8½-ton	20	1520	463	3300	1022	200	91
15	...	40	2297	700	3770	1167	100	45
15	Converted (1886).	35	1960	597	3000	929	112	50
15	(1880)	35	1970	600	2312	716	86	39
12	Skoda	40	2297	700	1922	595	52	24
12	'80 converted . .	35	1755	535	1210	372	57	26
12	'87 ,, . .	35	1970	600	1540	477	57	26
15	'87 ,, . .	26	1641	500	1358	420	73	33
9	Uchatius . . .	...	1470	448	?	?	14	6
7·5	Krupp	40	2379	725	500	154	12	5·5

the ships following the Austrian Navy also contains:—

rs, 450 tons, Szamos & Koros (1893), « armour, 10 kts. speed, 2E* guns. 2 other 300 tha (1871) (1895), 2E guns, 8 kts. speed. 2 frigates Landon & Radetzky, 3300 tons, 15E Donau & Saida, 2300 tons, 10E guns. 3 frigates Aurora, Zrinyi & Frundsberg, 1300 tons, vooden gunboats, 900 - 300 tons, 2 - 4E guns. 3 old ironclads for harbour service, Kaiser, Ferdinand Max. *circa* 5000 tons.

TORPEDO BOATS.

First Class. *trial mean.*

1 Yarrow (Viper) (1896), 130 tons, 147½ ft. long, speed 3 hours 26·6 kts., 2 tubes (18 in.)
1 Schichau (1896), 152 ,, 150 ,, ,, ,, ,, 24·4 ,, 3 ,, ,,
4 Yarrow (Cobra class) (1898), 135 ,, 152½ ,, ,, ,, ,, .. ,, 3 ,, ,,
— (These boats have 2 Yarrow boilers, engines *circa* 2000 I.H.P. Steam steering gear).
6

Second Class. *trial mean.*

9 boats (1891–87), 65 tons, 118 ft. long, speed 3 hours 18 kts., 1 tube (18 in.)
2 Yarrow boats (1886), 95 ,, 135 ,, ,, ,, ,, 22·5 ,, 2 tubes ,,
22 boats (various) (1886–9), 85 ,, 128 ,, ,, ,, ,, 18–21·5 ,, 2 ,, ,,
—
33

Third Class. *trial mean.*

22 boats (1887–3), 47 tons, 110 ft. long, speed 3 hours 17 kts. 1 tube (18 in.)
2 ,, (1881), 37 ,, 100 ,, ,, ,, ,, 17 ,, ,, ,,
7 ,, (1881–78), 27 ,, 87 ,, ,, ,, ,, 15 ,, ,, ,,
—
31

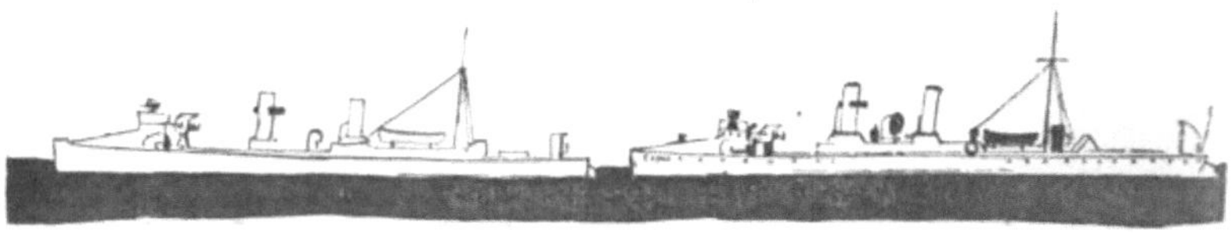

VIPER. COBRA, BOA, *etc.*

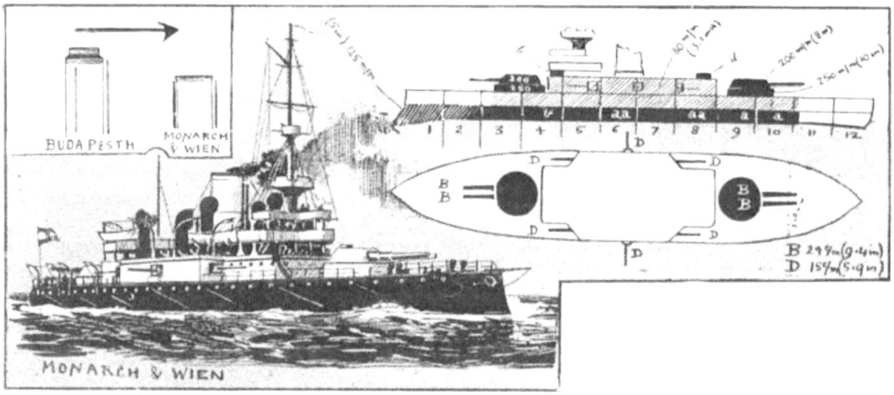

(II.) WIEN (1895) & MONARCH (1895) & BUDA

L.: 305 ft. = 93 m. B.: 56 ft. = 17 m. D.: 21 = 6·

5550 tons. Complement 469.

Guns: 4B (24 c/m 40 cal.) + 6D* (15 c/m 40 cal.) + 14 3 prs.' 4 Torpedo tubes.

Armour (Harvey): Belt from bow (255 × 7 ft. = 77·7 × 2 m.) deck aft of belt = *d*. Citadel redoubt and battery *d*. Barbettes *a*. head aft *b*. Fore C.T. *b* after C.T. *d*.

Machinery: 2 sets triple expansion, Maudslay. Boilers: Buda Other 3, 2-ended; 2, 1-ended cylindrical. 2 Screws. Budapesth Belle 6 hour *trial* 6608 I.H.P. = 17·1 kts. (*mean*). Wien, 6376 = 16·7 6110 = 16·2 kts. *Forced draught*—4 hours—B.: 9185 I.H.P. = W.: 8480 = 17·4 kts. M.: 8900 = 17·5 kts. Coal: 500 tons = c at 10 kts. Budapesth rather more.

(T.) SATELLIT (1893 Elbing).

L.: 220 ft. = 67 m.

Guns: 9*. Armour: 0. Sea speed: 18·5 kts.

&

(T.) PLANET (1889 Jarrow).

L.: 210 ft. = 64 m.

Guns: 2Z* + 8*. Armour: 0. Sea speed: 17·o kts.
Satellit 540 tons. Planet 500 tons.
Boilers, locomotive.
Trial speeds: S.: 21·9 kts. P.: 19·6 kts.

&

(T.) TRABANT (1890).

L.: 220 ft. = 67 m.

530 tons. Complement 78. D.: 9 ft. = 2·7 m. Guns: 10*. Machinery: 2 sets triple expansion, 4 loco. boilers. I.H.P. 3500 = 2 screws. Coal 150 tons.

KOMET & BLITZ (—1888 Elbing).

L.: 194 ft = 59 m.

0 tons. Complement 61. D.: 9 ft.= m. Guns: 9* ?. 4 torpedo tubes. P. 3500 = 21 kts. *trial*. 1 screw. 120 tons.

KOMET & BLITZ. SPALATO & ZARA.

SPALATO & ZARA (—1879).

Sea speed: *circa* 11 kts.

900 tons. D.: 13½ ft. = 4·2 m. Complement 142.

Guns: 4F (9 c/m) + 1F (7 c.m.) + 7*. 2 torpedo tubes (*but* Sebenico only 1). Armour: deck = *f*. 1 screw. Zara new boilers 1895. Coal 150 tons. I.H.P. 1200.

(T.) PELIKAN (1891 Elbing).

L: 279 ft. = 85 m. B.: 39½ ft. = 12 m. D.: 17 ft. = 5·2 m.

Guns: 2D (15 c/m) + 8*. Armour: 0. Sea speed · 15 kts.

2440 tons. Complement .

I.H.P. *natural draught* = 3832 = 17·8 kts. *trial* 6 hours. Coal 710 tons. rew. Torpedo depôt ship.

(IV.) PRINZ EUGEN (1877) & KAISER MAX (187 DON JUAN D'AUSTRIA (1875).

L.: 241 ft. = 73·5 m.

3566 tons. Complement 440. Sea speed: 9 kts.

Guns: 8C (21 c/m) + 4F + 31*. 4 torpedo tubes. **Armour**: Belt 2·8 m. broad) *d*—*e*. Battery and bulkheads to battery *e*.

Machinery: I.H.P. 2700 = 13 kts. (*trial*). Box boilers. 1 scre 380 tons = *circa* 1500 kts. at 8 kts.

(T.) METEOR (1887 Elbing).

350 tons. Complement 61.
L.: 187 ft. = 57 m. D.: 9 ft. = 2·7 m.

Guns: 9*. **Armour**: 0. **Sea speed**: 17 kts. 4 torpedo tubes.
Machinery: I.H.P. 3500 = 23 kts. (*trial*). 1 screw. 2 locomotive boilers. Triple expansion engines. Coal: 120 tons.

(T.) LUSSIN (1882.)

1011 tons. Complement 142.
L.: 200 ft. = 61 m. D.: 13¾ ft. = 4·2 m.

Guns: 2D + 1F + 7*. **Armour**: *prot.* **Sea speed**: 11 kts.
Deck = *f*. 2 torpedo tubes.
Machinery: I.H.P. 1830 = 14 kts. (*trial*). 2 sets compound. Dürr boilers. Coal: 200 tons.

&

(T.) SEBENICO (1882).

900 tons. Besatzung 142.
L.: 187 ft. = 57 m. Draught: 13¾ ft. = 4·2 m.

Guns: 4F (9 c.m.) + 1F (7 c/m) + 7*. 1 Torpedo tube. **Armour**: deck = *f*. **Sea speed**: 11 kts. 1 screw.

Lussin & Se

(AUTRICHE) OESTERREICH-UNGARN (AUSTRIA).

(IV.) ERZHERZOG ALBRECHT (1872).

5490 tons. Complement 535.

Guns: 8D (24 c.m.) + 8F (9 c.m.) + 11* (47 m/m). Torpedo tubes, 4.

Machinery: I.H.P. 3600 = 13 kts. (*first trials*). 1 Screw. Coal: 450 tons.

(T.) MAGNET (1896, Schichau).

510 tons.

Guns: 6* (47 m/m).
I.H.P. 5000 = 26 kts. (*trial*).
Boilers: Thornycroft. 2 screws.

Photo by Signor B. Circovitch.

(III.) KRONPRINZ ERZHERZOG RUDOLPH (1887).

6900 tons. Complement 492.

L. B. D.: 295 × 62½ × 26½ feet = 90 × 19 × 8 metres.

Guns (Krupp):
3A (30·5 c/m).
6E* (12 c/m, 35 cals.).
15* (47 m/m, 3 pdrs.).
Torpedo tubes:
4 *above water*.

Armour (compound):
12″–10″ Belt *a–b* (305–250 m/m)
Protection to vitals *a*.
2″ Deck (slopes) = *e* (50 m/m).
10″ Barbettes and bulkheads *b* (250 m/m).
Shields to battery guns *f*.
C.T. *a*.

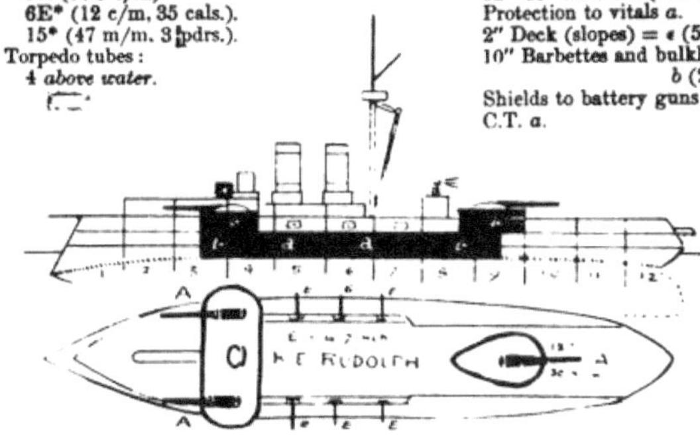

Machinery: 2 sets compound. 2 screws. 10 1-ended return-tube boilers. I.H.P. 5130 *natural* = 15 kts. (*trial*), 6500 *forced* = 16 kts. Sea speed, *circa* 13·5 kts. Coal, *normal* 600 tons, *max. capacity* 650 tons.
Note.—Belt is 9 feet (m. 2·70) wide.

(5.) **TIGER** (1887). 1675 tons. Complement 190. L. B. D.: 233 × 33 × 16½ feet = 71 × 10 × 5 metres. Guns: 4E (12 c/m) + 10*. Torpedo tubes, 4. Armour, 0. I.H.P. 4000. *Trial*, 5260 = 18·3 kts. 2 screws. Coal, 300 tons. Sea speed, 15 kts.

(5.) **LEOPARD** (1885) & **PANTHER** (1886). 1530 tons. Complement 148. L. B. D.: 226½ × 33 × 15 feet = 69 × 10 × 4·60 m. Guns: 2E* + 9*. Torpedo tubes, 4. Machinery: *natural d.* I.H.P. 4000; *forced d.* 6113 = 18·4 kts. (*trial*). 2 screws. Coal, 300 tons (*maximum capacity*).

(5.) **ZENTA** (1897) & **ASPERN** (1898) & *C* (building). 2300 tons. L. B. *D.* (*mean*): 302 × 39½ × *14½* feet = 92 × 12 × *4·30* m. Guns: 8E* (12 c/m, 40 calibres) + 10* (47 m/m) + 2*. Torpedo tubes, 7. Armour, 0. Machinery: I.H.P. 7000 = 20 kts. Boilers, Yarrow. 2 screws. Coal, 300 tons.

(II.) *New ships, I. & II.* (building).

8300 tons.

L. B. *D.* (*mean*): 354¼ × 65½ × *23* feet = 108 × 20 × *7·10* metres.

Guns (Krupp):
3B (24 c/m, 45 cals.).
12D* (15 c/m).
24*.
Torpedo tubes:
2 *submerged*.

Armour (Krupp):
8¾" Belt amidships *aa* (220 m/m).
2⅝" Deck = *b* (66 m/m).
Protection to vitals = *aaaa*.
4" Belt forward *c* (100 m/m).
4" Lower deck, side, *c* (100 m/m).
8¼" Barbettes *aa* (210 m/m).
8¼" Barbettes bases *aa* (210 m/m).
8¼" Hoods to B guns = *aaa*.
6" Casemates *a* (152 m/m).
8" Bulkheads *aa*.

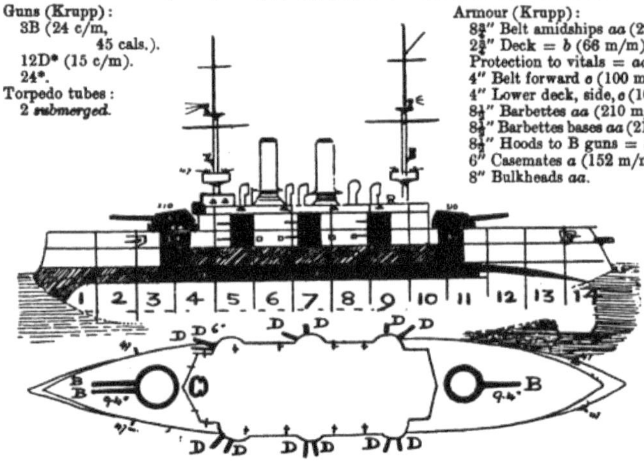

Machinery: I.H.P. 11,000 = 18 kts. (*estimated*). 2 screws. Boilers, Belleville.

OESTERREICH-UNGARN.

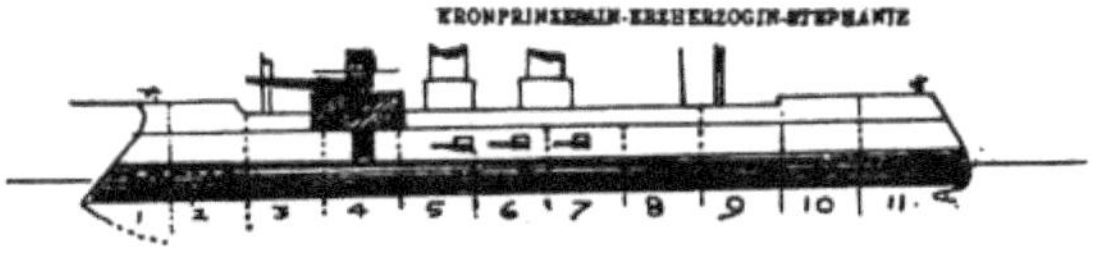

(IV.) KRONPRINZESSIN-ERZHERZOGIN-STEPHANIE (1887).

5150 tons. Complement 510.

L.: 279 ft. = 85 m. B.: $55\frac{2}{3}$ ft. = 17 m. D.: $23\frac{1}{4}$ ft. = 6·9 m.

Guns: 2A (30·5 c/m) + 6D (15 c/m) + 11*. 4 torpedo tubes. Sea speed: 14·5 kts.

Armour (compound): Belt (279 × 9 ft. = 85 × 2·75 m.) ($4\frac{1}{3}$ ft. = 1·30 m. belt under water) *c - d* (230 - 120 m/m). Barbettes *c* (200 m/m). C. T. *c*. Deck 25 m/m.

Machinery: Maudslay. 2 sets compound 5 cylinders. 10 return tube boilers. I.H.P. 6900 = 17 kts. (12 *hours trial*); 10500 (4 *hours forced draught*) = 17·5 kts. Coal: 500 tons. 2 screws.

(2.) KAISERIN-ELISABETH (1890) 4064 tons.
KAISER-FRANZ JOSEF (1889) 4030 tons.

Complement 367.

L. B. *D.* (*mean*): 321 × 49 × *19* feet = 98 × 15 × *5·79* m. (K. Elis.).
321 × 49 × 18⅝ feet = 98 × 15 × *5·60* m. (K. F. Josef).

Guns: 2B (24 c/m) + 6D (15 c/m) + 5* (47 m/m) + 10*. Torpedo tubes: 4.

Armour: 2" Deck = *ϵ* (50 m/m). Barbettes *ϵ* (100 m/m). C.T. *ϵ*.

Machinery: 2 sets horizontal triple expansion. 2 screws. Boilers: 4 2-ended cylindrical. I.H.P., *natural* 6400 = 18 kts. I.H.P., *forced* 9000 = 19 kts. On *trial* = 19·65 kts. Coal: 660 tons = *circa* 3000 miles at 10 kts.

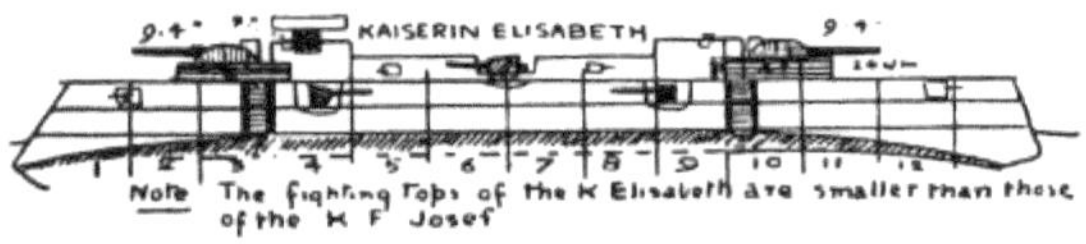

KAISER FRANZ JOSEF. *Photo, Le Y*

Note.—The tops of the Kaiserin-Elisabeth are smaller than those of the K. F. Jose
other difference.

(III.) TEGETTHOFF (1878) (1893).

7400 tons. Complement 578.

L. B. *D.* (*mean*): 285½ × 62⅓ × *25* feet = 87 × 19 × *7·60* metres.

Guns: 6B (24 c/m, 35 cals.) + 5D* (15 c/m, 35 cals.) + 15* (47 m/m). Torpedo tubes: 4.

Armour (iron): Belt *b–d* (370–225 m/m). Protection to vitals *b*. Bulkhead forward *c* (300 m/m). Battery redoubt *b* (370 m/m). Deck, flat on top of belt (76 m/m).

Machinery: 2 sets vertical triple expansion. 2 screws. Boilers: 8 return tube. I.H.P., *natural* 7000, *forced* 8800. Coal: 670 tons.

(1893) *Trials*, ? *hrs. natural d.* 7340 = 15·2 kts. 4 *hrs. forced d.* 8824 = 15·5 kts.

Note.—The belt is 9 feet (2·70 m.) wide.

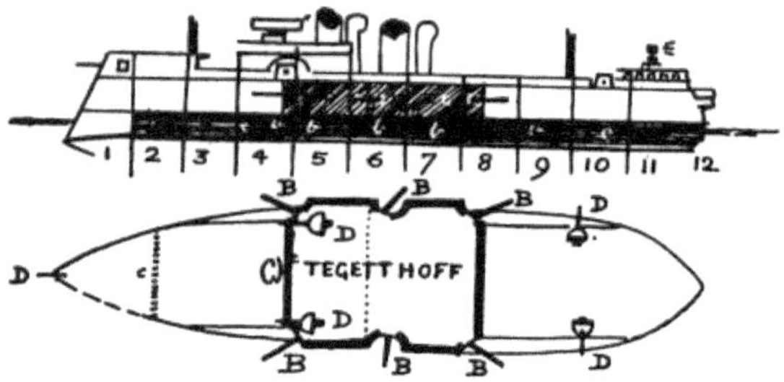

Photo by Signor B. Circ

OESTERREICH-UNGARN.

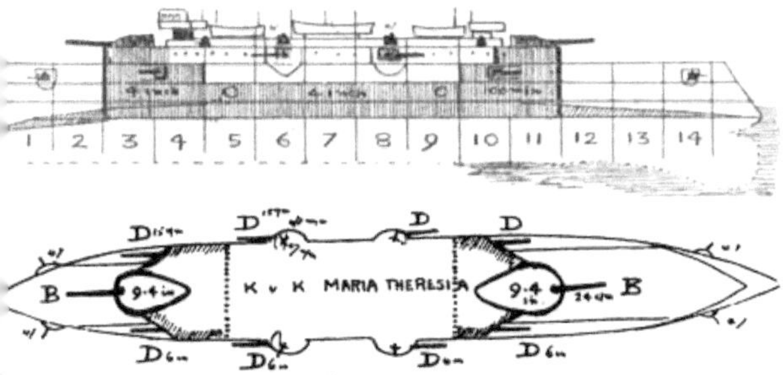

I.) KAISERIN-UND-KÖNIGIN-MARIA-THERESIA (1893).

5415 tons. Complement 450.

L.: 351 ft. = 107 m. B.: 52½ ft. = 16 m. D.: 22 ft. = 6·7 m.

Guns: 2B (24 c/m 35 cal.) + 8D* + 18*. Sea speed: 17·5 kts.

Armour: Deck = *e* Cofferdam. Citadel, bulkheads and barbettes *e*.

Machinery: Maudslay. Horizontal triple expansion. 4 2-ended boilers. I.H.P. *ours trial* 1895, 9755 = 19·4 kts. Coal 740 tons = *circa* 3000 miles at 11 kts.

Photo, Symonds & Co.

(IV.) CUSTOZZA (1872).

L.: 302 ft. = 92 m. B.: 58 ft. = 17·5 m. D.: 25½ ft. = 7·7 m.

Guns: 8C (26 c/m) + 8Z + 11*. 4 Torpedo tubes. Sea speed: 10 kts.

Armour: Belt *d* – *e*. Battery and bulkheads *e*.

Machinery: I.H.P. 4400 = 14 kts. (*trial*). 1 Screw. Coal 580 tons = *circa* 1000 miles at 8 kts.

(AUSTRIA) OESTERREICH-UNGARN.

(III.) KAISER KARL VI. (1898.)

6100 tons. Complement 450.

L. B. *D.* (*mean*): 367½ × 56 × 20½ feet = 112 × 17·20 × 6·20 metres.

Guns (Krupp):
2B (24 c/m, 40 cals.).
8D* (15 c/m, 40 cals.).
20* (47 m/m).
Torpedo tubes:
4 *above water* (behind *aa* armour).

Armour (Harvey nickel):
11" Belt amidships *aa* (230 m/m).
1½" Deck (on slopes) = *ε* (38 m/m).
Protection to vitals = *aaa*.
11" Lowerdeck side *aa* (230 m/m).
10" Bulkheads *aa* (200 m/m).
11" Barbettes *aa* (230 m/m).
11" Hoods to B guns = *aaaaa*.
6" Casemates *b* (152 m/m).

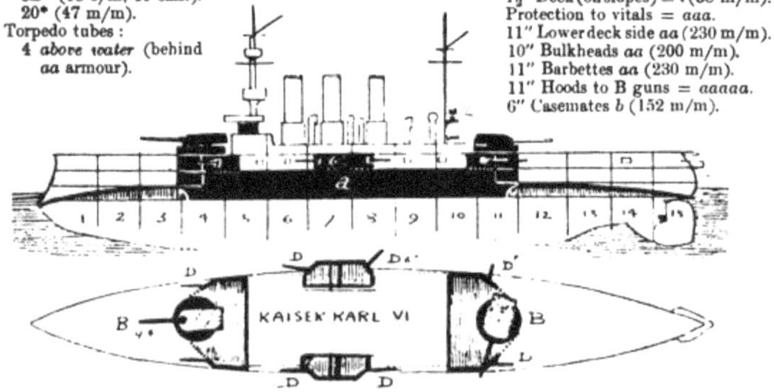

Machinery: 2 sets triple expansion. Boilers, Belleville. I.H.P. 12,000 = 20 kts. (*estimate*). Coal: *normal*, 800 tons.

PORTUGAL.

(IV.) *1 & 2* (projected). 2500 tons. Guns: 2B (24 c/m) + 4E* (12 c/m) + 4* (47 m/m). Torpedo tubes, 2. Armour (Krupp): Belt *aa* (200 m/m). Barbettes *aa*. (Type, Argentine Libertad.)

In addition to the ships enumerated, the Portuguese navy contains 6 corvettes, *circa* 1000 tons, and 21 gunboats (500–250 tons), 8 2nd class torpedo boats, and 4 3rd class (40–20 tons), and 1 submarine of old type.

PORTUGAL.

(IV.) VASCO-DI-GAMA (1875, Blackwall). 2500 tons. Guns: 2C (26 + 1E (15 c/m Krupp) + 2F* (65 m/m) + 2*. Armour (iron): Complet Battery *d*. Sea speed, *circa* 8 kts. Nominal speed, 13·2 kts.

(4.) SÃO GABRIEL & SÃO RAFAEL (Havre, 1898). 1800 tons. (15 c/m Canet) + 4E* (12 c/m) + 8* (47 m/m) + 4*. Torpedo tube, 1 Armour-deck = *ε*. Prot. to vitals (deck + coal) = *d*. I.H.P. *forced* 2650 = (2 *hrs. trial*); *n. d.* 15·75 kts.

(4.) ADAMASTOR (Leghorn, 1896). 1750 tons. Guns: 2D* (15 c/m) + 4F + 4* (57 m/m) + 4* (machine). Torpedo tubes, 3 (1 of them in bow). Armou Prot. to vitals = *d*. I.H.P. *natural* 3000 = 16 kts.; *forced* 4000 = 17·9 k 270 tons *normal*: *max.* 500 tons. 2 screws.

(4.) *Rainha Amelia* (building). 1860 tons. Guns: 4D* (15 c/m) + 4E* + 2* (47 m/m). Other details = São Gabriel.

SAÕ GABRIEL & SAÕ RAFAEL

VASCO DA GAMA

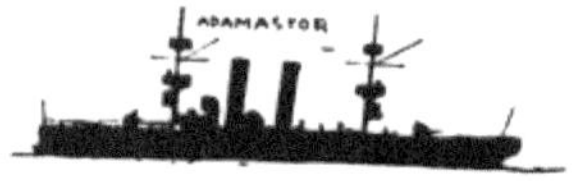

PORTUGAL.

(2). DOM CARLOS I. (Elswick, 1898).

4100 tons.

L. B. *D.* (*mean*): 360 × 46½ × 17½ ft. = 109·72 × 14·16 × 5·33 m.

Guns: 4D* (15 c/m, 45 cals.) + 8E* (12 c/m, 45 cals.) + 12* (47 m/m) + 10*. Torpedo tubes: 2 *submerged* + 3 *above water*.

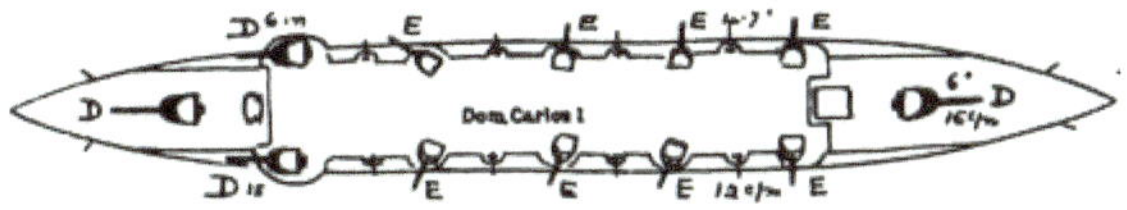

Armour: Deck = *b* (110 m/m) C.T. *d* (100 m/m). Gun shields = *c*. Cofferdam round machinery.

Machinery: 2 sets 4 cylinder triple expansion. I.H.P. *forced*, 12,500. 2 screws. Boilers, 12 Yarrow. Coal, 700 tons.

Trials.—6 *hrs. natural draught.* I.H.P. 8000 = 20·6 kts.
4 „ *forced* „ „ 12,690 = 22·15 kts.

Photo, Elswick.

ROUMANIA.

TORPEDO BOATS.

9 *2nd class*, 55 tons. 21 kts. speed (launched 1888–95).
4 *3rd class*, 32 tons. 4 of 15 tons.

(5.) ELISABETHA (1887, Elswick).

1325 tons. Complement 250.

L. B. D.: 222 × 33½ × 14 ft. = 67·50 × 10·20 × 4·30 m.

Guns: 4D (15 c/m) + 6*. Torpedo tubes, 4.

Armour: Deck = *d*.

Machinery: I.H.P., *natural draught* 4295 = 18 kts., *forced* 4714 = 18·5 kts.

(RUSSIE). РОССIЯ. (RUSSLAND).

(Revised by His Imperial Highness Grand Duke Alexander of Russia, Captain Imperial Russian Navy.)

—	Nominal Calibre.		—	Longeur.	Vitesse Initiale.		Énergie Initiale. circa.		Projectile: Obus Ordinaire.	
	in.	c/m		Cals.	ft. secs.	met. secs.	ft. tons.	met. tx.	lbs.	kgs.
A	12	30·5	Obuchoff, 56-ton	35	1942	592	25000	7750	732	332
A	12	30·5	„ 59-ton	40	2500	724	30000	9300	732	332
B	10	26	Obuchoff, 32-ton	45	2500	724	17000	5265	...	...
B	12	30·5	Obuchoff, 50-ton	30	1942	592	19300	6000	732	332
B	9·4	24	Canet	40	2500	724	15000	4645	...	...
C	12	30·5	(M. 1877) 40-ton	17	1705	448	10000	3100	990	448
C	11	28	(M. 1877)	19	1517	462	9000	2787	1020	462
C	11	28	(M. 1867)	19	1496	453	8000	2480	1000	452
C	9	23	...	...	...	...	10500	3252	...	...
C	8	20·3	...	35	1922	586	...	...	...	...
D	8	20·3	...	30	1794	547	4400	1362	192	87
D	9	23	(M. 1867)	20	1463	446	4100	1270	837	384
D	9	23	...	17	1260	384	3100	960	170	77
D	8	20·3	(M. 1867)	21	1352	412	2200	680	...	...
D✲	5·9	15	Obuchoff	45	2460	750	4000	1238	88	40
D✲	4·7	12	„	45	2460	750	2212	602	46·3	21
D✲	6	15	...	35	2080	633	3300	990	?	...
D & D✲	6	15	Obuchoff	28	1740	530	2700	836	119	54
D	6	15	„	24	1740	530	2200	680	86	39
E	6·03	15·3	„	21	1440	440	1500	464	86	39
E	6	15	„	19	1440	440	1100	340	86	39
F✲	3	7·5	Obuchoff	50	2658	810	648	200	13·2	6
F✲	2·4	6·3	Baranovsky	20	1246	380	...	...	27½	12·5
...	...	5·7	...	80	3149	960	...	...	6	2·7
Z	4 1	10·6	...	...	...	...	...	...	...	...

Volunteer fleet. *Moskva* and *Poltava*, 10225 tons, 20 kts.; Orel, St. Petersburg, Saratoff, circa 900 tons, 19 kts.; Kherson, 10255 tons, 19·5 kts.; and Chavarovsk, Orel, Kostroma, Tamboff, Vladimir, Yaroslav, Kief, Voronejtz, Ekaterinoslav, and Nijni Novgorod circa 7000 tons. Speeds circa 13 kts. Guns C and D. Also nine ships of Black Sea Steam Navigation Co.

DESTROYERS.

2 funnels.

4 Kit, Skat, Delphin, Kassatka (1898–99). 350 tons. 5700 I.H *Boilers: Schichau.* 27 kts.

+ *4* Perepel, Stchegol, Skvorets, & *unnamed*, *pro.* (= Skit).

4 funnels.

1 Sokol (1895). 240 tons. 4490 I.H.P. *Yarrow.* *Boilers: Yarrow*

2 Kretchet, Korshun (1898). 240 tons. 3800 I.H.P *Abo.* *Boil* 27·5 kts.

4 Berkout, Condor, Nyrok, Yastreb (1898). 240 tons. 3800 I. *Boilers: Yarrow.* 27·5 kts.

4 Filin, Gagara, Sova, Voron (1899–1900). 240 tons. 3800 I.H. *Yarrow.* 27·5 kts.

1 Som (1899). 370 tons. 6000 I.H.P. *Birkenhead.* *Boilers: Lair*

+ *3* Kefal, Osetr, Losos (*bldg.*). 312 tons. 4750 I.H.P. *La Seyn* *Normand.* 26 kts.

+ *2* Forel, Sterliad (*bldg.*) = Kefal. *Havre.* *Boilers: Normand.*

+ *9* Lebod, Pavlin, Phasan, Albatross, Pelikan, Bekass, Drozd, Diatol, B 312 tons = Kefal. *Normand boilers.*

+ *3* Gratch, Gorlitza, Kubik (*bldg.*). 240 tons = Sokol. *Yarrow boilers*

16 + *21* building.

[All the above carry 2 tubes. Guns 1F* + 3 - 5* *except* the new boa carry 2F* + 3*].

TORPEDO BOATS.

1st Class.

4	boats	(1890)	of 140	tons,	152	ft. long,	Speed *max.* 22	kts.
23	„	(1892–6)	of 118	„	138	„	.. 25	..
4	„	(1891)	of 100	„	152	..	.. 19	..
2	Adler type	(1890–3)	of 130	„	152	..	.. 27·4	..
4	boats	(1893–5)	of 99	„	127	..	.. 21	..
1	(Viborg)	(1886)	of 126	„	144	„	.. 20	..
1		(1877)	of 160	„	118	„	.. 16	„

3? + 6 *building* of 118 tons.

2nd Class.

21	{Schichau, etc. (Abo type)}	(1886)	87 tons.	128 ft. long.		Speed 20	kts.	2 t
2	Normand	(1892)	95 „	152	„	..	19·7	..	2
9	boats	(1890–1)	81 „	136	„	..	21	„	2
1	{Thornycroft, Sokhum}	(1883)	64 „	113	„	..	19·5	..	2
5	boats	(1891–5)	85 „	128	„	..	22	..	2
1	„	(1883)	78 „	120	„	..	18	..	2
1	„	(1883)	75 „	122	..	..	18	..	2
1	„	(1885)	67 „	124	„	..	16·5	..	2

41

3rd Class.

110 various (1877–90) 40—12 tons. Speeds *circa* 17 kts.

50 Submarines *projected.*

(3.) KHRABRY (1895).

Circa 1500 tons. Complement 135.

L. B. D.: 223 × 41⅔ × 13⅓ feet = 68 × 12·60 × 4 metres.

B (9·4 in.) + 1D* (6 in., 45 cals.) + 10*. Torpedo tubes, 1 (in bow). Armour (Harvey): ›ck forward *e*. Prot. to vitals (belt + coal) = *c*. Machinery: I.H.P. 2642 = 14·5 kts. (*trial*). ›ilers: 8 Niclausse.

ROZIASTCHY (1890) & GREMIASTCHY & OTVAJNY (—1892).

Circa 1500 tons. Complement 120.

L. B. D.: 223 × 41⅔ × 13⅓ feet = 68 × 12·60 × 4 metres.

C (9 in.) + 1D (6 in.) + 10* to 16*. Torpedo tubes, 2 (bow and stern). Armour (compound): ›ck forward *f*. Prot. to vitals = *d*. Machinery: I.H.P. 2000 = 13 kts. Sea speed, 12 kts. ›ilers, 6 Belleville. *Trials*, Groziastchy (1892) 6 *hours*, I.H.P. 2034 = 13·6 kts. (*mean*).

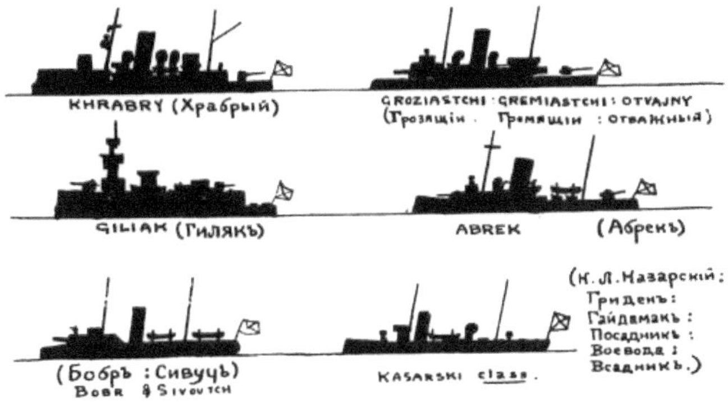

ITAN-LIEUTENANT KASARSKI (Schichau, 1890) & GRIDEN (1893) ›IDAMAK (1893) & POSADNIK (Schichau, 1892) & VOËVODA (Schichau,) & VSADNIK (1893).

400 tons. Complement 61. Guns: 6* (3 pdr., 47 m/m) (1 pdr., 37 m/m). Torpedo tubes, 2 (1 bow, 1 training). I.H.P. *natural* 1270; *forced* 3500 kts. (*trial*). 2 screws. Boilers: Locomotive (being replaced by Belleville and Yarrow).

EK (1896).

534 tons. Complement 109. Guns: 2E* (4·7 in.) + 6*. Torpedo tubes, 2. 5000 = 21 kts. (*trial, mean*).

(2 *others projected*.)

R (1885) & SIVOUTCH (1884, Stockholm).

950 tons. Guns: 1C (9 in.) + 1D + 4*. Machinery: I.H.P. 1150 = 12 kts. Sea speeds, 8 kts. (Flat bottoms.) 2 screws.

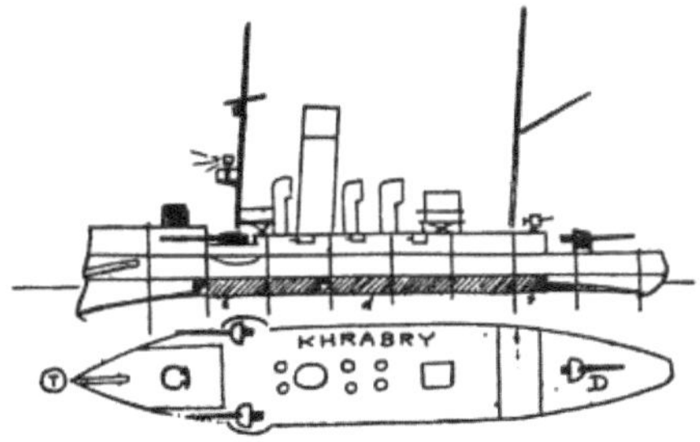

AK (1897) (river gunboat).

Circa 1400 tons. Guns: 1E* (4·7 in.) + 4F* (3 in.) (47 m/m, 3 pdrs.) + 2* (2½ pdr. Baranovsky) + 2 Maxims. Torpedo tubes, 1 (in bow). ›tion: deck ½ inch + coal = *f*. Machinery: I.H.P. 1000 = 12 kts. 2 screws. Boilers, Yarrow. ›num draught, 10 feet (3·04 m.).

(III.) ADMIRAL NAHIMOFF (1885) (1899).

Circa 10,000 tons. Complement 580.

L. B. D.: 333 × 61 × 28½ feet = 101·50 × 18·60 × 8·68 metres.

Guns: 18D* (6 in. 45 cals.) + . Torpedo tubes: 4 *above water*.

Armour (compound): 10″–8″ Belt *b–c* (250–200 m/m). Protection to vitals *b*. 8″ Barbettes *c* (200 m/m). Hoods *c* (Harvey, 152 m/m).

Machinery: Boilers: Belleville.

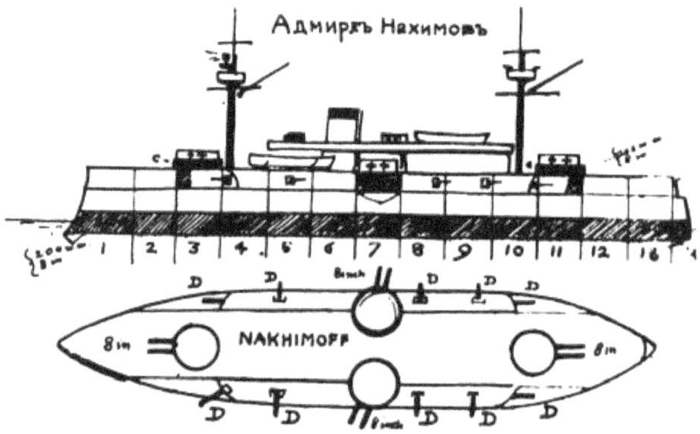

(III.) PETER VELIKY (1872) (1874) (1899–1900).

(Reconstructing.)

Circa 10,000 tons. Complement 436.

L. B. D.: 328 × 62 × 28 feet = 100 × 19 × 8·50 metres.

Guns (new armament): 4B (10 in. 45 cals.) + 4F* (3 in.) + . Torpedo tubes: 0.

Armour (as reconstructed): Belt *b–d*. Protection to vitals *b* Turrets *aa*. Redoubt below turret *d–e*.

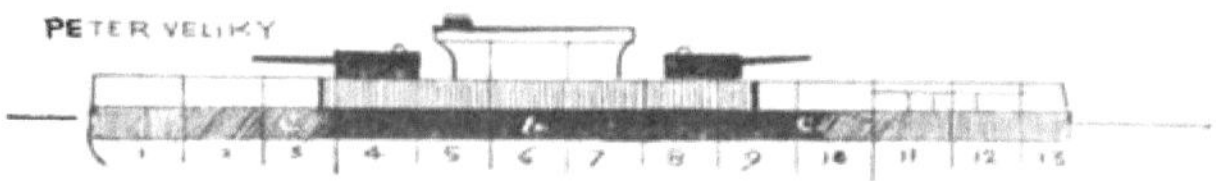

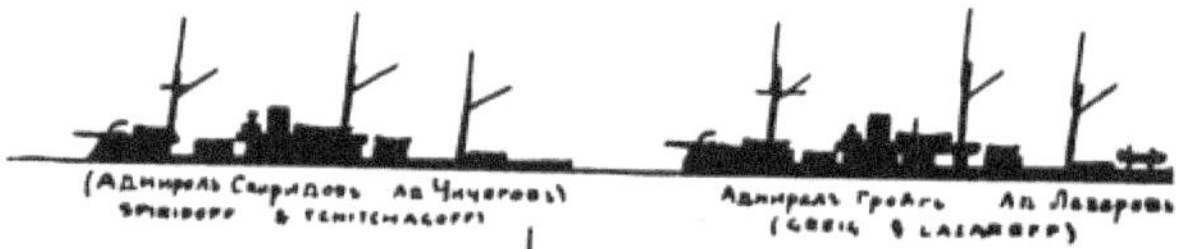

(V.) ADMIRALS—SPIRIDOFF & TCHITCHAGOFF (—1868): GREIG (1868) & LAZAREFF (1867).

3500 tons. Guns, *first two*, 2C (11 inch)+7*. *Greig & Lazareff*, 3C (11 in.) + 7*. Armour: Belt *e*. Turrets *d*. Sea speed, *circa* 9 kts.

ᴉERNOMORETZ & ZAPOROROJETZ & DONETZ (—1887):: KOUBANETZ URALETZ (—1887) & TERETZ (1888). 1230 tons. Guns: 2D (8 inch) +) (6 in.)+6*. Torpedo tubes: 2. Armour: deck over engines=*f*. 1 screw. Sea speed: 10 kts.

NDJOUR (Copenhagen, 1886) : : KOREITZ (Stockholm, 1886). 1500 tons. ıns: 2D (8 inch) + 1D (6 inch) + 4Z (4 inch) + 6*. Torpedo tubes: 2. 1 screw. Sea eeds: 11 kts.

ıIAT-MERKOURIA (Toulon, 1880). 3050 tons. Guns: 6D (6 in.) + 4Z (4 in.) 2*. Torpedo tubes: 2. Sea speed: 8 kts. 1 screw (*Euxine fleet*).

ᴚELOK (1879) & PLASTOUN (1879) & NAYEZDNIK (1878) & ESTNIK (1880) & RASBOINIK (1878) & DJIGIT (1876) & KREISER 875). *Circa* 1375 tons. Guns: 3E (6 inch). 1 screw. Speeds 8 kts.

KOREITZ.

(Euxine Fleet.)

(II.) GEORGI POBIEDONOSETS (1892).

10,500 tons. Complement 500.

L. B. D.: 339 × 69 × 29 feet = 103·50 × 21 × 8·84 metres.

Guns: 6A (12 in., 35 cals.) + 7D* (6 in., 45 cals.) + 8F* + 18*.
Torpedo tubes: 7 *above water*.
Armour (Creusot): Belt (16"–11") *aa–a* (400–275 m/m). Protection to vitals *aaa*. Redoubt above belt *aa*. Glacis = *aaaa*. Hoods to big guns fronts = *a*. Sides *c*. C. T. *a*. Flat deck on top of belt 3" (76 m/m).
Machinery: Vertical triple expansion. I.H.P. 13,468 *natural draught* = 16·5 kts. (12 *hours trial*). Boilers, cylindrical. 2 screws. Coal, 700 tons.

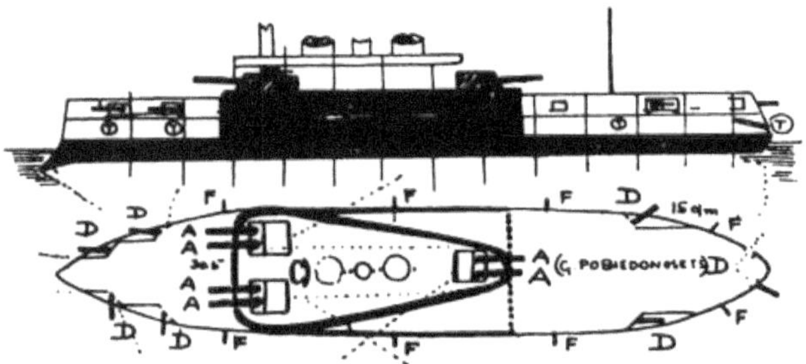

Georgi Pobiedonosets. *Photo by favour of C. de Grave Sells.*

(II.) EKATERINA II. (1886) (1900): TCHESMA (1886): SINOP (1887).

10,500 tons. Complement 500.

L. B. D.: 339 × 69 × 29 feet = 103·50 × 21 × 8·84 metres.

Guns: *Ekat. II.* 6B (12 in., 30 cals.) + 7D (6 in., 35 cals.) + 12*.
Tchesma 6C (12 in., Krupp) + 7D (6 in., 35 cals.) + 12*.
Sinop 6A (12 in., 40 cals.) + 7D* (6 in., 35 cals.) + 12*.
Torpedo tubes: 7 *above water*.
Armour (compound): Belt *aa–a*. Protection to vitals *aa*. Redoubt above belt *a*. Glacis = *aaa*. Hoods *e*.
Machinery: I.H.P. 11,000–13,000 *forced*. Boilers: Ekaterina II. (1900), Belleville. Others to be reboilered with Bellevilles. Coal, 870 tons.

Note.—Guns of Ekaterina II. and Tchesma are on disappearing mountings.

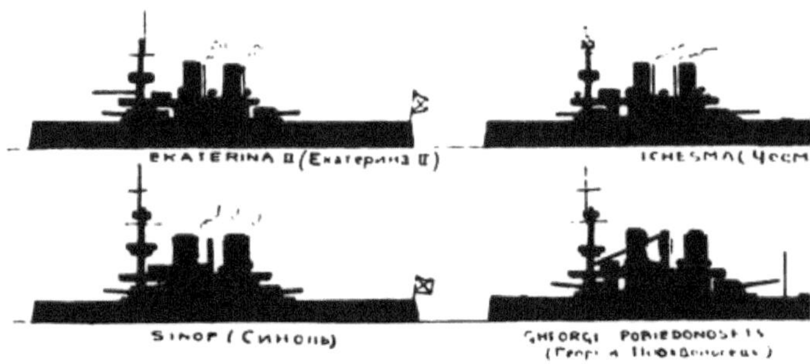

(T.)
LIEUT. ILYIN (1886).

L.: 230 ft. = 70·5 m.

Guns: 17*. Armour: ½ *prot.*

Sea speed: *circa* 17 kts.

714 tons. 2 screws. 7 tubes.

(T.)
KAPITAN SAAKEN (1889).

L.: 213 ft. = 65 m.

Guns: 10*. Armour: 0. Sea speed: *circa* 15 kts.

750 tons. 2 screws. 6 tubes.

(Reboilered 1897.)

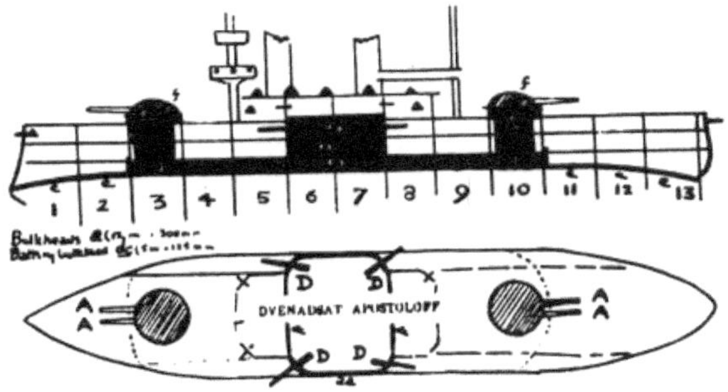

(II.) DVENADZAT APOSTOLOV (1890) (Euxine).

8500 tons. Complement, 530.

L. B. D.: 330 × 60 × ft. = 100·5 × 18·3 × m.

Guns: 4A (12 in., 30·5 c/m, 35 cals.) + 4D (15 c/m, 35 cals.) + 25* + 6 torpedo tubes.

Armour: Deck = *d*. Belt *aa* - *a*. Barbettes and barbette bases *a*. Battery *d*. C. T. *b*.

Machinery: I.H.P. 11,500 = 16·6 kts., *trial, natural* draught. 2 screws. Coal, — tons.

Photo by C. de Grave Sells, Esq.

РОССІЯ.

(II.) SISSOI VELIKY (1894) *circa* 9000 tons. Complement, 600.

L.B.D.: 345 × 69 × 25 ft. = 145 = 21 × 7·62 metres.

Guns:

- **4 A** (12 in.).
- **6 D*** (6 in.).
- **2 F***.
- **12*** (57 m/m).
- **24*** (37 m/m, etc.).

Torpedo tubes:

- 6 *above water*.

Note.—The * A guns are 40 cal., Obuchoff, the D, 45 cal. Canet, the F, Baronovsky.

Photo by M. Bar.

Armour (Creusot):

- 16″ Belt, *aaa* (400
- Prot. to vitals, *aaa*.
- 3″ Deck (slopes) =
- 5″ Lower deck, *d* (1
- 5″ Battery, *d* (127
- 5″ Bulkheads, *d*.
- 5″ Turret bases, *d*.
- 12″ Fore turret, *aa* (
- 10″ After turret, *a* (

Machinery: 2 sets vertical triple expansion. 2 screws. Boilers: 16 cylindrical. I.H.P. *forced* 8500 = 16 kts. Sea speed: 13·5 kts.

Coal: *Normal*, 500 tons: *maximum*, 800 tons.

Note.—The belt is 7 ft. (2·13 metres) wide.

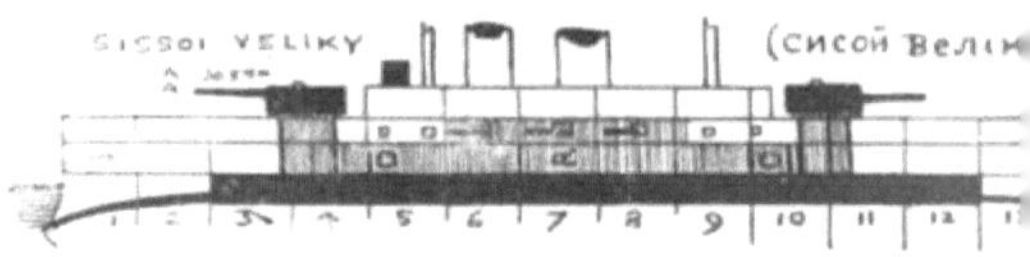

(I.) *TSAREVITCH* (Царевичъ) (bldg. La Seyne).

circa **13,000 tons. Complement 700.**

L. B. *D.* (*mean*): 387 × 75½ × *26* feet = 118 × 23 × *7·92* metres.

uns (Obuchoff):
4A (12 in., 40 cals.).
12D* (6 in., 45 cals.).
20F* (3 in.).
20* (3 pdr., 47 m/m).
8* (1 pdr., 37 m/m).
orpedo tubes:
2 *submerged*, forward.
1 *above water*, in bow.
1 *above water*, in stern.
2 *above water*, broadside.

Armour (Krupp):
9" Belt amidships *aa* (227 m/m).
4"—2½" Belt at ends
c—*e* (100—65).
4" Deck (slopes) = *a* (100 m/m).
2" Upper deck *flat* (50 m/m).
4" Hull under water (nickel) *d* (100).
4" Lateral bulkheads, internal, *e* (100).
Protection to vitals = *aaaaaa*.
6" Lower deck *a* (152 m/m).
11" A gun turrets *aaa* (280 m/m).
10" ,, bases *aaa* (253 m/m).
6" D gun turrets *a* (152 m/m).
5" ,, bases *b* (126 m/m).
Fore C. T. *aaa*.

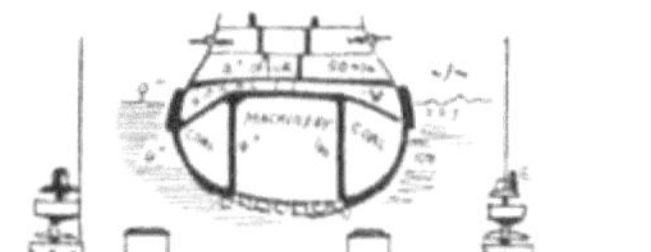

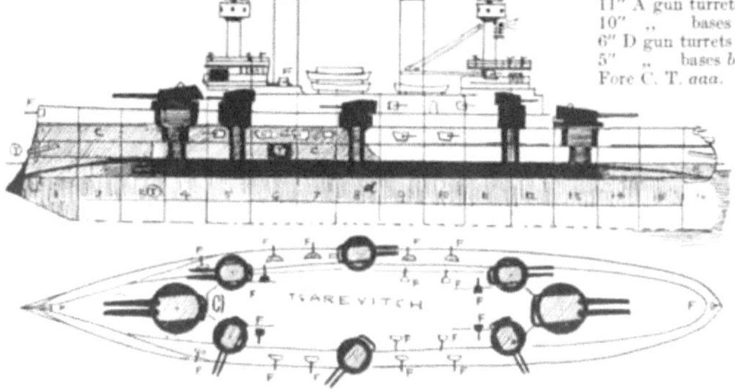

Machinery: 2 screws. I.H.P. 16,300 = 18 kts. (*contract*). Boilers: 20 Belleville with economisers. Coal capacity, 1250 tons.

(I.) *Borodino* (Бородинoъ) (bldg. at New Admiralty Yard) & *Orel* (Орелъ) (bldg. at Galernii Island).

13,566 tons. Complement 750.

L. (*p.p.*) B. *D.* (*mean*): 397 × 76 × *26* feet = 121 × 23·16 × *7·92* metres.

Guns, Armour, Machinery, Boilers, etc. = *Tsarevitch*.

(Imperator Alexander III., see page 218.)

Note.—These ships are armoured under water against torpedo attack. The hull under the belt is covered with 4 inch (100 m/m) nickel steel nearly to the double bottom. Internally two vertical bulkheads of the same maximum thickness run the whole length of the ship's side. These are of Krupp armour. Above the armour deck proper is a second deck 2 inches (50 m/m) thick. The entire space between these two decks is filled with coal.

In the Borodino, etc., the water-line belt forward will probably be 6 inches (152 m/m) thick (*a*).

(II.) ROSTISLAV (1897) (Euxine fleet).

circa 9000 tons. Complement 600. L. B. D.: 345 × 69 × 25 feet = 145 × 21 × 7·62 metres.

Guns: 4B (10 in. 45 cals.) + 8D* (6 in. 45 cal.) (Obuchoff) + 26*. Torpedo tubes: 6 *above water*.

Armour: (Harvey) Belt *aaa* (400 m/m). Deck = *c* (75 m/m). Protection to vitals = *aaaaa*. Lower deck *c* (152 m/m). D gun turrets and hoists *c*. B gun turrets *aa*. Bases *c*.

Machinery, etc. = *Sissoi Veliky*. But *Rostislav* also carries liquid fuel. On *trial* (Nov. 1898) *Rostislav* made 18 kts. Sea speed, 15 kts.

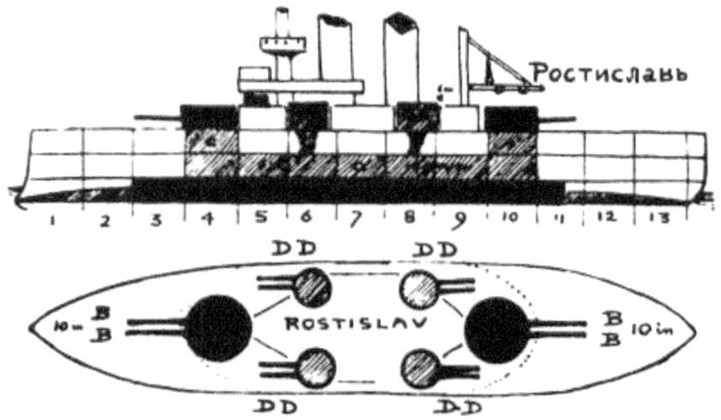

(I.) TRI-SVIATITELIA (1898) (Euxine fleet).

12,500 tons. Complement 582.

L. B. *D.* (*mean*): 371 × 72½ × 27 ft. = 113 × 22 × 8·28 metres.

Guns: 4A (12 inch, 40 cals.) + 8D* (6 in. 45 cals.) + 4E* (4·7 in.) · (47 m/m) + 40* (37 m/m). Torpedo tubes: 2 *submerged* + 4 *above water*.

Armour (special Creusot): Belt *aaa* (400 m/m). Prot. to vitals *aaa*. De Bulkheads *aa*. Redoubt, lower deck, *aaa*. Turrets *aaa* (400 m/m). C.' Battery *d*.

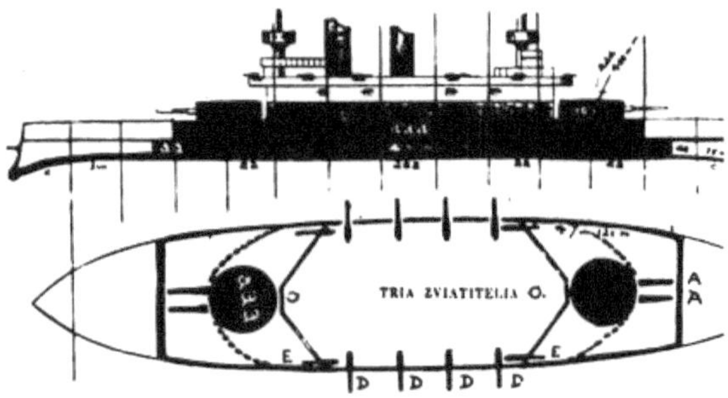

Machinery: 2 sets triple expansion. Humphrys & Tennant. I.H.P. 11, 17 kts (12 *hours trial*). 2 screws. Sea speed: 15 kts. Coal: 1000 tons

(IV.) ADMIRAL OUSHAKOFF (1893): ADMIRAL SENYAVIN (1894) & GENERAL-ADMIRAL-GRAF-APRAKSIN (1895).

4126 tons. Sea speed: 14 kts.

L. B. D.: 277½ × 52 × 17½ ft. = 84·5 × 15·8 × 5·3 m.

Machinery: Oushakoff, Maudslay; Senyavin, Humphrys and Tennant. 2 screws. I.H.P. *circa* 5700.

akoff (Oct. 1895), 12 *hours' trial natural draught* = 5769 H.P. = 16·1 kts. *mean speed.* Apraksin *trial* (1898), 7 *hours* natural draught = 5757 H.P.

Адмиралъ Ушаковъ. Адмиралъ Сенявинъ.
Генералъ Адмиралъ-Графъ Апраксинъ.

Guns:
4C (9 in.).
4D* (6 in.).
22*.
Torpedo tubes:
4 *above water.*
But *Apraksin* carries:—
3B (10 in.).
4D* (6 in.).
22*.
Torpedo tubes:
4 *above water.*

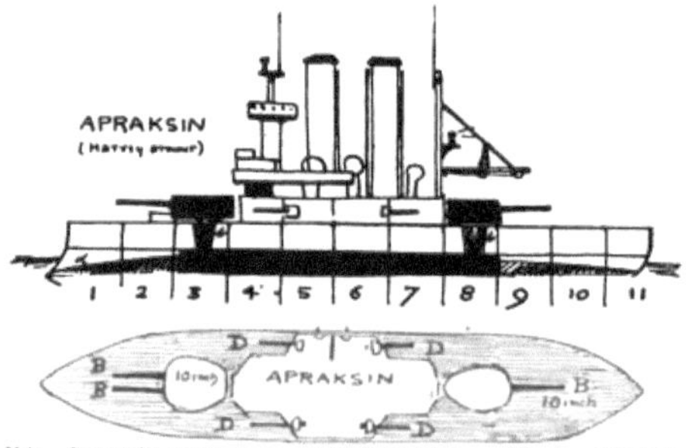

Note Oushakoff and Senyavin 4 B (9 inch) guns. Apraksin 3 B (10 inch)

Armou
10"
Prot.
3" A
Turr
Hoist
In *Apr*
10"
Prot.
3" A
Turr
Hoist

I.) POLTAVA : PETROPAVLOVSK (–1894)
SEVASTOPOL (1895).

10,960 tons. Complement: . Sea speed: 15 kts.

L. B. D.: 367 × 69 × 27 ft. = 112 × 21 × 8·2 m.

4A + 12D* + 36*. 4 torpedo tubes.

r: Deck = *b*. Belt, *aaa*. Bulkheads, *a*. Turrets, *aa*. D gun turrets
tes, *c*.

ıery: Poltāva, *natural draught* I.H.P. 11200 = 16·5 kts. (12 *hours trial*)
ısk, I.H.P. *natural draught* 11213 = 16·38 kts. (12 *hours trial*). 2 screws
sets triple expansion. Poltava, Humphrys and Tennant. Petropavlovsk
Leslie, Sevastopol, Baltic Works. Coal, 700 tons. Boilers, cylindrical.

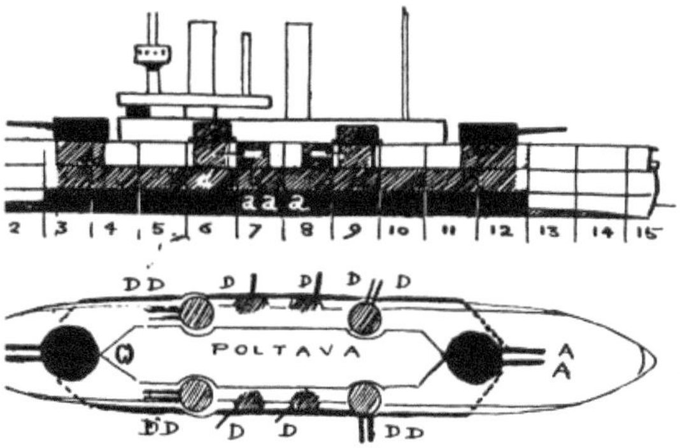

Petropavlovsk. *Photo by J. Geiser. Alger.*

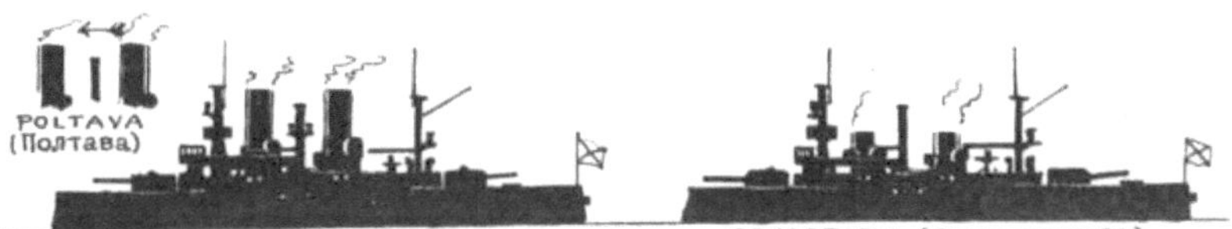

(III.) IMPERATOR NIKOLAI I. (1889): IMPERATOR ALEXANDER II. (1887).

L.: 326 ft. = 99·5 m. B.: ft. = m. D.: 25½ ft. = m.

Nikolai, 9700 tons. Alexander, 9900 tons. Complement, 600.

Sea speeds: Nikolai, 14·5 kts.; Alexander, 12·5 kts.

Guns: 2B (12 in., 30 c.m.) + 4C (9 in., 23 c.m.) + 8D (15 c.m.) + 18*. 6 torpedo tubes.

Armour: *a*. Belt *aa* - *d* flat deck on top 75 m.m. Protection to vitals, coal + side = *aaa*. Bulkhead (in Alexander II. only) *a*. In Alexander II.: Barbette *a*. shield *d*. On c guns in battery *d*. No conning tower. Hoist to big guns *d*. In Nikolai I.: Turret *b*. On c guns in battery and bulkhead to battery *c*. Conning tower *d*. Big gun hoist *d*.

Machinery: Nikolai, 16 Belleville boilers: 1898 with 11 boilers, speed = 14 kts.

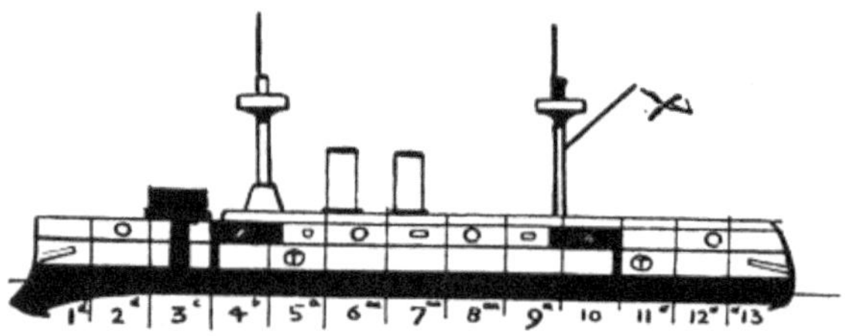

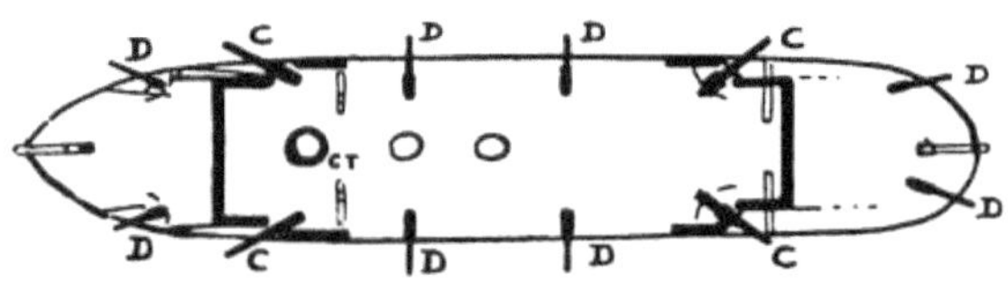

Императоръ Николай I.

Императоръ Александръ II.

Photos, Geiser.

(III.) VLADIMIR MONOMAAKH (1881) (1897).

L.: 295 ft. = 90 m.

Guns: 5D* (15 c.m., 45 cal.) + 6E* (12 c.m.) + 20*. Sea speed: 13 kts.

6000 tons. Complement 550. Armour: Belt *b* - *d*. On D guns in battery *c*.
2 screws. I.H.P. 7000 = 15·2 kts. (*trial*). 3 torpedo tubes. Coal: .

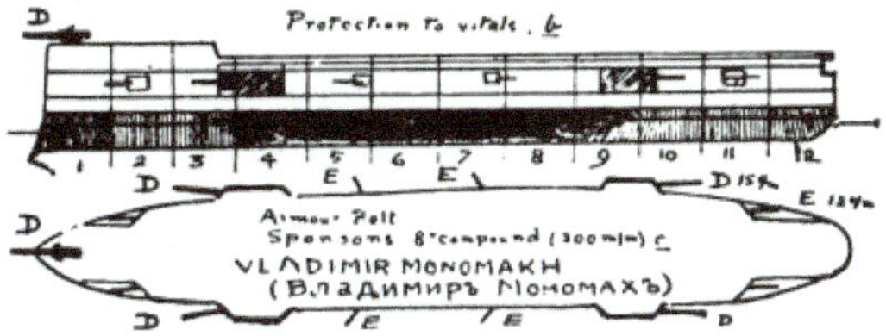

(III.) DMITRI DONSKOÏ (1883) (1895).

L.: 295 ft. = 90 m.

Guns: 6D* (15 c.m., 45 cal.) + 10E* (12 c.m.) + 28*. Sea speed: 14 kts.

5880 tons. Complement 550. Armour: Belt *d*. On D guns *b*. 4 torpedo tubes.
2 screws. I.H.P. 7000 = 16 kts. (*trial*). Coal: 400 tons.

(1.) RURIK (1892).

L.: 426 ft. = 130 m. B.: 67 ft. = 20·4 m. D.: 30 ft. = 9 m.

Guns: 4C (8 in.) + 16D* (Canet) + 6E* (12 c.m., 4·7 in.) + 22*. Sea speed: 17 kts. 10,950 tons. Complement 768. 6 Torpedo tubes.

Armour: Deck = *cd.* Belt (341 × 6½ ft. = 104 × 2 m. 10.), *b - c.* Bulkhead: fore *a*, aft *b*. Shields to C and E guns.

Machinery: I.H.P. 13,250 = 18 kts. (6 *hours trial*). 4 sets triple expansion engines. 48 furnaces.

Рюрикъ. *Photo, J. Geiser.*

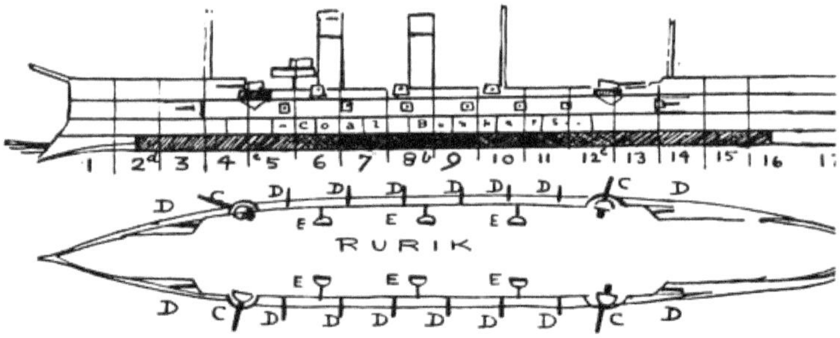

(2.) GERTZOG EDINBOURSKI (1875):
(2.) GENERAL ADMIRAL (1873).

L.: 285 ft. = 86·5 m.

Guns: {Herz. Edin.: 4D + 5E. / Gen. Ad.: 6D + 2E.} Armour: *c.* Sea speed: 8 kts.

4600 tons. 1 Screw.

(Training Service.)

(2.) RYNDA (1885).

L. : 269 ft. = 89 m.

Guns: 10D + 10. Armour: ½ *prot.* = *f*. Sea speed: 10 kts.

3506 tons.

(IV.) *MININE* (1869) (1878).

L. : 298 ft. = 91 m.

Guns :

6000 tons.

On transforme le Minine dans ce moment.

(IV.) KNIAZ POJARSKI (1867).

L. : 273 ft. = 83·5 m.

Guns: 2D + 2D + 2*. Armour: *e* Sea speed: 9 kts. (1894).

5000 tons.

Адмиралъ Корниловъ. *Photo by J. Geiser.*

(2.) ADMIRAL KORNILOFF
(Nazaire) (1895) (reboilered).

L. : 350 ft. = 106·5 m.

Guns: 14D* + 16*. Armour: *prot.* = *d*. Sea speed: 17 kts.

5000 tons. Complement, 450. 6 torpedo tubes.

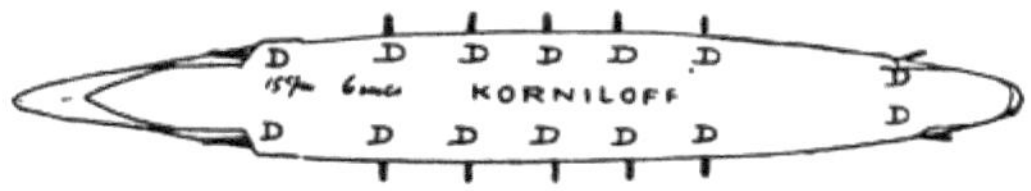

Photo by J. Geiser.

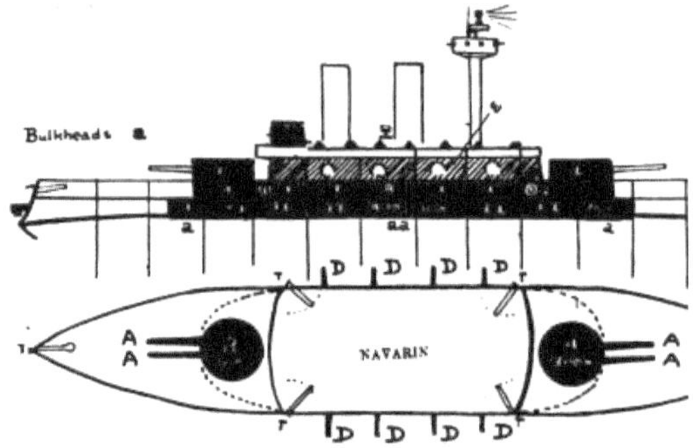

(II.) NAVARIN (1891).

L.: 338 ft. = 103·5 m.

Guns: 4A + 8D + 30*. Sea speed: 14 kts. 10000 tons. Compl
6 torpedo tubes (4 behind *a* armour).

Armour: Deck = *d*. Belt, *aa* - *a*. Redoubt, turrets, bulkhe
Battery *de*.

Machinery: I.H.P. 9000. 2 screws. Coal: *circa* 1000 tons.

(3.) SVIETLANA (1896, Havre).

L.: 331 ft. = 101 m. D.: 19 ft = 5·8 m.

Guns: 6D* + 12*. Armour: *prot.* = *d*. Speed: 20 kts. (*en essais*). 3900 tons.

Machinery: 2 sets vertical 4-cylinder triple expansion. 18 Belleville boilers. 2 screws. I.H.P. (*natural draught*) = 8500. Coal: 400 tons.

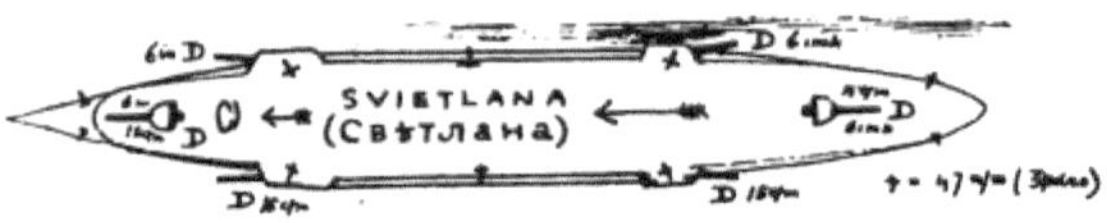

Свѣтлана.

) PALLADA & *DIANA* (1899) & *AURORA* (1900).

(Паллада: Діана: Аурора.)

6630 tons.

B. *D.* (*mean*): 406 × 55 × *21* feet = 123·70 × 16·76 × *6·40* m.

):
5 cals.).
30 cals.).

: 4 *above*

Armour:
$2\frac{1}{2}$" Deck = *c* (63 m/m).
Gun shields *c*.
Hoists *c*.
[All decks are steel, and no wood is used.]

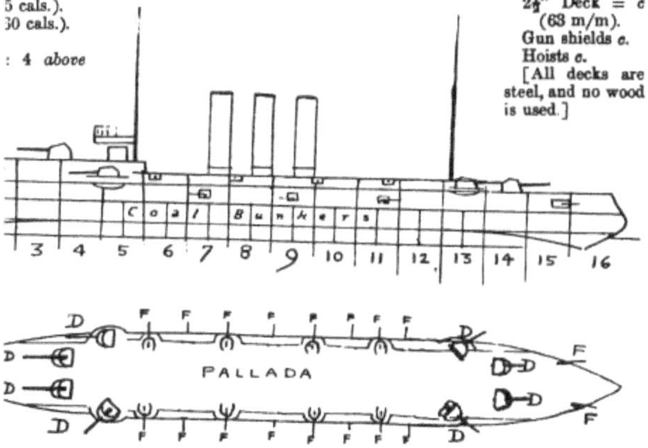

: I.H.P. 11,600 = 20 kts. Contract sea speed, 19 kts. 3 screws. Boilers, Coal, *normal* 900 tons; *maximum* 1430 tons + liquid fuel.

(I.) *KNIAZ POTEMKIN TAVRITCHESKY* (building) (Euxine fleet).

(Князь Потемкинъ Таврическій.)

12,500 tons.

L. B. D.: 371 × $72\frac{1}{3}$ × feet = 113 × 22 × m.

Guns (Obuchoff):
4A (12 in., 40 cals.).
16D* (6 in., 45 cals.).
14F* (3 in.).
14*.

Armour (Krupp):
9" Belt (amidships) *aa* (225 m/m).
3" Deck = *a* (76 m/m).
Protection to vitals = *aaaaa*.
9" Lower deck *aa* (225 m/m).
12" Turrets *aaaa* (400 m/m).
5" D gun turrets *b* (125 m/m).
5" Battery *b*.

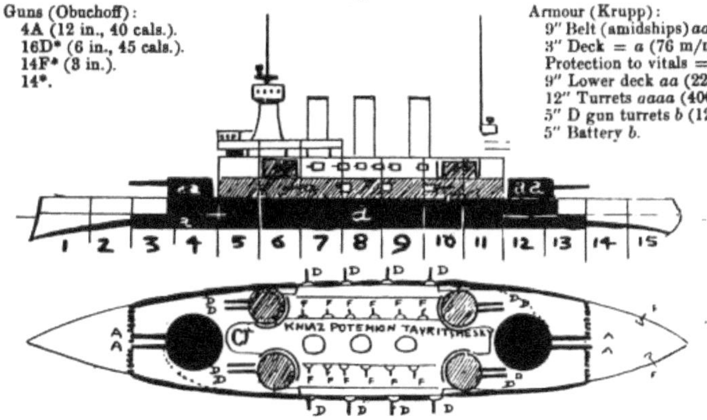

? some details

Machinery: I.H.P. 10,600 = 16 kts. (*est.*). 2 screws. Boilers, 27 Belleville.

(II.) PERESVIET & OSLIABIA (1898) & *POBIEDA* (1900).

(Пересвѣтъ: Ослябя: Побѣда.)

12,674 tons. Complement 732.

L. B. D.: 436 × 71½ × 27¼ ft. = 133 × 21·80 × 8·30 m.

Guns (Obuchoff):
4B (10 in. 45 cals.).
11D* (6 in. 45 cals.).
20F* (3 in.).
26*.
Torpedo tubes:
4 *submerged*
2 *above water*.

Armour (Harvey nickel made in Russia):
9" Belt *a* (227 m/m).
2¾" Deck = *c* (70 m/m).
Protection to vitals = *aaaa*.
6" Lower deck *b* (152 m/m).
6" Casemates *b* (152 m/m).
10" Turrets *aa* (250 m/m).
10" Turret bases *aa* (250m/m)
9" Bulkheads *a*.

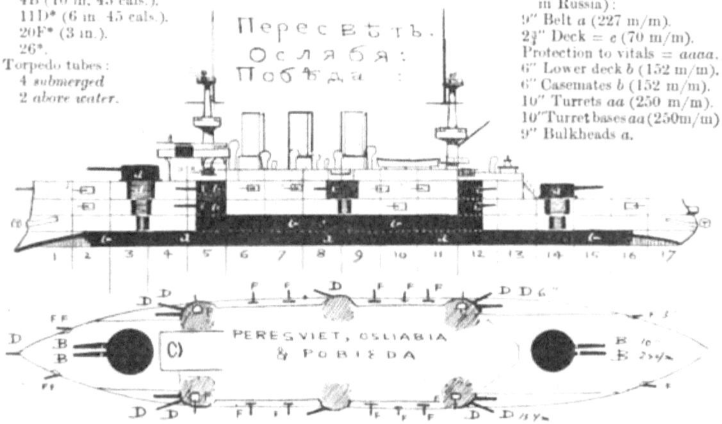

Machinery (made at Baltic Works): 3 sets vertical triple expansion. I.H.P. 14,500. 3 screws. Boilers: 30 Belleville in 6 groups. Sea speed (continuous): 16 knots. Coal, *normal* 1063 tons; *maximum* 2058 tons + liquid fuel

Trials.—**Peresviet**
Osliabia
Pobieda

Note.—**Pobieda, Krupp armour. Consequently armour values are:—Belt *aa*. Deck = *b*. Prot. to vitals = *aaaaa*. Lower deck *a*. Casemates *a*. Turrets and bases *aaa*. Bulkhead *aa*.**

Peresviet. *Photo by favour of H.I.H. Grand Duke Alex*

(I.) *RETVISAN* (bldg., Philadelphia) (Ретвизанъ).

12,700 tons. Complement 750.

L. B. *D.* (*mean*): 376 × 72 × *26* feet = 114·6 × 22 × *7·92* metres.

s.).
als.).

Armour (Krupp):
9" Belt amidships *aa* (225 m/m).
Belt, bow *e*.
3" Deck, on slopes = *a* (76 m/m).
Protection to vitals = *aaaaa*.
6" Lower deck *a* (152 m/m).
10" Turrets *aaa* (250 m/m).
9" Bulkheads and bases *aa* (225 m/m).
5" Casemates *b* (125 m/m).

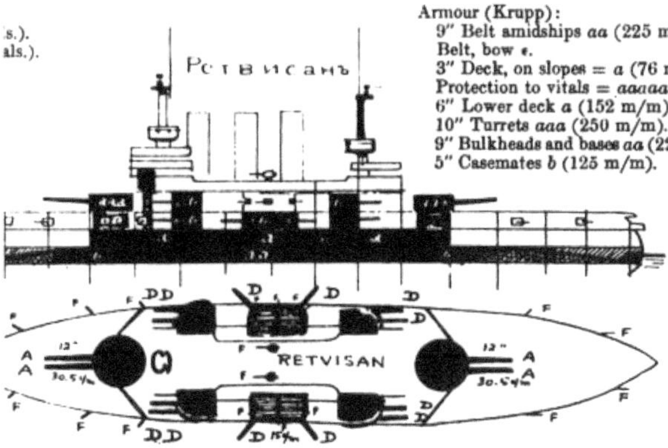

cylinder triple expansion. I.H.P. 16,000 *natural d.* = 18 kts. (*est.*). 2 screws.
se. Coal: *normal* tons; *max.* 2000. (12 *hours contract speed* = 18 kts.).

(I.) *Imperator Alexander III.* (building).

Императоръ Александръ III.

13,516 tons. Complement 750.

L. B. *D.* (*mean*): 397 × 75½ × *26* ft. = 121 × 23 × *7·90* m.

Guns, Armour, Machinery = Retvisan Boilers: 24 Niclausse.

(1.) *PAMIAT-AZOVA* (1888) (1900–01) (Память-Азова).

6700 tons. Complement 525. L. B. D.: 378 × 49 × 29½ feet = 115·50 × 15 × 9 metres.

Guns (new armament): 15D* (6 in., 45 cals.) + 15*. Torpedo tubes: 3 *above water* (1 of them in stern).

Armour (compound): Belt *b–c* (251–200 m/m). Deck = *e*. Prot. to vitals *b*. Barbettes on upper deck *c* (200 m/m).

Machinery: (To be re-engined, etc.). To have Belleville boilers. 2 screws. Coal, 1000 tons (*max.*).

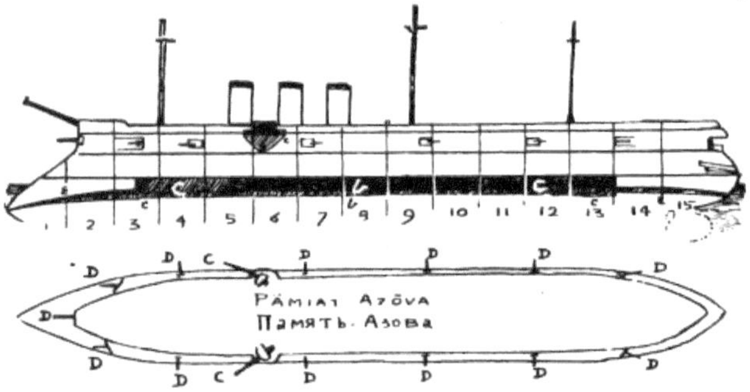

(To be reconstructed.)

Photo,

(2.) VARIAG (1899, Philadelphia) & *BÖIARIN* (1900, Copenhagen) & 1 projected. (Варягъ: Бояринъ)

Guns (Obuchoff):
12D* (6 in., 45 cals.).
12F* (3 in.).
6* (47m/m, 3 pdr.).
Torpedo tubes:
2 *submerged*.
2 *above water* (bow and stern).

6500 tons.
L. B. *D.* (*mean*): 400 × 32 × 19 feet = 122 × 9·75 × 5·90 metres.

Armour (Krupp):
3" Deck = *a* (76 m/m)
Protection to vitals = *aa*.
6" Gun shields *a* (152 m/m).

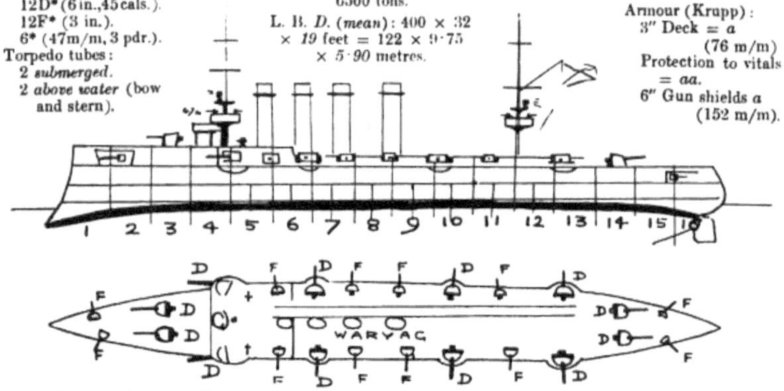

(III). *GROMOVOI* (1899).

14,367 tons. Complement, 800.

L. B. D. (*mean*): 472½ × 69 × 27¾ ft. = 144 × 21 × 8·40 m.

Guns: 4C* (8 in.) + 16D* (6 in., 45 cal.) + 20F* (3 in.) + 24*. Torpedo-tubes: 4 *submerged* + 2 *above water*.

Armour (Harvey Nickel): 6″ Belt *b* (152 m/m). Deck = *d* (51 m/m). Protection to vitals = *aa*. Casemates *b* (152 m/m). Lower deck, side *d* (100).

Machinery: 3 screws. Sea speed: 20 kts. (*est.*)

(1.) ROSSIYA (1896).

12,100 tons. Complement, 725. L.: *circa* 470 ft. = 143·5 m. Sea speed: **19 kts.**

Guns: 4C (8 in., 20·3 c.m.) + 16D* (6 in., 15 c.m.) + 12F* (75 m.m., 12 pdrs.) + 36*.

Armour: Belt *a*. Deck = *d*. Bulkheads fore *a*, aft *b*. Protection to vitals = *aa*. Lower deck amidships *d*. Coal on lower deck and water-line. Casemate to forward 8-in. guns *f*. Gun shields and battery screens *f*. Bulkheads to battery *e* (152 m/m).

Machinery: 3 screws (4-bladed). I.H.P. 18,446 = 20·25 kts. (*trial max.*) 19·74 kts. *mean* (10 *runs on mile*). 32 Belleville boilers. 3 sets vertical triple expansion engines. Coal, 2500 tons + liquid fuel.

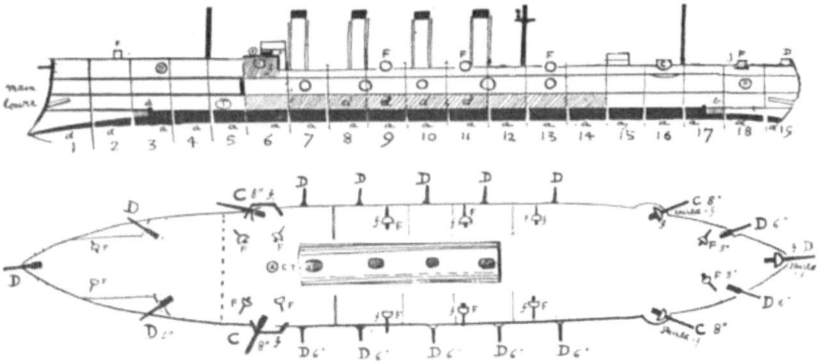

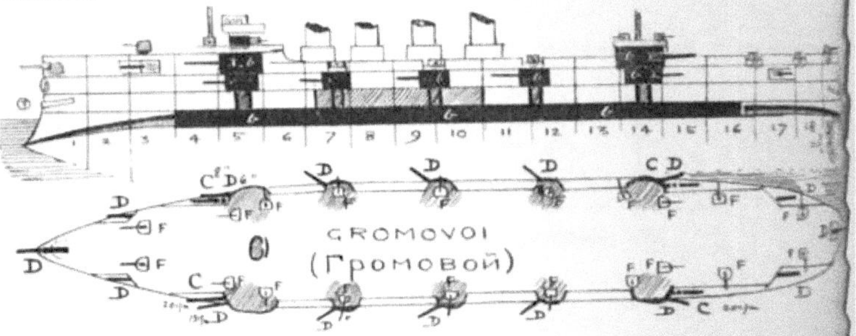

Россія.

Photo by W. ...

(Новикъ)

(4.) *NOVIK* (1900, Schichau).

3000 tons.

L. B. D. 361 × 39½ × 19 ft. = 110 × 12 × 5·80 m.

Guns (Obuchoff):
6E* (4·7 in. 45 cal.).
7* (47 m/m).
Torpedo tubes:
6 *above water*.

Armour (Krupp):
2″ Deck = *c* (51 m/m).

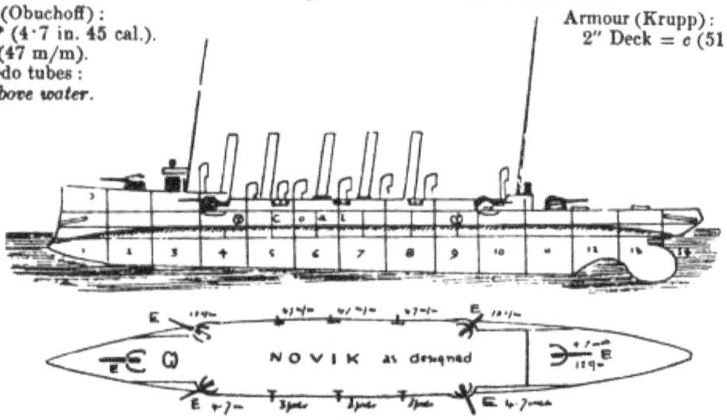

Machinery: I.H.P. 17,000 = 25 kts.
(Others of this class *projected*.)

(Баянъ) (Адмиралъ Бутаковъ)

(III.) *BAYAN* (La Seyne, 1900) & *Admiral Boutakoff* (bldg.).

7800 tons.

L. B. *D*. (*mean*): 443 × 55⅔ × *22* feet = 135 × 17 × *6·70* metres.

Guns (Obuchoff):
2C* (8 in.).
8D* (6 in. 45 cals.).
20F* (3 in.).
7* (47 m/m, 3 pdr.).
Torpedo tubes:
2 *submerged* + 3 *above water*.

Armour (Krupp):
8″–4″ Belt *aa–c*.
2″ Deck = *c* (50 m/m).
Prot. to vitals = *aaaaa*.
3″ Lower deck side *d* (80 m/m).
3″ Redoubts *d* (80 m/m).
7″ Big gun turrets *a* (170 m/m).
6¼″ C. T. *a* (160 m/m).
3″ Tube *d* (80 m/m).

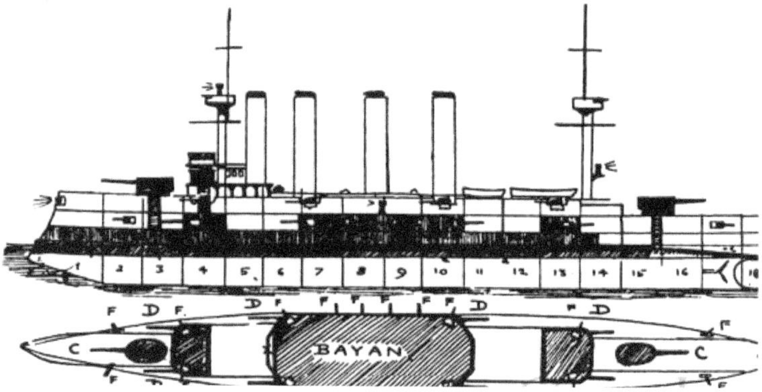

Machinery: I.H.P. 17,000 = 21 kts. (*est.*). 2 screws. Coal, *normal* 750 tons; *max.* 1100 tons. Boilers, 26 Belleville, with economisers, in 4 groups.

Note.—The contract sea speed is 21 kts. for 24 hours. The ammunition carried is as

RUSSIE.

(Аскольдъ) (Богатырь)

III. *ASKOLD* (1900, Krupp's) & *BOGATYR* (bldg., Stettin).

7500 tons.

L. B. D. (mean): (Askold) 426½ × 49 × 19¾ ft. = 130 × 15 × 6 metres.
(Bogatyr) 423 × 52¼ × 21⅛ ft. = 129 × 16 × 6·60 m.

uchoff):
6 in., 45 cal.).
3 in.).

ubes:
erged.

Armour (Krupp):
? 6" Belt *a* (152 m/m)
2" Deck = *c* (51 m/m).
Protection to vitals = *aaa*.
4" Main deck casemates *c* (100 m/m)
Hoists

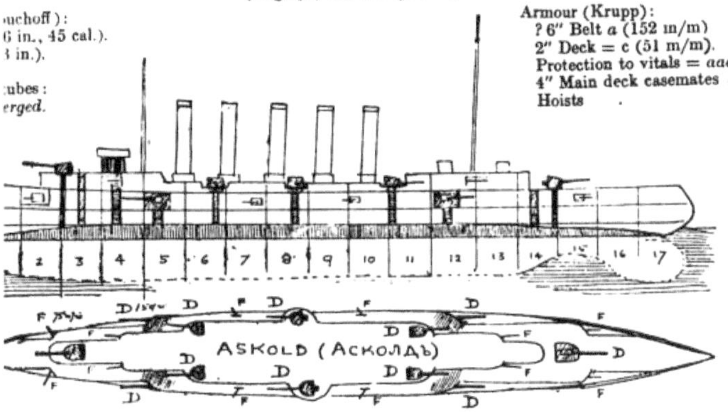

hinery: I.H.P. 19,500 = 23 kts. (*Est.*).
:—The bow will possibly be covered with thin armour. *See* Bayan.

SIAM.

(4.) MAHA CHAKRKRI (1892 Leith).

L.: 290 ft. = 88·5 m.

Guns: 4E* + 10*. Armour: 0. Sea speed: 13·5 kts.

2500 tons. Coal 280 tons. 2 Screws.

Siam also possesses (6.) MAHUT RAJAKUMAR 500 tons. 3E guns. 1 torpedo tube.

SVERIGE och NORGE.

(Edited by Herr Ingeniör C. G. Björkman.) (Drawings by Lieut. Erik Hägg.) (Photographs, unless otherwise noted, all by favour of Herr Björkman.)

(I.) SVERIGE (Suéde).

GUNS AFLOAT.

Notation.	Nominal Calibre.		—	Length in Calibres.	Initial Velocity.		Muzzle Energy.		Obus de Rupture. A.P. Shell.	
	in.	c.m.		cals.	ft. secs.	met. secs.	ft. tons.	met. tx.	lbs.	kgs.
B	10	25	Canet (M. 1894)[1] Bofors (M. 1894)	42	2362	720	17380	5386	450	204
BC	10	25	(Elswick)[2]	34	2099	640	13750	4258	450	204
C	8	21	(1898 M) Bofors	45	2297	700	11260	3494	308½	140
C	11	27	(1881) Elswick[3]	26·5	1788	545	10550	3267	476	216 a.p.
C	11	27	(1876) Finspong	20	1474	449	5970	1848	396	180 com.
D	9·4	24	(1876) Finspong	21	1355	413	4043	1252	317 273	144 124 com.
D	6	15	Bofors (1889)	34	2067	630	2963	918	100	45·4
D*	6	15	„ (1898)	45	2460	750	4228	1301	100	45·4
E	6	15	(1883) Elswick	28	1660	506	1915	592	100	45·4
E*	4·7	12	Bofors (1894)	45	2427	740	1892	586	46½	21
E	4·7	12	„ (1881)	26	1640	500	903	280	48½	22 com.
	2·2	5·7	Nordenfelt (1889)	55	2310	704	222	68·7	6	2·72
		4·7	Bofors	55	2428	740	135	41·8	3¼	1·5

[1] Oden, Thor (Canet), Niord (Bofors). [2] Svea, Göta, Thule. [3] Edda.

TORPEDO BOATS.

Second Class.

7 boats	(1896–'99)	Komet class	*circa.* 92 tons,	39 metres long,	*Max.* speed 24 kts.	2 tubes.
6 „	(1884–1894)	(1, 3, 5, 7, 9, 11),	65-86 „	39-35 „	„ 20-18 „	1-2 „
13						

Third Class.

9 boats (1882-1891) (61—77) 40-58 tons, 30 metres long, *Max.* speed 20-18 kts. 1-2 tubes.

9

2 boats Komet class and 2 third class building.

Note.—The Swedish navy also contains the gunboat Svensksund, 280 tons. All Swedish warships are built in Sweden.

(V.) BERSERK (1874) (1898) & ULF (1873) (1896) & SÖLVE (1

460 tons. Complement, 48.

L. B. D.: 130½ × 26 × 9 ft. = 39·7 × 8 × 2·7 m.

Engines: I.H.P. 155 = 6 kts. (sea speed). 2 screws.
Coal capacity (*max.*), 25 tons = kts. at full speed.
Guns: 1E* + 2*. Armour: Belt, *f.* Casemate, b (418 m.m. iron).
(Gerda, Hildur, Björn & Folke to be reconstructed as Berserk).

Saga.

(Training service).

(6.) SAGA (1878).

L.: 200 ft. = 61·1 m. 1530 tons. Complement, 208.

Guns: 1E (15 c.m.) + 6F (12 c.m.) + 1Z + 2* (38 m/m) + 1*. Speed: 11
1 screw. I.H.P., 900.

J. ERICSSON. *Photo by favour of Herr Björkman.*

(V.) JOHN ERICSSON & THORDÖN & TIRFING (1865–7) (1897–1900).

1500 tons. Complement 103.

L. B. D.: 200 × 45 × 12⅓ ft. = 60·9 × 13·8 × 3·7 m.

Guns (J. Ericsson) 2D (15 c.m.) + 2 6 prs.* (57 m.m.) + 2*. (Others) 2E* (12 c.m.) + 8* (47 m.m.).

Armour: Belt *f*. Barbette *d* (270 m.m.) Shield *f*.

Machinery: 1 Screw. I.H.P. 380 = 8 kts. (during 6 *hours* trial). Coal capacity: 150 tons.

& (V.) LOKE, to be re-constructed.

Guns at present 2C (27 c.m.)

(6.) BALDER (1870) (wood).

L.: 203 ft. = 61·8 m. 1880 tons. Speed: 12 kts. (*trial*). Complement, 218.
Guns: 1E (15 c.m.) + 8F (12 c.m.) + 1Z (65 m.m.) + 2* (38 m.m.) + 5*.
I.H.P., 1380. 1 screw.

(6.) FREJA (1885).

L.: 214 ft. = 65 m.
Guns: 4D + 8E + 2Z + 9*. Armour: 0.
Sea speed: 10 kts.
2000 tons. 1750 H.P. 1 Screw. 14 kts. *trial speed.*
Complement 284. School ship.

BJÖRN
FOLKE
GERDA
HILDUR
(1870–3).

460 tons. Complement 48. L.: 131 ft. = 40 m.
Guns: 1C + 2*. Armour: belt, *f*; Casemate, *b* (418 m.m. iron.)
Speed: 8 kts.
(To be reconstructed as *Berserk*.)

SUÈDE.

GÖTA

(IV). SVEA (1886): *GÖTA* (1891) (1900): THULE (1893).

SVEA (1886)	*GÖTA* (1891) (1900)	THULE (1893)
3100 tons.	3290 tons.	3300 tons.
Complement, 232.	Complement, 237.	Complement, 237.

. B. D.: Svea $\left\{\frac{248 \times 49 \times 17 \text{ ft.}}{75\cdot7 \times 14\cdot8 \times 5\cdot2\text{m.}}\right\}$: Göta $\left\{\frac{258 \times 48 \times 17 \text{ ft.}}{78\cdot8 \times 14\cdot6 \times 5\cdot1 \text{ m.}}\right\}$: Thule $\left\{\frac{260\frac{1}{2} \times 48 \times 16\frac{1}{2} \text{ ft.}}{79\cdot5 \times 14\cdot6 \times 5\cdot1 \text{ m.}}\right.$

Guns: Svea, 2BC + 4E* + 6 6-pdrs.* (57 m.m.) + 8*: Göta and Thule, 2BC + 4D (15 c.m., 34 cal.) - 5 6-pdrs.* + 8*. Torpedo tubes: Göta, 1 submerged: Svea, 1 submerged, 2 above water: Thule, above water.

Armour: (Creusot) deck = *d* (49 m.m.) Belt *a* - *b* (293 - 198 m.m.) Turret *a* (293 - 243 m.m.) T. *b* (268 m.m.)

Machinery: 2 horizontal compound triple expansion, surface condensation.

Boilers: 6 Admiralty in 2 groups. 2 screws. Svea reboilered 1897.

Speed ***trial:*** Svea, **natural draught**, 14·5; ***forced draught***, 15·9 kts., 3640 I.H.P. Göta, natural, ·6 kts.; ***forced***, 15·9 kts.; max. forced, 4750 H.P. = 16·3 kts. Thule, 14·5 natural; forced, 15·9; ax. forced, 5007 H.P. = 16·9 kts. Coal endurance: 4 days full speed.

Details of Göta as to be reconstructed. Thule and Svea will undergo the same reconstruction.

ıns (Bofors):
1C* (21 c/m, 45 cals.).
7D* (15 c/m, 45 cals.).
11* (57 m/m, 6 pdrs.).
2* (37 m/m).
rpedo tubes:
submerged (bow).

Armour (Creusot):
11⅝" Belt amidships *a* (293 m/m).
Protection to vitals *a*.
2" Deck = *d* (49 m/m). C. T. *b*.
Armour (Krupp):
7½" C turret, front = *aaaa* (190 m/m).
5" „ sides *b* (130 m/m).
6" D gun turrets *a* (152 m/m).
Hoists to C *a*, to D gun *e* (100 m/m).

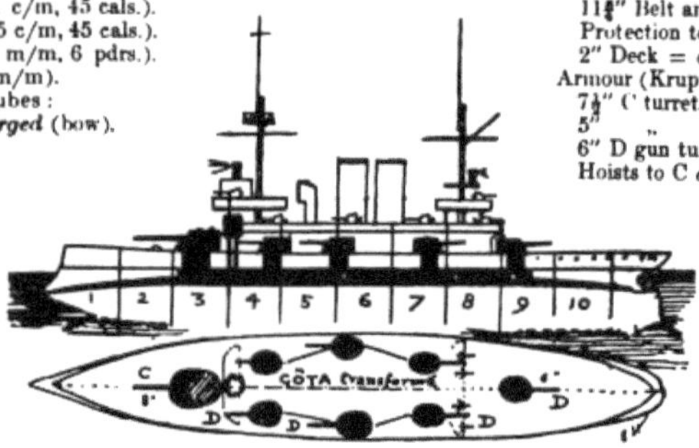

SVEA. *Photo by favour of Herr Björkma*

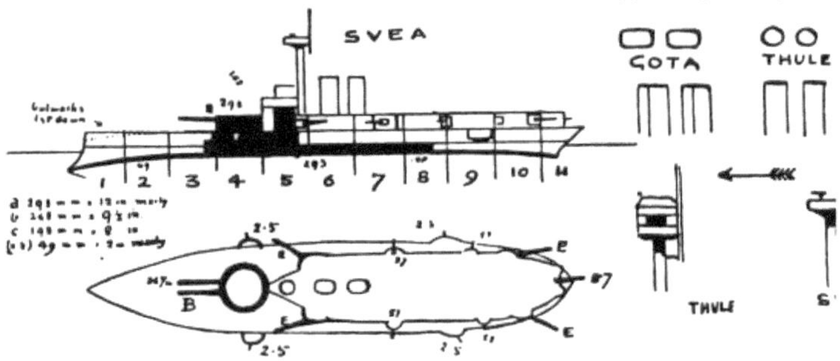

(T.) ÖRNEN (1897): JAKOB BAGGE & CLAS HORN (— 1898)

L. B. D.: 223 × 27½ × 10 ft. = 68 × 8·2 × 3·1 m.

& (T.) PSILANDER & CLAS UGGLA (— 1899).

L. B. D.: 233 × 27⅛ × 9 ft. = 71 × 8·3 × 2·7 m.

700 tons. Complement 99. Guns (Bofors): 2E* + 4* (57 m/m). Torpedo tubes: 1 *submerged* (bow). Armour: ¾" Deck (19 m/m). 2" C.T. (50 m/m). Gun shields *f*. Machinery: I.H.P. 4000 = 20 kts. (*forced d., maximum*); ***natural draught*** = 17·5 knots. 2 screws. Boilers: 2 Yarrow (*except* Örnen which has 2 Admiralty).

JAKOB BAGGE. *Photo by favour of Herr Björkman.*

(5.) ROTA, SKÄGGALD, SKULD, VERDANDE (1877–80).

500 - 540 tons. Complement, 86.

L.: 171 ft. = 52 m.

Guns: 1C (27 c.m.) + 1E + 2*. Armour: 0. Speed: 13 kts.

(5.) EDDA (1885).

L.: 181 ft. = 56 m. 640 Tons. Complement, 86.

Guns: 1C (27 c.m.) + 1D + 4*. Armour: 0. Speed: 13·6 kts.

(5.) URD & SKAGUL (1877) (1896–8).

L.: 171 ft. = 52 m. 500 – 540 tons.

Guns: 1D + 1E + 2 6 pdrs.* (57 m.m.) + 2*. Armour: C.T. 51 – 45 m.m.; Gunshields (Bofors) *f*. Speed, 13 kts. Complement, 86.

URD & SKAGUL.

EDDA

(IV.) *DRISTIGHETEN* (1900).

3500 tons. Complement 250. L. B. D.: 285 × 48½ × 16 ft. = 86·9 × 14·8 × 4·9 metres.

Guns (Bofors):
  2C* (21 c/m, 45 cals.).
  6D* (15 c/m, 45 cals.).
  10* (57 m/m, 6 pdr.).
  2* (37 m/m).
Torpedo tubes:
  2 *submerged* (Elswick).

Armour (St. Chamond):
  8" Belt amidships *a* (200 m/m)
  Prot. to vitals *a*.
  2" Deck = *e* (48 m/m).
  8" Turrets *a* (200 m/m).
  3½" Battery *d* (91 m/m).
  3½" Hoists *d*.
  8" C. T. *a*.
  Turret supports *a*.

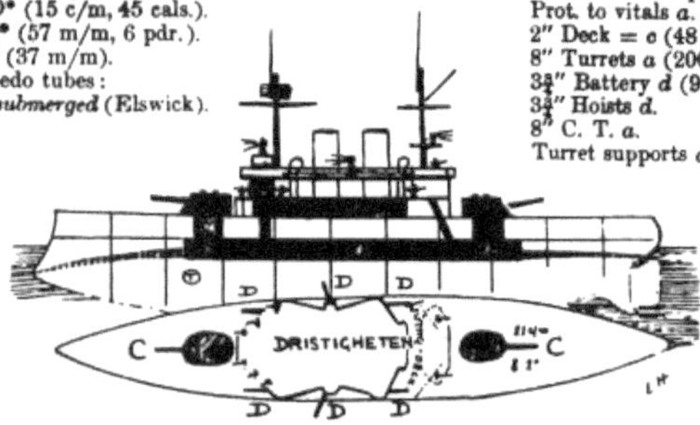

DRISTIGHETEN as designed.

Machinery: 2 sets triple expansion. I.H.P. 5000 = 16·5 kts. *natural draught* (*est.*).
2 screws. Boilers: 8 Yarrow. Coal: *normal* 250 tons. *Maximum* 400 tons = 6 days full speed.

(IV.) *ÄRA & BRODERSKAP & HELSA* (building).

3650 tons. Complement 250. L. B. D.: 287 × 49¼ × 16 ft. = 87·5 × 15 × 4·9 m.

Guns (Bofors):
  2C* (21 c/m, 45 cals.).
  6D* (15 c/m, 45 cals.).
  10* (57 m/m).
  2* (37 m/m).
Torpedo tubes:
  2 *submerged* (Elswick).

Armour (Krupp):
  7" Belt amidships *a* (175 m/m).
  2" Deck, on slopes = *e* (48 m/m).
  Protection to vitals = *aaa*.
  8" fronts, C turrets = *aaaa* (200 m/m).
  5" sides, C gun turrets *b* (130 m/m).
  D gun turrets, fronts = *aaa*, sides *e*.
  8" Supports C turrets *a*.
  4" D gun hoists, etc. *e*.

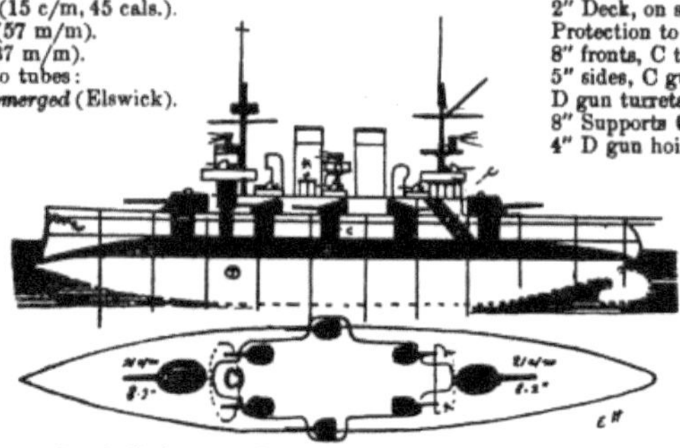

A. B & C. as designed

Machinery, Speed, Coal, etc. = Dristigheten.

(IV.) ODEN (1897): THOR (1898) & NIORD (1899).

3400 tons. Complement 210.¶
L. B. D.: 279 × 40 × 17 feet = 84·8 × 15 × 5·2 metres.

Oden—Guns:
2B (25 c/m, 42 cals.).
4E* (12 c/m, 45 cals.).
10* (57 m/m).
4* (37 m/m).
Torpedo tubes:
1 *submerged* (bow).

Armour (Harvey nickel):
10″ Belt (amidships) *aa* (250 m/m).
Protection to vitals *aa*.
2″ Deck = *d* (48 m/m).
10″ Turrets (fronts) = *aaaaa*.
10″ Supports and hoists *aa* (250 m/m).
3¾″ Battery *d* (91 m/m).

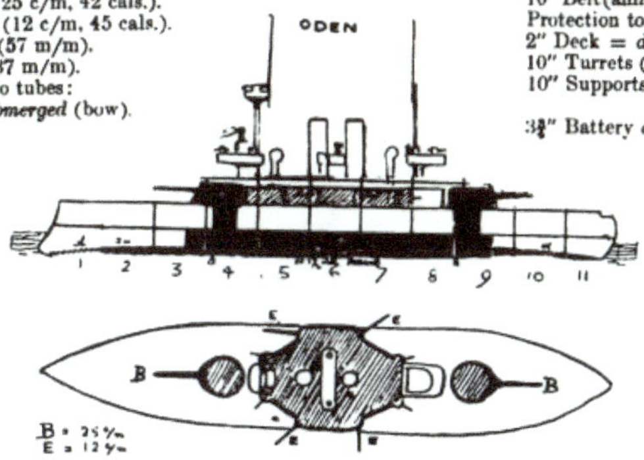

ODEN. *Photo by Kadett J. Krook.*

Thor & Niord—Guns:
2B (25 c/m, 45 cals.).
6E* (12 c/m, 45 cals.).
10* (57 m/m, 6 pdrs.).
4*.
Torpedo tubes:
1 *submerged* (bow).

Armour (Harvey nickel):
10″ Belt *aa* (250 m/m).
Prot. to vitals *aa*.
2″ Deck = *d* (48 m/m).
8″ Turrets (fronts) = *aaaa* (200 m/m).
8″ Supports and hoists *aa* (200 m/m).
3¾″ Battery *d* (91 m/m).

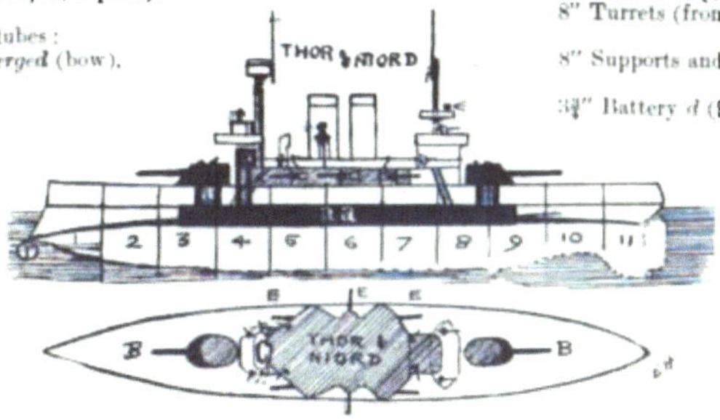

(The 25 c/m guns of *Oden* and *Thor* are Canet, those of *Niord* are Bofors.)

Machinery: 2 screws. Boilers, 6 Admiralty.

Trials—Oden *natural* I.H.P. 3700 = 15 kts.: 6 *hours forced* 5350 = 16·82 kts.
Thor „ = 15·5 „: 6 „ „ = 16·7 „
Niord „ = 15·5 „: 6 „ „ = 16·5 „

THOR. (NIORD *identical.*) *Photo by favour of Herr Björkman.*

SVERIGE OCH NORGE.

(1. SVERIGE).

(5.) BLENDA (1875) & DISA (1877).

L.: 167 ft. = 51 m.

Guns: Blenda: 1D (27 c.m.) + 1E + 2*. / Disa: 1D (15 c.m.) + 1E + 2*(57m/m) + 2*. Armour: 0. Sea speed: 11 kts.

590 tons.

(2. NORGE).

NORVÈGE.

GUNS.

Elswick guns, see list for Japon. Krupp guns, see list for Oesterreich-Ungarn.

TORPEDO BOATS.

SECOND CLASS.

2 boats (1894), 43 tons, 112 ft. long, *max. speed* 20·5 kts., 2 tubes.
3 ,, (1899), 65 ,, ,, ,, 19 ,,

THIRD CLASS.

8 boats (1887–82). 40 tons, 100 feet long, *max. speed* 18 - 20 kts., 2 tubes.
1 Thornycroft (1875), 16 ,, 60 ,, ,, ,, ,, 18 ,, 2 ,,
3 boats (1878), 16 ,, 50 ,, ,, ,, ,, 12–9 ,, spar torpedo.

12

4 second class boats building.
2 third ,, ,, ,,

Norwegian Navy includes:—NORDSTJERNEN, ELLIDA, NORNEN, all *circa* 1000 tons, and several small gunboats, all of no fighting value.

(T.) VALKYRIEN (Elbing, 1896).

380 tons.

190 × 24 × 13 ft. = 58 × 7·4 × 4 m.

Guns: 2F* (Krupp) + 2*. 2 torpedo tubes (1 of them in bow).

Machinery: I.H.P. 3300 = 23 kts. (short *trial*).

NORVÈGE.

(V.) THOR (1872) & SKORPIONEN (1866) & MJÖLNER (1868) & THRUDVANG (1869) (1898).

L : 205 ft. = 62·5 m.

Guns: 2E* (4·7 in. Elswick) + 2F* (65 m/m) + 2*. Armour: *b* - *e*
Speed: 8 kts. (*trial*)

2000—1500 tons.

THRUDVANG. ÆGIR class.

(T.) VIKING (1891).

L.: 197 ft. = 60 m.

Guns: 2D + 4F* (65 m/m) + 4* (37 m/m) + 2*.
Armour: deck = *f*. Sea speed: 14 kts.

1110 tons. Complement 130. 3 torpedo tubes (1 submerged).

I.H.P. 2000 = 15 kts. *trial*. Coal: 140 tons.

(5.) AEGIR (1892), TYR (1888), GOR (1885), BRAGA (1878), NOR (1878), VIDAR (1882), ULLÆR (1876), VALE (1874)

L.: 108—90 ft. = 33—27·5 m.

Guns: 1C+3*. Armour: 0. Speed: *circa* 10–8 kts.

250 - 200 tons. C guns = 26 c/m Krupp, or 27 c/m Armstrong.

(*One building.*)

(IV.) HARALD HAARFAGRE & TORDENSKJOLD (1897, Elswick).

3400 tons. Complement 220.

L. B. D.: 279 × 48½ × 19 feet = 85 × 14·7 × 5·8 metres.

Guns (Elswick):
2C* (21 c/m).
6E* (12 c/m, 45 cals.).
6F* (7·5 c/m).
6* (Bofors, 37 m/m).
Torpedo tubes:
2 *submerged*.

Armour (Harvey):
7" Belt *b* (177 m/m) (*d* at ends)
2" Deck = *d* (51 m/m).
Protection to vitals = *aa*.
8" Fronts C turrets
= *aaa* (200 m/m).
5" Bases of turrets *d* (125 m/m).
C. T. *a*. Hoists *e*.

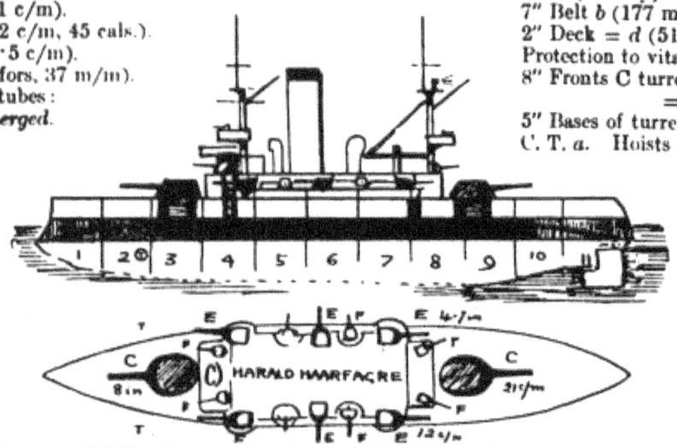

Machinery: I.H.P. 3700 = 17·2 kts. (*trial* 1897). Sea speed: *circa* 14 kts. 2 screws.
Coal: 200 tons, *normal*; *max.* 500 tons.

TORDENSKJOLD. *Both exactly alike.* *Photo, Elswick.*

(IV.) *NORGE* & *EIDSVOLD* (1900, Elswick).

3800 tons. Complement 250.

L. B. *D.* (*mean*): 290 × 50 × 16½ feet = 88·40 × 15·24 × 5 metres.

Guns (Elswick):
2C* (21 c/m).
6D* (15 c/m).
8F* (76 m/m).
6* (47 m/m).
Torpedo tubes:
2 *submerged*.

Armour (Krupp):
7" Belt *a* (177 m/m).
2" Deck = *c* (100 m/m).
Protection to vitals = *aaa*.
6" Turrets (inclined) = *aaaa*.
6" Bases *a* (200 m/m).
5" Casemates (to 4 of the D guns)
b (125 m/m).

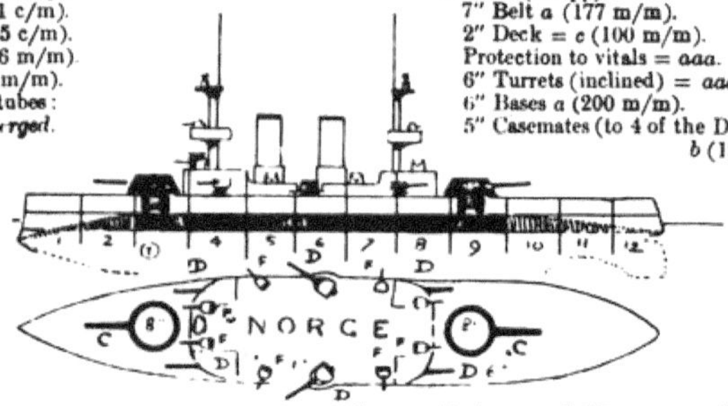

Machinery: I.H.P. 4830 = 17 kts. Boilers: 8 Yarrow. Coal: *normal*, 250 tons; *maximum*, 400 tons.
Note.—The forward and after D guns are in casemates.

NORVÈGE.

(T.) FRITHJOF (1896).

L.: 216·5 ft. = 66 m

Guns: 2E* (4·7 in.) + 4F* + 4*. Armour: 0. Sea speed: *circa* 14 kts.

1370 tons. 3 torpedo tubes. I.H.P. 3000. 2 screws.

SLEIPNER (1877).

L.: 175 ft. = 53·5 m.

Guns: C (26 c.m. Krupp) + 1E (15 c.m. old). Armour: 0. Sea speed: 8 kts.

580 tons. Complement 87. 2 screws.

TURQUIE.

ZIZIEH (1864) (1894) & MAHMUDIEH (1864) (1894) & MANIEH (1864) (1897) & ORKANIEH (1865) (1898).

L.: 292 ft. = 89 m. 6400 tons. 3735 I H.P. Complement, 600.

C + 8D (6 in.) + 6E*. Armour: Belt *b* - *e*; battery *d*. Sea speed: 8 kts.

The Turkish navy possesses 2 destroyers, 270 tons, speed ?. 2 1st class torpedo boats, *circa* 120 tons. 11 2nd class, *circa* 85 tons. 11 3rd class, *circa* 40 tons. Speeds of all, ? ?. There are also some ironclads in addition to those enumerated here of no fighting value, and a number of miscellaneous craft.

Guns: The C guns of Azizieh are short Krupp, 24 c.m.; all other guns D and E are very old pieces (10–8 in.) 25–19 c.m. The E* are Krupp 10 c.m. The reconstructed (1900–01) ships Vickers guns, for *data* see *England*.

(III.) ASSAR-I-TEWFIK (1868, La Seyne)

5600 tons.

(Converting into an "armoured cruiser" at Stettin.)

L. B. D.: 276 × 52 × 25 feet = 84 × × m.

Guns (Vickers):
2B (9·2 in.).
6D* (6 in., 45 cals.).
10F* (3 in.).
12* (57 m/m)

Armour (Krupp + iron):
+ 8" Belt.
+ 6" Battery.
Barbettes.

Machinery: Being re-engined, re-boilered, etc. Niclausse boilers.

(III.) MESSUDIYEH (1874, Blackwall) (1900, Genoa).

L. B. D.: 331 × 59 × ? feet = 101 × × ? m.

Guns (Vickers):
2B (9·2 in.).
12D* (6 in., 45 cals.).
16F* (3 in.).
10* (57 m/m).

Armour (Krupp + iron):
+ 12" Belt.
Prot. to vitals.
+ 8" Battery.
Barbettes.

Machinery: Being re-engined, etc. Niclausse boilers.

PELENK-I-DERIA & NAMET

(T.) PELENK-I-DERIA & NAMET (Gaarden, 1890).

900 tons.

L. B. *D.* (*mean*): 230 × 31 × 16½ feet = 70 × 9·45 × 5 metres.

Guns: 2E* (10 c/m, 4 inch Krupp) + 6* (57 m/m). **Torpedo tubes:** 3 (1 of them in b

I.H.P. 4500 = 19 kts. Boilers, locomotive

(IV.) HAMIDIEH (*ex* Nusretyeh) (1885).

L.: 292 ft. = 89 m.

Guns: 10C + 2D. **Armour:** *d.* **Sea speed:** 6 kts.

6700 tons. I.H.P 4000.

GUNS IN THE U. S. FLEET.

Nominal Calibre.		—	Length in Calibres.	Muzzle Velocity.		Muzzle Energy. *circa.*		Projectile.	
in.	c.m.			ft. secs.	met. secs.	ft. tons.	met. tnx. *circa*	lbs.	kgs.
13	33	Indiana, etc.	35	2100	640	33627	10415	1100	499
12	30·5	Iowa, Texas, Monterey, etc.	35	2100	640	25985	8048	850	385·5
12	30·5	model '99	40	2800	854	46246	14322	850	385·5
10	25	Monterey only	35	2100	640	15285	4734	500	226·7
10	25	...	35	2060	628	14710	4556	500	226·7
10	25	...	30	2000	609	13864	4294	500	226·7
10	25	model '99	40	2800	854	27200	8425	500	226·7
8	20·3	...	40	2150	655	8000	2478	250	113·3
8	20·3	...	35	2080	634	7500	2286	250	113·3
8	20·3	...	30	2000	609	6932	2147	250	113·3
8	20·3	...	45	2800	854	13600	4212	250	113·3
		model '99	50	2900	884	5840	1809	100	45·3
...	...	(Elswick)	50	2600	792	4840	1499	100	45·3
6	15	...	40	2150	655	3200	991	100	45·3
6	15	...	35	2080	634	2990	925	100	45·3
6	15	...	30	2000	609	2773	858	100	45·3
4·7	12	(Elswick)	50	2600	792	2150	666	45	20·4
5	12·5	model '99	50	2900	884	3503	1085	60	27·2
5	12·5	...	40	2000	609	1834	568	50	22·6
5	12·5	...	30	2000	609	1660	514	60	27·2
4	10	model '99	50	3000	915	2000	619	32	14·5
4	10	...	40	2000	609	915	283	33	15
3	7·5	model '99	50	3000	915	874	270	14	6·3

Note.—Figures for the '99 models are not service velocities.

MERCHANT AUXILIARY CRUISERS.

· of New York and Paris). 3 fun. 3 masts. Sea speed: 19 kts. 20000 I.H.P. 2 (St. Louis
d). 2 fun. 2 masts. Sea speed: 19 kts. 20000 I.H.P. 535 ft. long. And 19 others, speeds
. Also 8 for the Pacific, average speed 14 kts.

or first four 8D + 6*, for the rest 6E + 4* or else 8F + 6*.

. S. navy also contains gunboats Annapolis, Newport, Princeton and Vicksburg, for police duties,
farietta, etc., but 3 masts and swan bow.

DESTROYERS.

(*Two funnels.*) (*All* carry 2 tubes.)

1 Farragut (1898), 273 tons. I.H.P. 5600 = 31·76 kts. Guns: 6* (57 m/m).
1 Goldsborough (1899), 247 tons. I.H.P. 5600 = 30 kts. Guns: 4* (57 m/m, 6 pdr.).
1 Stringham (1899), 340 tons. I.H.P. 7200 = 30 kts. Guns: 7* (57 m/m).
1 Bailey (1899), 235 tons. I.H.P. 5600 = 30 kts. Guns: 4* (57 m/m).

(*Four funnels.*)

2 Lawrence (1899) & Macdonough (1900), 400 tons. I.H.P. 8400 = 30 kts. Guns: 2F* + 5* (57 m/m).
+ *9* Barry & Bainbridge & Chauncey & Dale & Decatur & Paul Jones & Perry & Preble & Stewart (1900), 420 tons. I.H.P. 8000 = 29 kts. Guns: 2F* + 5* (57 m/m).
+ *2* Hopkins & Hull (1900), 408 tons. I.H.P. 7200 = 29 kts. Guns: = *Barry*.
+ *3* Truxton & Whipple & Worden (1900), 433 tons. I.H.P. 8300 = 30 kts. Guns: = *Barry*.

TORPEDO BOATS.

1st Class.

3 Porter class	(1896),	185 tons,	170 ft. long,	*Trial* speed	28-26 kts.	3 tubes.
3 Winslow class	(1897),	142 „	160 „	„	24 „	3 „
4 Dahlgren „	(1898),	130 „	147 „	„	30 „	3 „
1 Ericsson	(1894),	120 „	150 „	„	23 „	3 „
1 Cushing	(1890),	105 „	139 „	„	22 „	3 „
+ *7*	(1900),	165 „	175 „	„	*est.* 26 „	3 „
+ *3*	(1900),	167 „	157 „	„	*est.* 28 „	3 „
+ *2*	(1900),	174 „	174 „	„	*est.* 26 „	3 „
12 + *12*						

3rd Class.

2	(1897),	46 tons,	98 ft. long,	*Trial* speed	20 kts.	2 tubes.
1 Stiletto	(1886),	31 „	89 „	„	18 „	2 „
3						

Submarine.

					above.	below.	
1 Plunger	(1897),	168 tons,	80 ft. long,	*Trial* speed	13	8	2 tubes.
1 Holland	(1897),	138 „	75 „	„	13	16	2 „

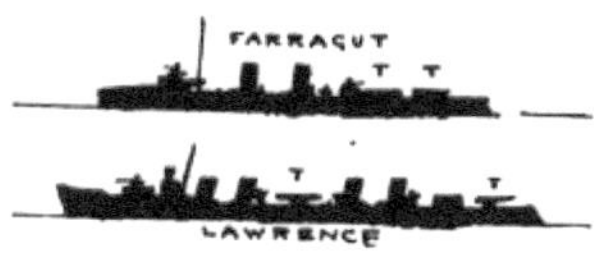

(X.) KATAHDIN (1893).

L.: 251 ft. = 76·5 m.

2183 tons. Complement, 100. D. (*moyen*) 15 ft. = 4·5 m.

Guns: 4*. Sea speed: *circa* 12 kts.

Armour: All upper deck and sides (sloping) = *aa* (or *a*). C. T. *aaa* (18" Harvey). Base of funnel *c*. Glacis = *a*. **Machinery**: 2 sets horizontal triple expansion engines. 2 2-ended Scotch boilers. 1 1-ended. 15 furnaces. I.H.P. 4800 = (*trial 4 hours*) 16 kts. Coal, 175 tons., *max*. 192 tons. 2 screws.

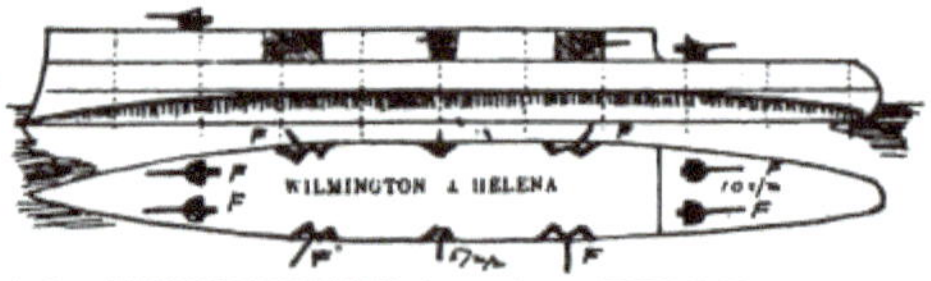

(4.) WILMINGTON (1895) & HELENA (1896).

L.: 250 ft. = 76 m.

1392 tons. Complement, 175. D.: 10 ft. = 3 m.

Guns: 8F* (4 in.) + 6*. **Armour**: *prot.* Sea speed: *circa* 12 kts.

Armour: Deck = *d*. Sponsons for F guns *e*, for 2 6-pdrs. *, amidships *f*.

Machinery: 2 screws. 2 rudders. I.H.P., Wilmington, 1894 = 15 kts. (*4 hours trial*); Helena, 1998 = 15·7 kts. (4 *hours trial*) ship light at the time. Coal, 100 tons, *max*. 275 tons.

(V.) *ARKANSAS* & *CONNECTICUT* & *FLORIDA* & *WYOMING* (building).

2755 tons. Complement 130.

L B. D.: 228 × 50 × 12½ feet = 69 6 × 15·23 × 3·8 m.

Guns: 2A (12 in. mod. '99) + 4E* (4 in.) + 3* (57) × 4* (37). Torpedo tubes: **2** *above water*.

Armour (Krupp) 11"-5" Belt *aaa–b* (279–127 m/m). Deck = *d* (37 m/m). Prot. to vitals = *aaaaa*. Turret and base *aaa* (279 m/m). C.T. *aa* (200 m/m).

Machinery: I.H.P. *natural* 2400 = 12 kts. (*est.*). 2 Screws. Boilers: *Arkansas*, 4 Thornycroft; *Connecticut*, 4 Niclausse; *Florida*, 4 Normand; Wyoming, 4 Babcock and Wilcox. Coal: 400 tons.

Note.—It is proposed to mount a 16 inch (40·5 c/m) Gatham gun, firing a shell with a charge of 500 lbs. (226·7 kgs.) of high explosive in some of these monitors.

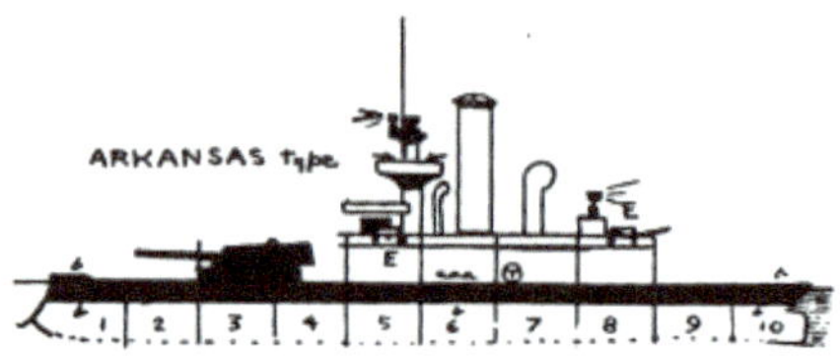

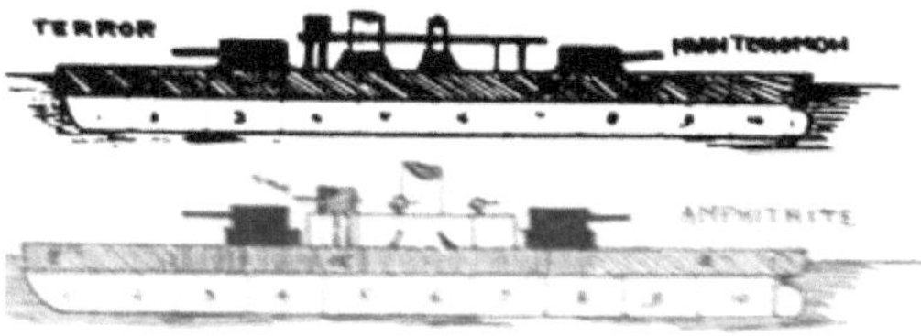

MIANTONOMOH : TERROR : AMPHITRITE (—1883).

[illegible] tons. Complement, 160 (Terror, 190).

L. B. D.: 250 × 56 × 16 ft. = 79 × 17 × 4·9 m.

[illegible] (35 cal. in Terror, 30 cal. in others): 4 6-pdrs.*: Amphitrite [illegible] Q.F.) + 4 6-pdrs.* (57 m.m.) Torpedo tubes, nil.

[illegible] (Mian. and Terror) belt e. B gun turrets h. Flat deck 2" steel

[illegible] Belt d. Barbettes h. Hoods d. [illegible] steel deck. C. T. d.

[illegible] I.H.P. [illegible] = 12 kts. (trial). All 2 screws. Coal capacity, circa [illegible] Sea speed, circa 8 kts.

Photo. W. Rau. AMPHITRITE. Photo. W. Rau.

(V.) MONTEREY (1891).

4138 tons. Complement 203.

L. B. D.: 256 × 59 × 16 ft. = 78 × 18 × 4·9 m.

Guns: 2A (12,in.) + 2B (10 in. 35 cals.) + 6 6 prs.* + 4*. Torpedo tubes, 0.

Armour: Complete belt, *a - c*. *Prot.* to vitals = *a*. Fore barbette, *a*; after barbette, *b*. Barbette shields, *c*. C.T., *b*.

Machinery: I.H.P. 5070 = 14·2 kts. (trial). Sea speed, 9 kts. Coal: 200 tons = *circa* 1000 miles. 2 Screws.

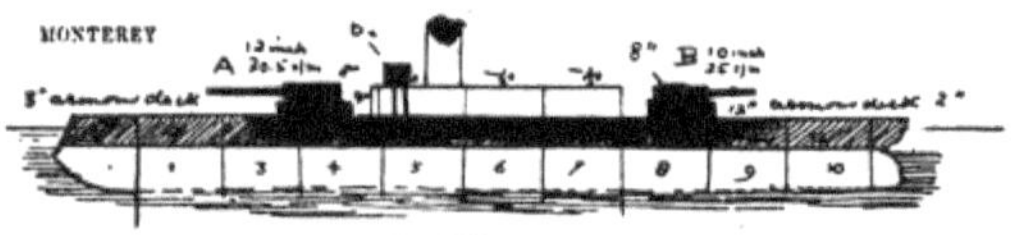

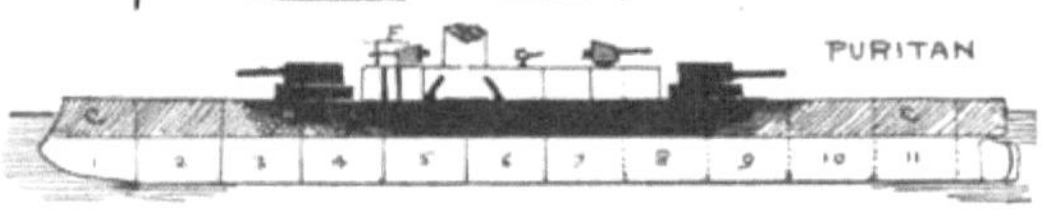

(V.) PURITAN (1883).

6060 tons. Complement 230.

L. B. D.: 289 × 60 × 20 ft. = 88 × 18 × 6 m.

Guns: 4A (12 in.) + 4D* (6 in Q.F.) + 4 3 prs.* (37 m.m.) + 8*. Torpedo tubes, nil.

Armour: Belt, *a*. On A guns, *a*. Hoods to guns, *c*.

Machinery: I.H.P. = 3700 = 12·4 kts. (trial). Coal: 580 tons = 2000 kts. at 10 kts. 2 Screws.

Photo, W. Rau.

U. S. A.

(V.) MONADNOCK (1883).

L.: 259 ft = 79 m.

Guns: 4B + 2E* + 4*. Armour: *b* (guns) - *d*. Speed: *circa* 12 kts.

Details = Amphitrite, p. 247.

(T.) VESUVIUS (1888).

930 tons. Complement, 70.

L. B. D.: 252½ × 26½ × 13 ft. = 77 × 8 × 4 m.

Guns: 3 - 15 in. (38 c.m.) Dynamite guns (velocity *mean* 185 ft. secs. = 56·3 metre secs.), shell 600 lbs. (272 kilogrammes) of explosive. Range 2000 - 3000 yards.

Machinery: I.H. P. (*forced draught*) = 3794 = 21 kts. (*short trial, ship very light*). 2 screws. Coal, 150 tons.

(6.) BUFFALO. (*Ex* Nitheroy. *Ex* El Cid.)

7080 tons.

Guns: 1 - 15 in. (38 c.m.) Dynamite gun + 1E* (5 in.) + 2F* (4 in.) + 8* (6 pdr. 57 m.m.) + 12*. 4 torpedo tubes + 1 Halpine "dirigible torpedo" appliance. Armour: 0.

Machinery: I.H.P. 3500 = 18 kts. 1 screw.

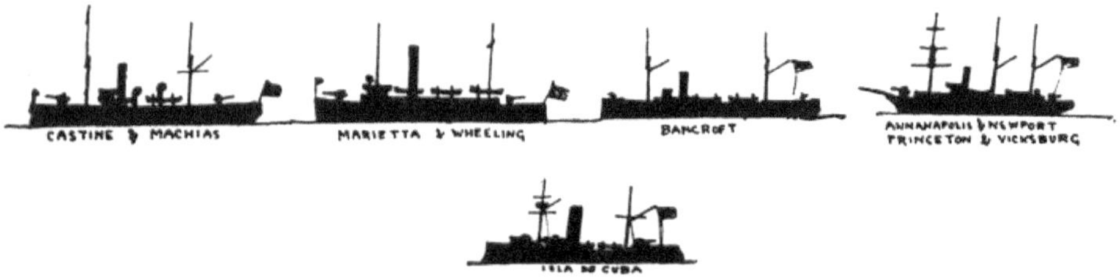

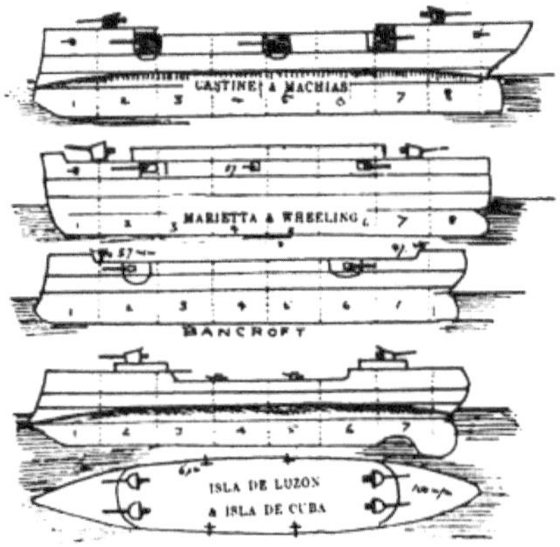

(6.) CASTINE (1891) & MACHIAS (1892)

L.: 204 ft. = 62 m. D.: 15 ft. = 4·5 m

Guns: 8F* + 6*. Armour: 0. Sea speed: 13 kts.

1200 tons. Complement 150.

Coal: 125 tons: *max.* 250 tons I.H.P. 2199 = 16 kts. (*trial*).
(Lengthened after completion, to correct instability).

&

MARIETTA & WHEELING, (—— 1897).

Guns: 6F* (4 in.) + 4* (6 pdr.) + 2* + 1 howitzer (3 in.).
Speed 12 kts. Other details (nearly) = Castine.

(5.) ISLA DE LUZON & ISLA DE CUBA (*ex* Spa

1040 tons. Complement 164.

L.: 185 × 30 × 15 ft. = 56·5 × 9 × 4·5 m.

Guns: 4E* (4 in.) + 4*. Armour: deck *ε*. Sea speed: *circa* 12 kt
tubes. Coal: 100 tons.

(6.) ANNANAPOLIS & NEWPORT & PRINCET
& VICKSBURG (1896–97).

1000 tons. All details = Marietta.

(6.) BANCROFT (1892).

840 tons.

Guns: 4F* (4 in.) + 2* (6 pdr.) + 2* (3 pdr.) + 2*. 2 torpedo t
1218 = 14·3 kts. (*forced, trial*).

U. S. A.

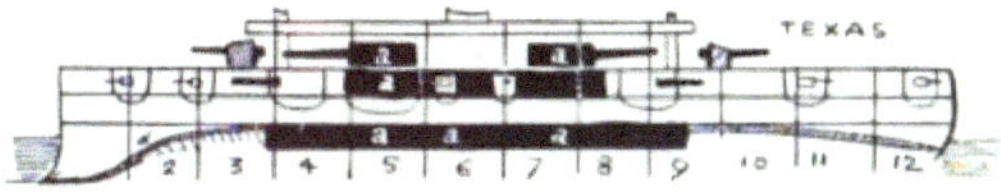

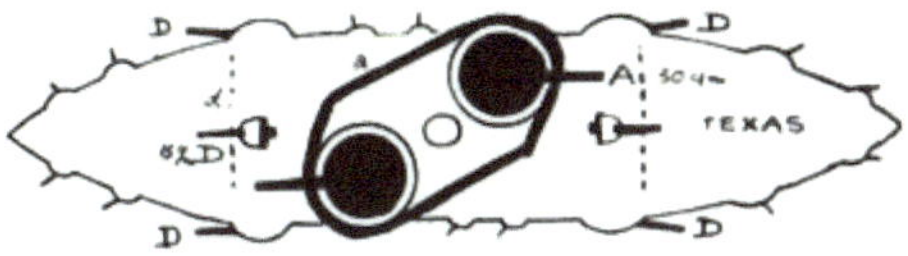

(III.) TEXAS (1892).

6300 tons. Complement, 380.

L. B. D.: 301 × 64 × 24 ft. = 92 × 19·5 × 7·3 m.

Guns: 2A (12 in.) + 6D (6 in.) + 12 6 pdrs.* + 12*. 6 above water torpedo tubes.

Armour: Deck = *d*. Belt, turrets and redoubt *a*. C. T. *a*. Bulkheads *d*. Hoists *d*. Prot. to vitals, belt + coal = *aa*.

Machinery: I.H.P. forced = 8000 = 17 kts. (trial). Speed at Santiago, *circa* 13 kts. Coal capacity, 500, *max.* 950 tons = 6000 miles at 10 kts.

Photo. W. Rau, Philadelphia.

(6.) BENNINGTON (1890)
CONCORD (1890)
YORKTOWN (1888)

L.: 230 ft. = 70 m. D.: 16½ ft. = 5 m.

Guns: 6D + 4*. 1700 tons. Complement, 150. Coal, 400 tons. Sea speed: 14 kts.

(6.) DON JUAN D'AUSTRIA.

1152 tons. Complement, 150.

L.: 210 ft. = 64 m.

Guns: 4E* (5 in.) + 4*. 1 screw.

I.H.P. 1500 = 13 kts.

(6.) PETREL (1888).

L.: 177 ft = 54 m.

Guns: 4D + 3*. Sea speed: *circa* 12 kts. 800 tons. Complement, 100. I.H.P. 1513.

(6.) DOLPHIN (1884).

L.: 240 ft. = 73 m.

Guns: 2F* + 2*. **Armour**: 0. Sea speed: 14·5 kts. 1485 tons. Complement, 115. I.H.P. *circa* 2250.

(I.) ALABAMA & ILLINOIS & WISCONSIN (— 1898).

11,525 tons. Complement 490.

L. B. *D.* (*mean*): 368 + 72 × 23½ feet = 112 × 22 × *7·16* m.

Guns:
- 4A (13 in. 35 cals.).
- 14D* (6 in. 40 cals.).
- 17* (57 m/m. 6 pdrs.).
- 6* (37 m/m, 1 pdr.).

Torpedo tubes:
- 4 *above water*, behind *c* armour.

Armour (Harvey nickel):
- 16½" Belt (amidships) *aaaa*.
- 4" Deck (slopes) = *a* (100 m/m).
- Protection to vitals = *aaaaa*.
- 4" Belt at bow *d* (100 m/m).
- 5½" Lower deck *c* (129 m/m).
- 5½" Battery (129 m/m).
- 6" Casemates *b* (152 m/m).
- 16½" Turrets (412 m/m).
- Turret bases *c* and *aaa*.
- 12" Bulkheads *aaa* (300 m/m).
- 10" C. T. *aa*.

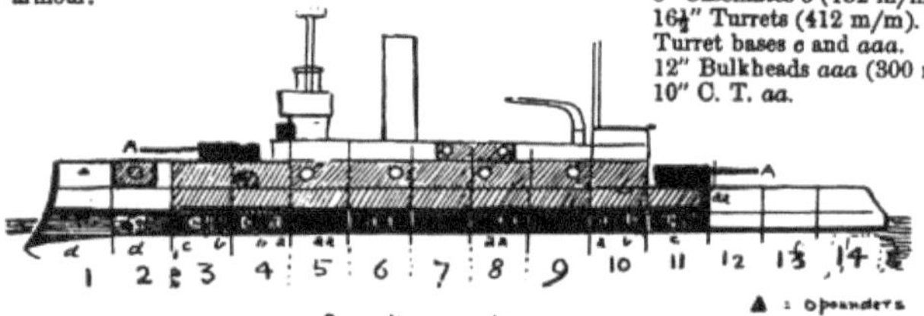

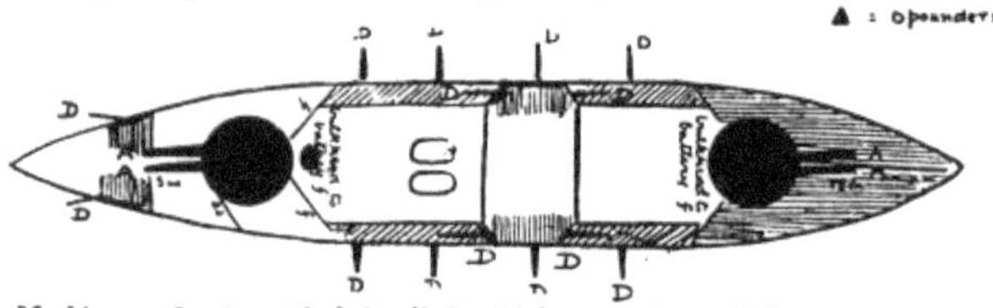

Machinery: 2 sets vertical 3-cylinder triple expansion. Boilers: 8, 1-ended. 2 screws. Nominal I.H.P. 10,000 = 16 kts. (*forced d.*). Coal: *normal* 800 tons: *max.* 1200 tons.

Trials.

(1900) *Alabama* ? *hours*. ? I.H.P. = 16·23 kts.
Illinois
Wisconsin

Note.—The total weight of armour is 2800 tons. There are 1½" screens (33 m/m) in the battery between the D* guns. There is a cofferdam behind the water-line belt.

ALABAMA, ILLINOIS & WISCONSIN.

(II.) IOWA (1896).

11,296 tons. Complement, 486.

L. B. D.: 360 × 72 × 28 ft. = 109·7 × 30 × 8·53 m.

Guns: 4A (12 in.) + 8C (8 in.) + 6E* (4 in.) + 20 6 pdr.* + 6*. 6 above water torpedo tubes.

Armour (Harvey): Deck = *c*. Belt (7½ ft. (2·3 m.) deep), *aa*. Bulkheads, *aa*. Lower deck side, *d*. Turrets (A guns), *aa* (350 m.m.) Barbettes (to C guns), *b* (200 m.m.). Hoods to C guns, *c* (140 m.m.) C.T., *a*. Hoists, *e*. Shields to E guns = *d*.

Machinery: I.H.P. *natural* 10,000, *forced* 11,000 = 16·5 kts. (trial). Coal capacity *normal*, 625, *max*. 1780 = 6000 miles at 10 kts.

Photo, W. Rau.

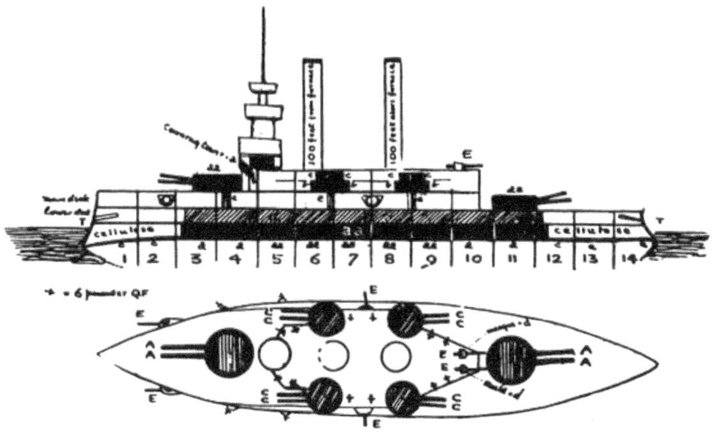

(II.) INDIANA & MASSACHUSETTS : OREGON (—1893).

10400 tons. Complement, 470.

L. : 348 ft. = 106 m.

Guns : 4A + 8C + 4D + 20 6 pdrs.* **Sea speed** : Oregon, 15 kts. ; Mass. & Indiana, 14 kts. (?).

Armour (Harvey) : Deck = *c*. Belt (200 × 7½ ft. = 61 × 2·25 m.) *aaa* (18" Harvey). Bulkheads and barbettes *aaa*. A gun shields *b*. Lower deck and bulkheads above belt *d*. C gun barbettes *a*. Gun hoods *e*. Hoists *e*. D gun sponsons *e*. C. T. *a*. 7 torpedo tubes (several of these removed recently.)

D. : Tirant d'eau, 28 ft. = 8·5 m.

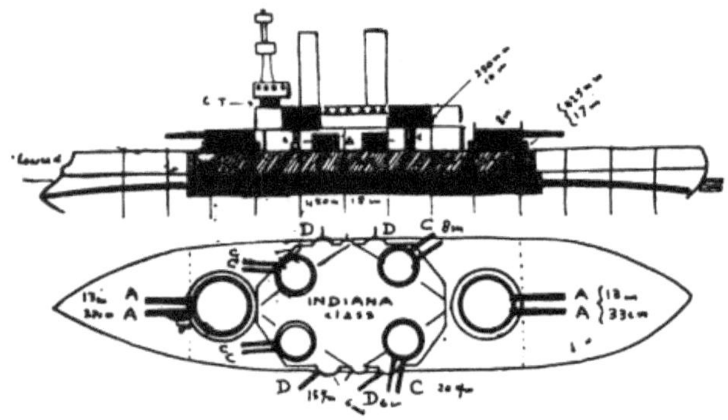

Machinery : 4 boilers, horizontal return fire, tubular. Also 2 auxiliary boilers. I.H.P. 8000 forced ; Indiana, 11,800 = 15·61 (*4 hours trial*) : Massachusetts (*4 hours*) 16·15 kts. ; Oregon, 16·78 kts. (*4 hours*) ; Oregon at Santiago did 16 kts. for a short while. 2 screws. Coal, 400 tons, *max. circa* 1000 tons.

INDIANA. *Photo,*

OREGON (1900)

INDIANA *class.*

(3.) *Chattanooga* & *Cleveland* & *Denver*, *Desmoines* & *Galverston* & *Tacoma* (building).

(Wood sheathed and coppered.) 3200 tons. Complement 293.
L. B. D.: 292 × 44 × 17¼ feet = 89 × 13·4 × 5·22 metres.

Guns ('99 model):
10D* (5 in., 50 cals.).
8* (57 m/m, 6 pdr.).
2* (37 m/m).
2 Colts.
Torpedo tubes: 0.

Armour (Harvey nickel):
2" Deck (on engines) = *d*
(50 m/m).
½" elsewhere.

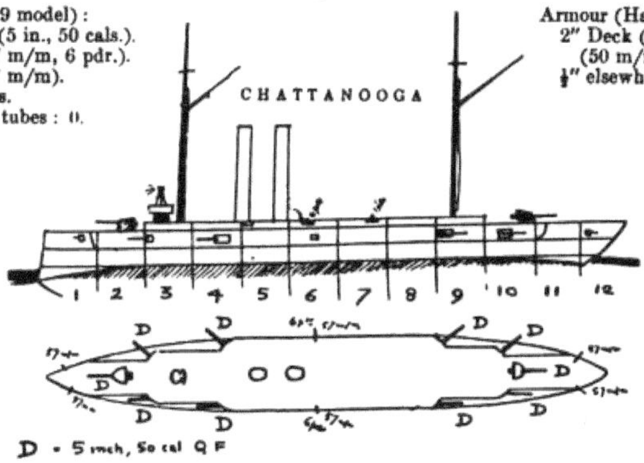

Machinery: 2 sets 4 cylinder triple expansion. I.H.P. 4500 = 16½ kts. (*est.*). 2 screws. Boilers, 6 watertube. Coal, *normal* 467 tons: *max.* 700 = *nominal* 7000 miles at 10 kts.

(4.) NASHVILLE (1895).

1371 tons. Complement 150.

L. B. D.: 220 × 36 × 15 ft. = 67 × 11 × 4·57 m.

Guns: 8F* (4 in.) + 4 6 prs.* + 2*. 1 bow torpedo tube.
Armour: Deck = *d*. 4 sponsons for F guns = *f*. Shields to others, *f*.
Machinery: 2 sets vertical quadruple expansion. I.H.P. 1790 = 14 kts. *trial*. Sea speed, *circa* 12 kts. 2 screws. Coal 150 tons.

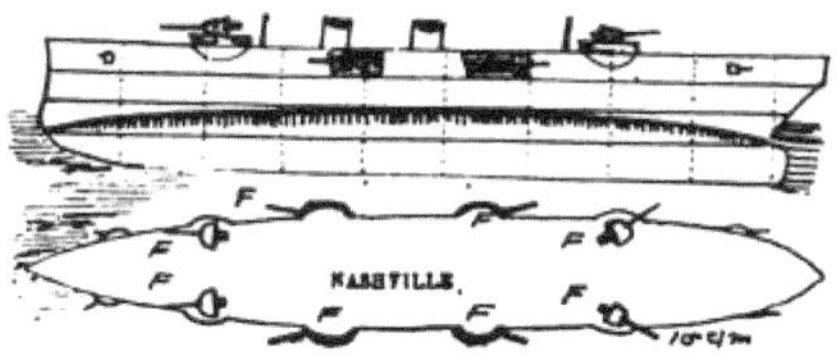

(3.) CINCINNATI (1892).

3180 tons. Complement, 300.

L.: 300 ft. = 91·5 m. B. 42 ft. = 12·8 m. D.: 20 ft. = 6 m.

Guns: 1D* + 10E* + 8*. Armour: *prot.* Sea speed: 17 kts. 6 torpedo

I.H.P. 10,000 = 19 kts. (4 *hours trial*). 2 screws. Coal, 500 tons. Ar Deck = *d*. Sponsons for E guns = *d*. Shields to others, *e*. Cofferdam.

(4.) DETROIT & MONTGOMERY & MARBLE HEAD (—1

2000 tons. Complement, 250.

L.: 257 ft. = 78·5 m. B.: 37 ft. = 11·3 m. D.: 16½ ft. = 5 m.

Guns: 9E* + 6*. Sea speed: *circa* 16·5 kts. 6 torpedo tubes.

I.H.P. *nat.* 2588 = 16·4 kts., *forced* 4096 = 19·2 kts. (2 *hours trial*). 2 Cellulose belt. Deck ⅜ in. (10 m.m. steel).

MARBLEHEAD. (All 3 alike.) *Photos, Copyright, Symonds & Co.*

(2.) PHILADELPHIA (1889).

Complement 326. 4324 tons. L. B. D. = $\frac{328 \times 48\frac{1}{2} \times 23 \text{ ft.}}{100 \times 14\cdot8 \times 7 \text{ m.}}$

12D + 4 6-pdrs.* + 4 3-pdrs.* + 7*. 5 torpedo tubes. Sea speed: 18 kts.
: Deck (4" steel) = *dc*. I.H.P. 8800 = 19·6 kts.
Coal: 400 tons (*normal*). 2 screws. 4 2-ended boilers.

Photo, copyright, Rau.

(2.) MINNEAPOLIS (1893).

mplement 500. 7450 tons. L. B. D. = $\frac{412 \times 58 \times 25\frac{1}{2} \text{ ft.}}{125 \times 17\cdot7 \times 7\cdot8 \text{ m.}}$

lC + 2D + 8F* (4-in.) + 12*. 6 torpedo tubes. Sea speed: 18·5 kts.

(Harvey): Sponsons ("fixed shields") *de* (100 m.m.) on F. guns: D gun) m.m.): on 8" C gun aft shield *e* (75 m.m.). Deck 100 m.m. = *b* oists *e*.

ry: vertical inverted triple-expansion. 3 screws. Boilers: 8 2-ended 58 furnaces.

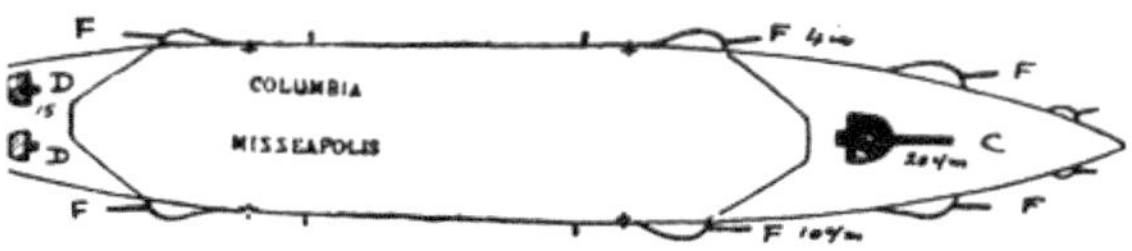

Photo, W. Rau, Philadelphia.

1896

(3.) RALEIGH (1892).

3180 tons. Complement 300.

L : 300 ft. = 91·5 m. B.: 42 ft. = 12·8 m. D.: 20 ft. = 6

Guns: 1D* + 10E* + 12*. Sea speed: *circa* 16 kts.

Armour: Deck = *d*. E gun sponsons *d*. D gun shields *e*. Coffer

1898

(4.) ATLANTA (1884) & BOSTON (1884).

3189 tons. Complement 272. L. B. D. = $\frac{276 \times 42 \times circa\ 18}{84 \times 12 \cdot 8 \times circa\ 5 \cdot 5}$

Guns: 2C (8-in. 25 calibres) + 6D + 6*. Sea speed: 11 kts.

Armour: Deck over engines only = *f*. I.H.P. *circa* 4000. 2 screws.

(To be re-boilered and re-constructed.)

(2.) NEWARK (1890) (1898) & SAN FRANCISCO (1889), (1899).

4098 tons. Complement 345.

L. B. D. (*mean*): $311\frac{1}{2} \times 49 \times 18\frac{3}{4}$ ft. $= 95 \times 14{\cdot}9 \times 5{\cdot}7$ m.

Guns: 12D* (6-in. 40 cal.) + 4* (6-pdrs.) + 4* (3-pdrs.) +8*. 6 torpedo tubes.

Armour: Deck (75 m.m.) = *ed*. Shields 2-in steel (50 m.m.).

Machinery: 2 sets horizontal triple expansion. 4 2-ended boilers. 2 screws.

Trial speeds: Newark, 4 *hours forced draught* 8869 = 19 kts. San Francisco, 4 *hours forced draught* 9913 = 19·5 kts. Sea speeds: *circa* 15 kts.

Coal: *normal*, *circa* 300 tons; *max.* 850 tons = *circa* 6000 miles.

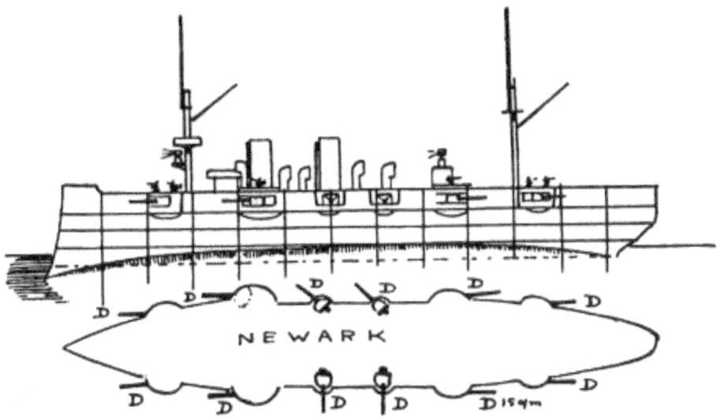

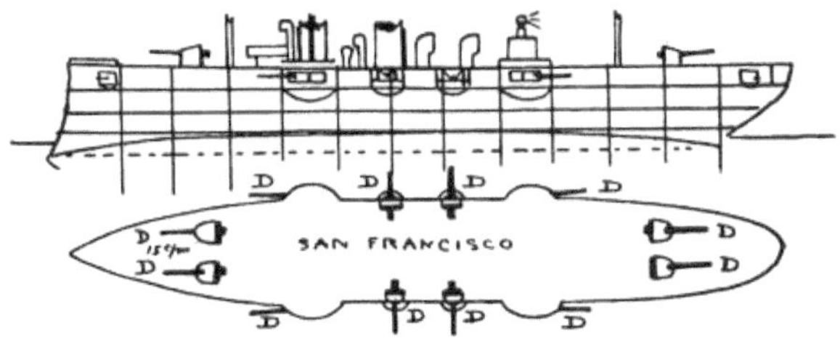

(2.) CHICAGO (1885) (1897–9).

Circa 4500 tons. Complement .

L. B. D.: 328 × 48 × ft. = 100 × 14·7 × m.

Guns: 4C* (8 in. converted) + 14D* (5-in. 50 cals.) + 9* (6-pdrs.) + 6*. Torpedo tubes: ?. Armour: Deck = *f*.

Machinery: 2 sets triple expansion. 6 Babcock boilers + 4 cylindrical. 2 screws. Coal: *max.* 940 tons.

Chicago is fitted with automatic watertight doors.

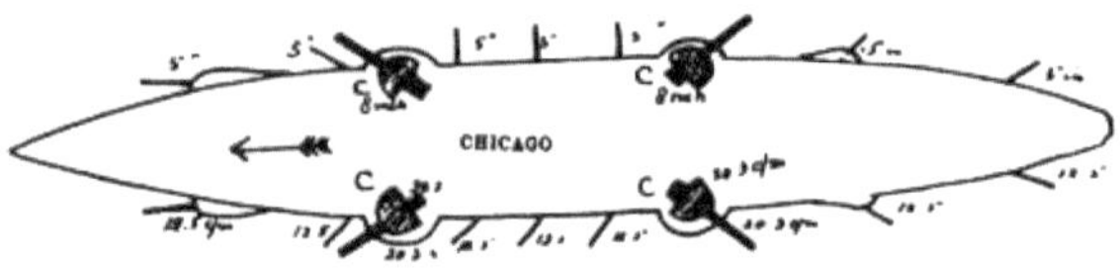

(2.) BALTIMORE (1888).

4600 tons. Complement, 395.

L.: 328 ft. = 100 m. B.: 48½ ft. = 14·8 m. D.: 23 ft. = 7 m.

Guns: 4C (8 in. = 20·3 c.m. of 25 cals.) + 6D (6 in.) + 8*. Sea speed: *circa* 18 kts. 5 torpedo tubes.

Armour: Deck (100 m.m.) = *d*. Gun shields *e*. Cofferdam and cellulose belt.

I.H.P. *circa* 10,000 = 20 kts. (*trial*).

Machinery: Horizontal triple expansion. 2 screws. 4 2-ended Scotch boilers. 32 furnaces. Coal, 400 tons (*normal*), *max.* = 900 tons = *circa* 6000 kts. at 12 kts.

Trial. 4 *hours forced draught* (1890), I.H.P. 10,064 = 20·1 kts.

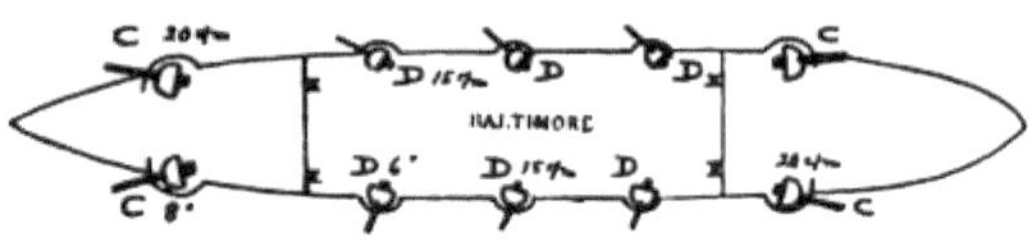

Photo by Loeffler.

Photo, copyright, Rau, Philadelphi

(I.) KEARSARGE & KENTUCKY (— 1898).

11,500 tons. Complement 520.

L. B. *D.* (*mean*): 368 × 72 × 23½ feet = 112 × 22 × 7·2 m.

s:
A (13 in.).
C (8 in. 40 cals.).
4E* (5 in. 40 cals.).
D* (57 m/m, 6 pdr.).
* (37 m/m, 1 pdr.).
Colts.
pedo tubes:
above water (4 behind
c armour).

Armour (Harvey nickel):
16½" Belt (amidships) *aaaa* (418 m/m).
2¾" Deck slopes (63 m/m).
Protection to vitals = *aaaaa*.
4" Belt forward *d* (100 m/m).
4" Deck aft = *a* (100 m/m).
5½" Lower deck and battery *c* (138 m/m)
15" A turrets and bases *aaaa* (381 m/m).
9" C turrets *a* (228 m/m).
10" Fore C. T. *aa*. After C. T. *b*.

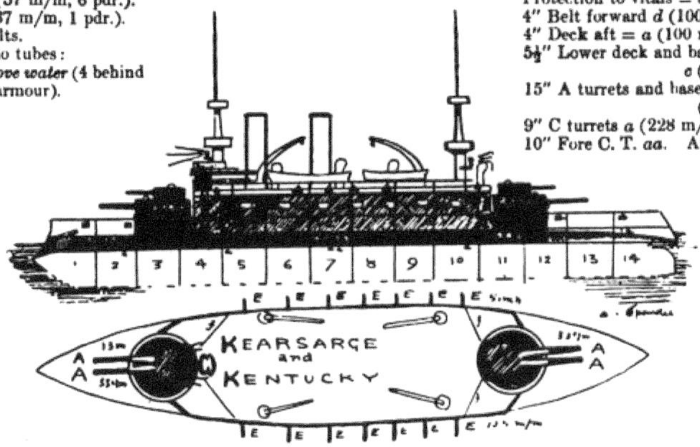

Machinery: 2 sets vertical triple expansion. 2 screws. Boilers: cylindrical, 6 2-ended + 4 1-ended. I.H.P. (*est.*) 10,500 = 16 kts. (*f. d.*). Coal: *normal* 410 tons; *maximum capacity* 1210 tons.

Trials (1899).

Kearsarge	40 *hrs. natural d.* I.H P.	= 14·1.	4 *hrs. forced d.* I.H.P.?	= 16·84.
Kentucky	,, ,, ,, ,,	= 14·93.	,, ,, ,, ,,	= 16·89.

KENTUCKY. *Photo, copyright, by Loefeller, Tompkinsville.*

(2.) OLYMPIA (1892).

5870 tons. Complement, 466.

L. B. D.: 344½ × 53 × 25 (?) ft. = 105 × 16·2 × 7·61 m.

Guns: 4C (8 in., cals.) + 10E* (5 in., cals.) + 14* (6 pdr. 57 m.m (37 m.m. 1 pdr.) + 4*. 6 torpedo tubes.

Cofferdam {3½ ft. wide / 1·20 m. „}. Belt, cellulose. Armour (Harvey): Dec (120 m.m.) Turrets *dc* (120 m.m., 4·7"). Hoists *e* (80 m.m.) Gun shields *e*.

Machinery: 2 sets vertical triple expansion. Boilers, 4 2-ended + 2 1 2 screws. Coal ?

Speed: *Trial, 4 hours forced draught* (1894) I.H.P. = 17,363 = 21·7 kt light). Sea speed: *circa* 18 kts.

(2.) NEW ORLEANS (*ex* Amazonas) & ALBANY (*ex* Almirante Abreu) (1897–98, Elswick).

3450 tons. Complement, 300.

L. B. D.: (*mean*) = 330 × 43¾ × 17 ft. = 100·5 × 13·5 × 5·20 m.

Guns: 6D* (6 in., 50 cals.) + 4E* (4·7 in., 50 cal.) + 10* (6 pdr. 57 m.m.) + 4* (1 pdr.) + 4*. 3 torpedo tubes (1 of them in bow).

Armour: Deck (Harvey) (90 m.m.) = *b*. D gun shields fronts = *aa*, sides *e*.

Machinery: 2 sets 4-cylinder triple expansion. Boilers, 4 2-ended Scotch.

Sea speed: . Coal, *normal* tons.

Trials. New Orleans, *nat. d.* = 19·8 kts.: 4 *hours forced* = 20·5.
Albany „ „ = 19·1 kts.; 4 „ „ = 20·5.

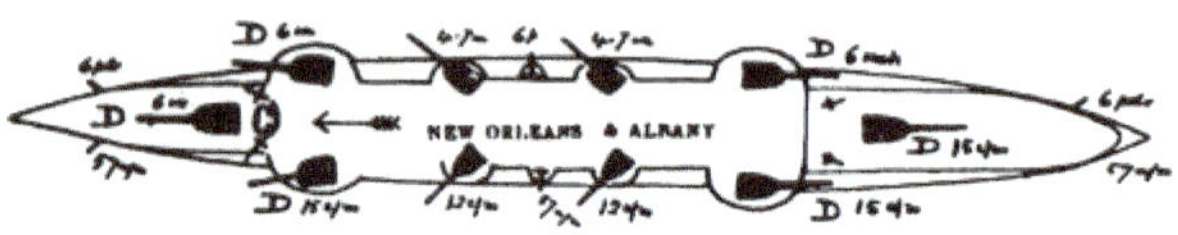

ALBANY. *Photo, Sir W. G. Armstrong, Whitworth & Co.*
(New Orleans similar, but ventilators less conspicuous.)

Photo, Rau, Philadelphia.

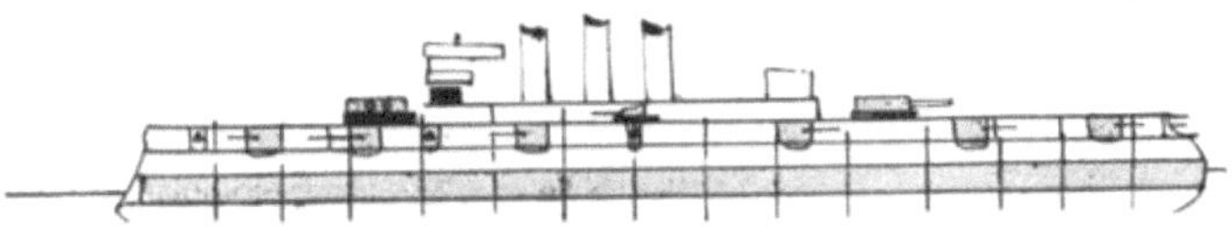

(1.) NEW YORK (1891).

8480 tons. Complement, 566.

L. B. D.: 381 × 64 × 28 ft. = 116 × 19·5 × 8·5 m.

ns: 6C (8 in., 35 cal.) + 12F* (4 in. Q.F.) + 8 6 pdrs.* + 8*. 6 above water o tubes.

mour (Harvey): Belt *d* (100 m.m.) Curved deck behind belt = *aa* (6" = .m.) Cellulose behind armour belt (6¼ feet = 2 metres wide). Barbettes *a* n.m.) Shields to C guns *c* (150 m.m.) C. T. *c*. Shields to F guns *ed* (100 m.m.) *d* (125 m.m.)

achinery: I.H.P. 17,400 = 21 kts. (*trial*). 2 screws. 4 sets of engines. 8

al capacity, *normal* 750, *max.* 1150 = 10,000 miles at 10 kts.

(1.) BROOKLYN (1895).

9250 tons
Complement, 500.

L. B. D.: 400 + 65 × 28 ft. = 122 × 20 × 8·5 m.

Guns: 4C (8 in., 35 cal.) + 12E* (5 in.) + 12 6-pdrs.* + 4 1-pdrs.* 6 torpedo tubes.

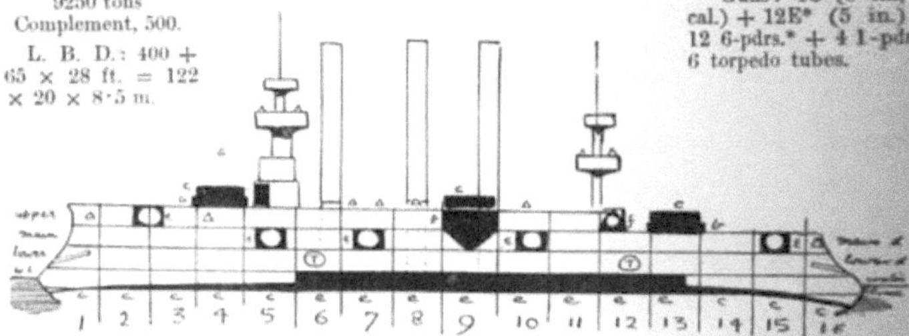

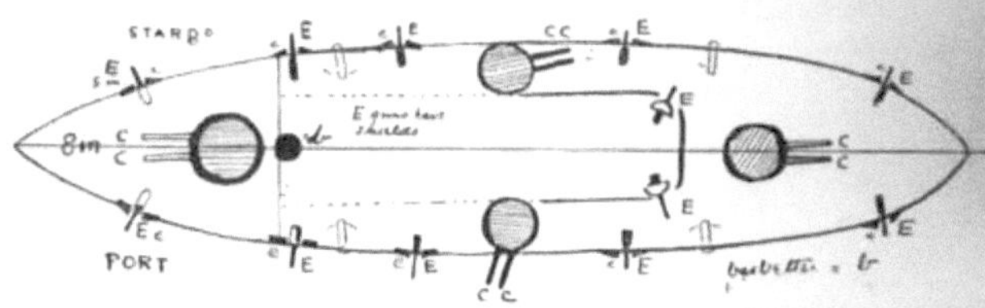

Armour (Harvey). Deck = *aa* (152 m.m.) Belt (partial) *e* (75 m.m.) Deck inclined behind belt. Protection to vitals = *aaa*. Cofferdam. Barbettes *b* (?). Shields *c*. Fixed shields to E* guns *e*. No bulkheads.

Machinery: 4 sets vertical triple expansion engines. 20 double-ended boilers, 8 single-ended. I.H.P. 18,769 = 22 kts. (trial—ship very light). At Santiago, without forward engines, Brooklyn made 16 kts. Coal capacity, *normal* 900; *max.* 1650 tons = 10,000 miles at 10 kts.

BROOKLYN. *Photo Copyright. Le Yacht.*

(II.) *A, B & C (projected).*

13,500 tons.

Copies of the Russian Peresviet type (*q.v.*)

Guns (1899 model):
4A* (12 in. R.L.).
16D* (6 in., 50 cals.).
20* (57 m/m, 6 pdr.).
6 Colts, 4 machine.
Torpedo tubes:
2 *submerged*.

Armour (Kr
16" Belt (
4" Deck (
Prot. to v
5½" Lower
4" Belt fo
14" Turret
10" Bulkh
2" Screen
Fore C. T.

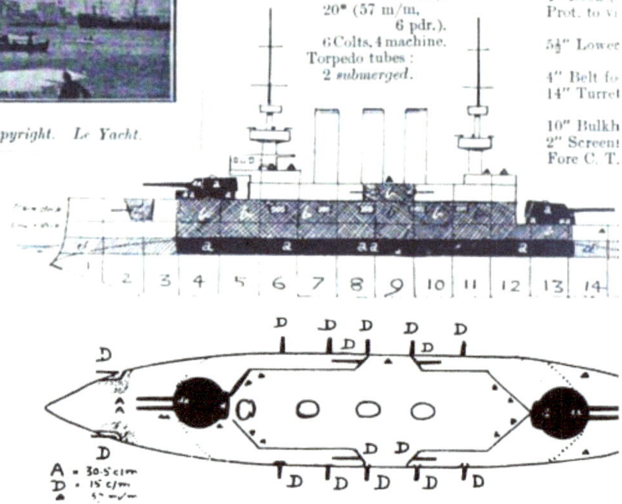

(I.) *MAINE* (1899) & *MISSOURI* (1900) & *OHIO* (1900).

12,500 tons.

L. B. *D.* (*mean*): 388 × 72 × 24 feet = 118 × 22 × 7·31 metres.

(Length over all, 393 feet (120 m.).)

Machinery: 2 sets vertical inverted triple expansion 3 cylinder, *except* in Missouri, which is 4 cylinder. 2 screws. Boilers: *Maine* and *Ohio*, 24 Niclausse; *Missouri*, Thornycroft. I.H.P. 16,000 *forced* = 18 kts. (*estimated*).

(I.) *New Jersey* & *Georgia* & *Pennsylvania* (to be built).

Displacement 14,650 tons. Complement 703.

L. B. D. (*max.*): 435 × 76 × 26 feet = 132·58 × 23·16 × 7·92 metres.

Guns (model 1899):
4A* (12 inch, 40 cals.).
8C* (8 inch, 45 cals.).
12D* (6 inch, 50 cals.).
12F* (3 inch, 40 pdr.).
12* (3 pdrs., 47 m/m).
4* (1 pdr., 37 m/m).
4* (1 pdr., automatic).
2 Gatlings.
6 automatic, ·03 calibre.
Torpedo tubes:
2 *submerged* forward.

Armour (Krupp and Harvey nickel):
11" Belt amidships *aaa* (275 m/m)
4" Belt at ends (H.N.) *d* (100 m/m).
3" Armour deck (flat amidships) = *a* (76 m/m).
Prot. to vitals = *aaa*.
6" Lower deck side *a* (152 **m/m**).
6" Bulkheads forward, lower **deck** = *aaaa*.
6" Bulkheads, main deck, *a* (152 **m/m**).
10" A gun turrets *aaa* (250 **m/m**).
6" C gun turrets (H.N.) *b* (152 **m/m**).
9" Conning tower *aa* (226 **m/m**).
5" Signal tower aft (H.N.) *c* (125 **m/m**)
2" Side to F guns *e* (50 m/m).

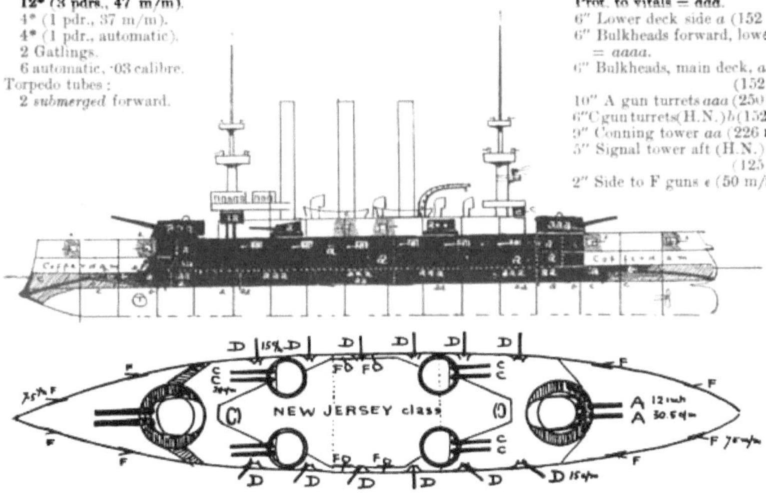

Machinery: 2 sets 4 cylinder triple expansion. 2 screws. I.H.P. *maximum* 19,000 = 19 kts. (*estimated*). Boilers: 24 straight tube water-tube. Coal: *normal* 900 tons; *maximum capacity* 1900 tons.

Notes.—At the normal displacement the upper edge of the thick belt will be 1 foot (30 c/m) above the water. This belt is 8 feet (m. 2·44) in total width. Its thickness is everywhere, except at bow and stern, about 25 per cent. less at the bottom than at the top edge. The thinning usually begins 3 feet (1 m.) from the bottom. All fire mains are under the protective deck. The protective deck is flat on top of the belt amidships, but reinforces the thin belt at the ends. The flat deck is 1½ inches thick (38 m/m). The lower deck bulkhead forward is inclined at an angle of 45° against the turret base. The upper deck is planked, fireproof wood laid on a complete metal deck—all other decks are linoleum on steel. These ships were approved March 1899, but, owing to delay in settling the plans, were not finally passed till May 1900. They are to be ready for sea 1st January 1902.

Arcs of fire: 12 in. guns, 270°. 8 in., 110°, from right ahead or astern. 6 in., 110°, *i.e.*, 55° before or abaft the beam. Ammunition hoists (electric) serve at the following rates:—12 in., 1 in 90 seconds; 8 in., 1 in 50 seconds; 6 in., 3 per minute. There is a complete cellulose belt along the lower deck.

Photo, A. Renard, Kiel.

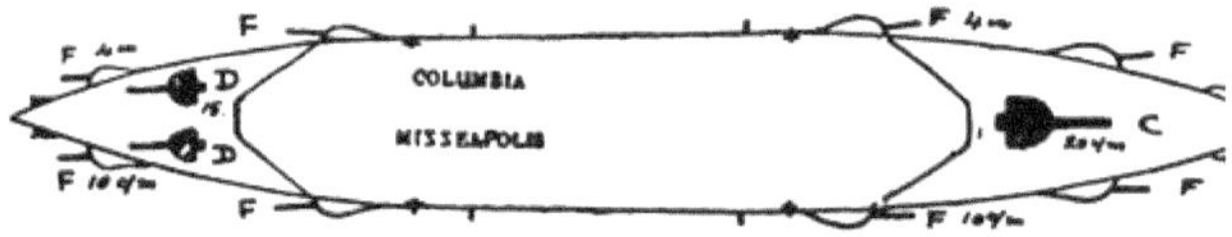

(2.) COLUMBIA (1892).

7475 tons. Complement, 524.

L. B. D. (*mean*): 413⅓ × 58⅓ × 24 ft. = 126 × 17·9 × 7·3 m.

Guns: 1C (8-in. 35 cals.) aft + 2D* (6-in. 40 cals.) + 8F* (4-in.) + 12* (6- 57 m.m.) + 8*. 5 torpedo tubes.

Armour (Harvey): Deck = *b* (100 m.m.). Cofferdam. Cellulose belt 4 (1·50 m. high). Gun shields to C gun *e* (75 m.m.), to D gun *f* (50 m.m.). Ve armour on F* guns *ed* (100 m.m.), on the 6-pdrs. *f* (50 m.m.).

Machinery: 3 sets vertical triple expansion. 8 2-ended + 2 1-ended bo 3 screws. Speed *trial, light, 4 hours forced draught*, I.H.P. *circa* 22000 = 22·8 7 *days* (*sept jours*) full speed sea trial = *moyen de* 18·4 kts. (*le dernier jour il ne marcher au tirage force*).

(III.) *California* & *Nebraska* & *West Virginia*.

(To be built).

Sheathed and coppered.

12,000 tons. Complement,

L. × B. × D.:

Guns (model '99): 4C* (8-in. Q.F.) + 14D* (6-in. Q.F.) + 10 F* (3-in. Q.F.) + 10* (6-pdr., 57 m/m) + 6* (1-pdr., 37 m/m) + 6 Colts.

Torpedo tubes: 2 *submerged*.

Armour:

Machinery: Speed, *est.* 22 kts.

SUMMARY OF THE WORLD'S FIGHTING SHIPS,

ACCORDING TO THE CLASSIFICATION USED IN THIS BOOK (*see* p. 17).

(The numbers of ships building are put in italics, and preceded by a +.)

	Armoured Ships. *Class* I.	II.	III.	IV.	1.	V.	2.	3.	4.	5 & 6, including T.	Destroyers.	Torpedo boats.	Submarines.	Total.
entine .	...	...	4+ *0*	3+ *0*	...	2+ *0*	3+ *0*	...	1+ *0*	8+ *0*	4+ *0*	22+ *0*	...	47+ *0*
sil . . .	...	...	2+ *0*	2+ *0*	...	...	2+ *0*	1+ *0*	1+ *0*	18+ *0*	0+ *8*	20+ *0*	...	54+ *0*
li . . .	...	...	2+ *0*	...	1+ *0*	2+ *0*	2+ *0*	...	2+ *0*	5+ *0*	6+ *0*	23+ *0*	...	43+ *0*
ne . . .	...	...	...	...	...	...	5+ *0*	...	...	15+ *0*	4+ *0*	14+ *0*	...	38+ *0*
emarck .	...	...	1+ *1*	2+ *0*	...	4+ *0*	1 (T.X.)+*0*	1+ *0*	1+ *0*	13+ *0*	0+ *4*	27+ *2*	...	50+ *7*
tschland .	...	7+ *8*	0+ *2*	16+ *0*	...	11+ *0*	6+ *0*	2+ *0*	2+ *2*	27+ *1*	11+ *1*	149+ *1*	...	231+*15*
land . .	14+*15*	13+*10*	11+*10*	17+ *0*	29+ *0*	14+ *0*	24+ *1*	21+ *0*	33+ *0*	104+ *6*	99+*14*	186+ *2*	...	565+*58*
agne . .	...	1+ *0*	...	5+ *0*	1+ *0*	...	2+ *1*	...	1+ *1*	22+ *0*	4+ *0*	16+ *0*	1+ *0*	51+ *2*
nce . .	5+ *8*	4+ *7*	19+*12*	9+ *0*	3+ *0*	14+ *0*	17+ *1*	4+ *0*	8+ *0*	27+ *0*	6+ *6*	238+ *49*	4+ *6*	358+*89*
ce . . .	...	...	3+ *0*	1+ *0*	...	1+ *0*	...	...	...	13+ *0*	...	36+ *0*	...	54+ *0*
lande . .	...	...	0+ *2*	4+ *0*	1+ *0*	15+ *6*	6+ *3*	1+ *0*	...	62+ *5*	...	45+ *17*	...	134+*33*
a . . .	0+ *3*	6+ *3*	10+ *6*	4+ *0*	3+ *0*	1+ *0*	1+ *0*	12+ *1*	3+ *0*	35+ *0*	1+ *8*	184+ *0*	1+ *0*	261+*21*
on . . .	4+ *2*	...	5+ *2*	...	...	2+ *0*	10+ *0*	5+ *0*	2+ *0*	16+ *3*	12+ *0*	89+ *12*	...	95+*19*
-Ungarn	...	3+ *2*	3+ *1*	6+ *0*	...	...	2+ *0*	...	...	36+ *1*	...	70+ *0*	...	120+ *4*
tugal . .	...	...	...	1+ *2*	...	...	1+ *0*	...	3+ *1*	27+ *0*	...	12+ *0*	1+ *0*	45+ *3*
sie . .	1+ *6*	12+ *2*	6+ *5*	3+ *0*	3+ *0*	4+ *0*	6+ *4*	5+ *0*	0+ *1*	28+ *0*	16+*21*	181+ *6*	0+*50*(p.)	265+*95* (50 p. [illegible]
rige *och*	...	...	...	6+ *4*	...	13+ *0*	0+ *3*	...	...	19+ *0*	...	22+ *4*	...	60+*11*
orge . .	...	...	...	2+ *2*	...	4+ *0*	...	...	...	20+ *0*	...	14+ *4*	...	40+ *6*
quie . .	...	...	0+ *2*(r.)	4+ *0*	...	...	...	...	...	2+ *0*	2+ *0*	24+ *0*	2+ *0*	34+ *2*
S. A. . .	5+ *6*	4+ *3*	1+ *3*	...	2+ *0*	7+ *4*	10+ *0*	2+ *6*	8+ *0*	18+ *0*	6+*14*	15+ *12*	2+ ? (2)	80+*48*
als . .	29+*40*	50+*35*	67+*46*	85+ *8*	43+ *0*	94+*10*	98+*13*	63+ *7*	65+ *5*	515+*16*	171+*76*	1837+*113*	11+*56*	2630+*413*

Note.—The classification of all ships has been revised this year. In the body of the book the following misprints occur:—p. 129, Bouvines, Jémappes, etc., are (III.) *not* (II.); p. 168, Lombardia, Liguria, etc. (3) *not* (2). They are correctly counted in this summary.

(T.X.) includes Tordenskjöld (torpedo ram). (r.) includes 2 ships that are to be rebuilt. (p.) these submarines are all "projected"; and none are yet building. (2.) From 4 to 30 submarines, projected, according to report at moment of going to press.

THE BELLEISLE EXPERIMENT.

By the courtesy of the proprietors of *The Engineer*, London, which dealt exhaustively with these experiments in its issue of May 24th, June 1st, 8th and 15th, 1900, the following condensed account of their report is published.

The following diagram represents the course and ranges :—

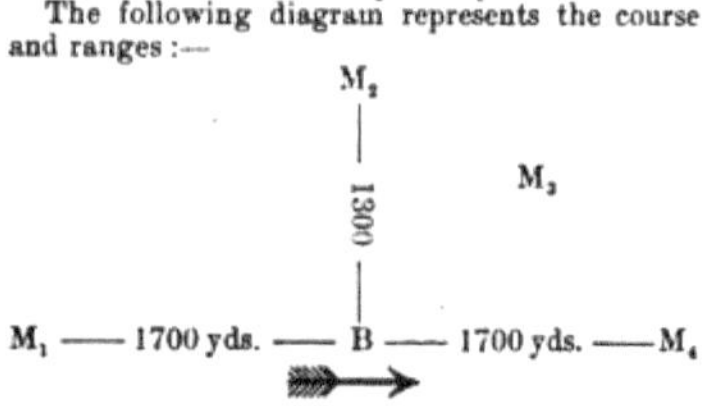

The Majestic steaming firing from M_1 to M_4, the Belleisle (B) being moored with her bow in the direction of the arrow. All the serious hits appear to have taken place between M_1 and M_2. The Majestic steamed at six knots.

There being considerable doubt as to the sequence of hits, saving that the second 12in. shot hit the belt amidships, and that the funnel was knocked out when the Majestic was at M_2 or thereabouts, we propose simply to take them in the sequence in which they registered themselves upon the hull.

So far as can be ascertained, the Majestic fired :—

8	rounds	12in. common shell
7	„	12in. A.P. shot
About 100	„	6in. lyddite (at bow and battery)
About 100	„	6in. common (at stern)
400	„	3in. various
750	„	3-pounder shell

And an indefinite number of Maxims. No torpedoes were discharged. Full charges were used throughout for all guns.

Roughly speaking, about 30 to 40 per cent. of the projectiles fired were effective, the rest went over or just short, or else went clean through and burst beyond. Only about 10 per cent. of the projectiles were clear misses outside the black-and-white haystack of smoke that shrouded the target.

Here, perhaps, we should recapitulate the data of the Belleisle. She is of about 5000 tons displacement, and the iron armour is disposed as follows :—

Belt	12in.	6in.
Battery	10in. — 8in.	
Redoubt below battery...	9in. — 8in.	
Conning tower	12in.	

The belt rises about 2ft. to 2½ft. above the water line, but at the time of the trial about 4ft. were shown. The redoubt is octagonal like the battery. The conning tower is situated under the bridge on top of the battery ; it is very low, and the hood stood at least 18in. above the walls.

Taking the 12in. hits first, the demolition of the erection in the bow is variously attributed to a 12in. and a 6in. We think, however, that it was due to one of the latter. On the belt, below where the funnel used to be, there are two very large stars where shell have burst. One or both these are presumably 12in. hits, there being in one case—the after one of the two—a 12in. penetration of about 2in. deep. The armour belt just below is cracked laterally for some 7ft., and a vertical seam has also opened. The lateral crack is about 6in. wide. There are the marks of a 6in. shell in the aperture. The other hit has no mark beyond the large star caused by the burst.

On the water-line, directly under the battery, there is a large hole made by a 12in. shot. This hit was that by which the ship eventually sank ; though not, as erroneously supposed, the direct cause of her sinking. As we have said above, it was the second 12in. projectile fired. It is probably the prime cause of the crack a little more forward, as this hit blew the side right in, sending one fragment of armour right up through the flat armoured deck above it. It did not, however, send anything down into the engine-room. We have not been able to learn where the shot went after going through the belt, which here offered a resistance equal to from 9in. to 12in. Harvey, because it was struck at a considerable angle—one, indeed, in which we might conceivably have expected the shot to glance off.

Immediately abaft the casemate a 12in. shell has hit. This shell has blown away some 10ft. square of side, which here looks something like the ruins of a recessed port. In the armoured bulkhead there is a penetration, made either by the head of this shell or by a 6in. shell that did not burst on impact.

Well aft another 12in. shell has hit. Here, too, a large piece of side is blown in, the deck above has disappeared, and the iron deck beams are curled up into shapeless remains more like shavings than anything else. The external area of damage is about 10ft. ; internally the whole of the cabins hereabouts were reduced to matchwood.

There remain two more hits that may be 12in. As we observed at the beginning of this article, there are so many hits, so many close together, and such complete destruction about their immediate areas, that the projectile cannot always be named to a certainty. But at the after end of the unarmoured top of the battery, and towards the after end of what little remains of the funnel base, there are huge irregular holes, averaging perhaps 5ft. to 7ft. each. Most of the edges are bent inwards.

We now come to the most mysterious hit of any. High up on the port side after porthole is a clean, nearly exactly circular penetration of exactly 10in. in diameter. There is around it an irregular depression, with a maximum depth of
2½in., and some 2ft. across at the widest.
made this hole no one seems inclined
positive about ; and, for the present at an
we prefer to leave it to conjecture. The
a 12in. projectile, a 6in. lyddite, and a fr
from the hole in the side below, have all
into the conjectures of those who have se
hole, but not been allowed to inspect it
And here we may mention that on the
and outer edge of the after port, on the st
side of the battery, another projectile
not very traceable has chipped out a p
armour, and burst, judging by the marks.
is the total of the big projectiles ; three o
certain, and five of varying degrees of dou

The bow and battery were fired at
lyddite, the 6in. common shell were direct
Between the damage done by one and th
there is no comparison. Both have mad
in the unarmoured ends ; both have done n
at all to armour, both have made matchw
the woodwork inside. But while th
common has destroyed as one might
a wooden box with an axe, lyddite has
ised completely. Practically there are no r
beyond a dust. The lyddite has act
a moment much as dry rot acts in a score
of years. The deck in places is simply go
the edges have just the crumbled appear
dry rot. In other places, too—to use a
homely simile—the wood looks like the en
deal board that has been heavily hit by a ha
This exactly gives the curiously com
appearance. It is noteworthy that there i
single sign of charring. It is not merely t
burnt wood is visible, there are no si
scorching either, though possibly a week's b
ing under the sun may account to some
for this.

Where 6in. common shell have burst be
decks, the deck above shows no sign of it,
is quite another story with lyddite. No

B

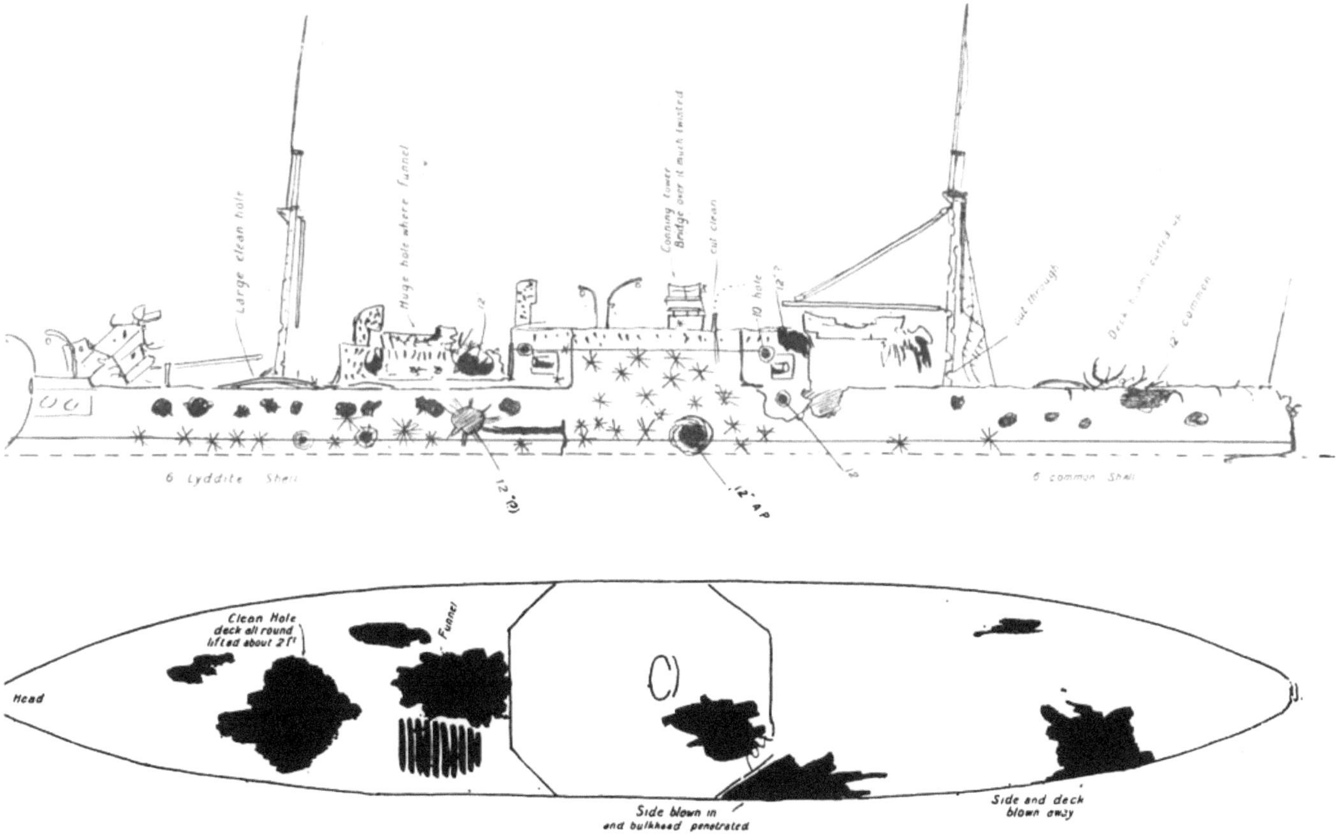

THE BELLEISLE—DIAGRAM SHOWING THE DAMAGE DONE.

(Reproduced by kind permission of the proprietors of "The Engineer.")

are huge holes blown upwards, but the entire deck is bulged up. The cross beams were thick and plentiful, but their resisting power has been nil.

The 6in. holes of entry and exit along the main deck side are all plugged, so no description of them is yet possible. In the diagram they are all indicated by black spots, armour penetrations by a spot surrounded by a circle, shell bursts against armour by a star. These stars along the water-line are very numerous, and effectually dispose of the theory that a water-line can hardly be hit. The extreme bow is absolutely untouched, then comes a series of stars innocent of the slightest dent. Abreast of the funnel—curiously enough just where the armour is thickest—there are several dents where shells have gone in an inch or two. Many shells have burst against the battery, and this, being painted white, looks from a little distance as though it had been covered with Japanese naval flags. There is a great sameness about these shell marks, many of which are almost absolutely identical, and most are symmetrical in shape.

The masts are in a bad way. The foremast is the least damaged. The yard now on it was placed for signalling purposes after the firing, the yards on both masts were shot away, and both peaks were dropped. The mainmast is especially in a very bad way, and had to be shored up. It was cut in two all but a tiny piece, low down, and remained standing only because all the shrouds were not destroyed.

The only boat left is the steam pinnace carried between the battery and mainmast. A 6in. common shell entered this boat and blew the starboard side of it out into a number of shapeless streamers. It is absolutely clear that no ordinary boats will float after an action; it is doubtful if they will exist.

The damage to the funnel cannot well be described, as there is nothing there but an enormous hole, with strips of iron twisted over it.

The bridge, though much bent and twisted, survived; but the searchlight was blown away. The conning tower was not hit at all. It had inside it some thin wooden dummies—mere laths of wood—so badly put together that the concussion of shells bursting on the topside shook them to pieces. The story about a shell having hit the tower and done this has no foundation in fact.

The upper works were blown to pieces, half the 6-pounders either knocked clean away or else blown sideways with some of their fittings melted. Ventilators and so forth were pitted with innumerable small holes from splinters and fragments.

Inside the battery, in which at least one shell that had entered a porthole had burst, all the gun sights were destroyed. The guns themselves are absolutely unhurt; but, owing to damages to sights and gear, would have been of no use at all. The dummies around them were all burned.

It is worth noting that two davits ha cut clean in two by projectiles.

Directly firing ceased the ship was bo fire parties and by the officers told off to her. She was then full of dense smoke by the exploded shell, by the pierced base, by a few small bonfires of dummie lyddite shell or two that had failed properly and were burning. Practically was to be seen for the smoke and fume.

It was taken for granted that the ship fire, and water was pumped into her lowered her freeboard, and the sea then r at the hole on the water-line made by th shot. The ship canted heavily, and s after the last of the exploring party had

Some of the Belleisle's hoses were still when the ship was boarded, others wer cut to pieces. Most of the valves and had disappeared.

ДОПОЛНЕНІЯ.

Дополненія къ книгѣ „Всѣ Военныя суда на Свѣтѣ" издаются каждыя 10 недѣль и ихъ можно Достать подпиской по 6 рублей въ годъ, съ пересылкой, у любого книгопродавца.

THE NAMES OF ALL THE WORLD'S FIGHTING SHIPS
MADE BY

THOMAS STEVENS,

Stevengraph Works, Coventry, England.

INVENTOR AND MANUFACTURER OF

HAT RIBBONS

With Ships' Names Woven in Gold Wire.

CONTRACTOR TO HER MAJESTY'S GOVERNMENT.

W. THACKER & CO.'S NAVAL PUBLICATIONS.

THE NAVAL POCKET BOOK. (Founded by W. LAIRD CLOWES.)
Fifth Year of Issue. Edited by L. G. CARR LAUGHTON.
The most valuable work of reference now available, containing a full list of Battleships, Ironclads, Gunboats, Cruisers, Torpedo Boats, a list of Dry Docks, and other valuable information concerning ALL THE NAVIES IN THE WORLD. 5s. net (6 francs).

THE CAPTAIN OF THE "MARY ROSE." By W. LAIRD CLOWES.
Illustrated by the CHEVALIER DE MARTINO and FRED. T. JANE. Seventh Edition. Cloth, 6s. (7 francs, 20).

WITH SAMPSON THROUGH THE WAR. By W. A. M. GOODE.
An account of the Naval Operations during the Spanish-American War. With Chapters by Rear-Admiral SAMPSON, Captain ROBLEY D. EVANS, Commander C. C. TODD, U.S.N. 10s. 6d. (12 francs, 60).

WORKS BY FRED. T. JANE, Author of "All the World's Fighting Ships," etc.

THE IMPERIAL RUSSIAN NAVY. Its Past, Present, and Future.
Dedicated to H.I.H. the Grand Duke Alexander of Russia. 180 Illustrations. In One Handsome Volume. 750 pages. 30s. net (38 francs).

THE TORPEDO IN PEACE AND WAR. Oblong folio. Full of large Sketches of Torpedo Operations, made at sea, by the Author. 10s. 6d. (12 francs, 60).

BLAKE OF THE "RATTLESNAKE." Illustrated by the Author.
3s. 6d. (4 francs, 20).

W. THACKER & CO., 2, CREED LANE, LONDON, E.C., ENGLAND.

NAVAL AND MILITARY SONGS AND BALLADS.

SEVENTEENTH EDITION. Fcap. 8vo. 1s. net.
"ADMIRALS ALL," and other Verses.
By HENRY NEWBOLT.
"Genuinely inspired patriotic verse. . . . There are about a dozen pieces in this shillingsworth, but there is no dross among them."—*St. James's Gazette.*

BY THE AUTHOR OF "ADMIRALS ALL."
THIRD THOUSAND. Crown 8vo. 5s. net.
THE ISLAND RACE. By HENRY NEWBOLT.
Of the Forty Poems in this Volume, Twelve appeared in "Admirals All."
"Mr. Newbolt's ballads have the right trumpet tone. . . . Everywhere is seen the evidence of a delicate ear and a delicate hand."—*Spectator.*

Vigo Cabinet Series.
Roy. 16mo. 1s. net.
Nos. 1 and 2 now ready.

THE QUEEN'S HIGHWAY, and other Lyrics of the War, 1899-1900. By JOHN HUNTLEY SKRINE, Warden of Trinity College, Glenalmond, Perthshire. [*Second Edition.*

HOME IN WAR TIME. Poems. By SYDNEY DOBELL. Selected and Edited with an Introduction, by W. G. HUTCHISON.

London: ELKIN MATHEWS, Vigo Street, W.

ADVERTISEMENTS.

Advertisements should be Pictorial if possible, and embody original ideas. The Manager will gladly submit ideas FREE of cost to Advertisers.

You see this? . . .

Aim to infuse individuality in your advertisements. Whether an advertisement be large or small, it ought to possess some distinctive characteristic. **Illustrate** your advertisements whenever possible. *While* the attention is attracted the story is told and the impression made.

ADVERTISEMENT SCALE AND FULL PARTICULARS FROM
C. GILBERT WOOD, Granville House, Arundel Street, LONDON, W.C.

LEAN'S

ROYAL NAVY LIST.

SUPPORTED BY THE ADMIRALTY.

Containing the War and Meritorious Services of all Officers on the Active and Retired Lists of the Royal Navy and Royal Marine Forces, with all the Royal Orders, Medals, Decorations, and Honours, given for Distinguished Services.

The Dates of all Commissions and of all Retirements, with Dates of Birth of Flag Officers, Captains and Officers of the Royal Marines, showing at a glance their progressive rank, from entrance in the Navy to the present time.

A LIST OF SHIPS ON ACTIVE SERVICE, GIVING THE TONNAGE, STEAM POWER, AND CLASS, WITH THEIR OFFICERS ARRANGED IN ORDER OF SENIORITY.

The War Services of all the Famous Battle Ships of the Royal Navy, showing the Glorious Records of our "First Line of Defence."

The Officers of the Royal Indian Marine, the Australasian Naval Defence Forces, and of the Royal Naval Reserve. A complete Alphabetical List of all Officers, both active and retired, in the Royal Navy and Royal Marines, and of all Officers in the Royal Indian Marine and Royal Naval Reserve. An Exhaustive List of all the Civil Departments of the British Admiralty.

PUBLISHED QUARTERLY (500 PAGES), PRICE 7s. 6d., POST FREE.
Annual Subscriptions (payable in advance), £1 7s. 6d. post free.

"What 'Whitaker' is to the general public, so is 'Lean's Royal Navy List' to everyone interested in the Naval Service."—*Nautical Magazine.*

LONDON: WITHERBY & CO., PUBLISHERS, 326, HIGH HOLBORN, W.C.
Advt. Dept.—GRANVILLE HOUSE, ARUNDEL ST., STRAND, W.C. (Mr. C. GILBERT WOOD, Sole Agent.)
☛ Officers are earnestly requested to send at once their special services to the Editor.

MARINE PHOTOGRAPH.

ARTHUR RENARD,

KIEL, BRUNSWIEKERSTRASSE

PHOTOGRAPHIEN SÄMTLICHEN DEUTSCHEN KRIEGSSCHIFFEN.

MEHRFACH PRÄMIERT.

SOWIE AUFNAHMEN VIELER SEGELYACHTEN DES KAISERLICHEN YACHT-CLUBS.

Auserdem empfehle meine hochinteressanten Aufnahmen vom eiben der Deutschen Marine an Bord, und Stereoscopaufnahmen Tausend Dessins.

PHOTOGRAPHS OF ALL WARSHIPS OF THE GERMAN N

SEVERAL DIPLOMAS AND PRIZE MEDALS.

Photographs of most Sailing Yachts of the Ge Imperial Yacht Club.

N.B.—I recommend my highly interesting Photographs of Life on Shipboar German Navy; also my Stereoscopic Pictures, of which I have over a Thousand Var

ARTHUR RENARD, BRUNSWIEKERSTRASSE 30, KIEL, GERMANY.

G. WEST & SON'S

Magnificent Marine Studies.

AWARDED 46 MEDALS FOR EXCELLENCE IN PHOTOGRAPHY.

Photographs of H. M. Battleships, Cruisers, Torpedo Warfare, Torpedo Boats and Destroyers, Mercantile Steamers, Yachts, Foreign Ships of War, and other interesting Naval Subjects.

YACHTS AND SHIPS PHOTOGRAPHED TO ORDER.
LISTS POST FREE.

G. WEST & SON, Marine Photographers,
SOUTHSEA, PORTSMOUTH, ENGLAND.

英國軍艦并ニ商船
外国軍艦并ニ珍奇ナ
ル海軍寫真
右賣捌申候艦船
写真ハ御好ニヨリ撮影
可仕候
代價表ハ御申越次第郵
送可仕候

G. WEST & FI

46 MÉDAILLES POUR L'EXCELLI

Photographies des navires de guerre de Reine d'Angleterre, cuirassés d'escadre, gar cuirassés, croiseurs, torpilleurs et contre-to 'destroyers,' les grands paquet-boats, yachts, de guerre étrangers, et plusieurs autres suj intéressants.

Catalogues franco.

COLLECTION UNIQUE DES YACHT

G. WEST et FILS, Photographies du
PALMERSTON ROAD, À SOUTHSEA, PORTSMOUTH, ANGLE

GUTERMAN & CAMPBELL'S

Patent Belt for Saving Life at Sea

The cost of the Belt is not excessive, so that now, at a nominal cost, a thoroughly reliable life-belt can be obtained by any ordinary passenger on board ship.

The practical uses of the Belt were amply demonstrated at the VICTORIA BATHS, PECKHAM, recently, on the occasion of the Swimming Contest of the Public Elementary Schools. There seemed to be a concensus of opinion that the result was most satisfactory, and that there is a great future for this invaluable invention.

The Belt is made so portable and compact that it can be worn unobserved under the coat during the voyage, or can be folded up and packed away in a small compass when not required for use.

The patentees, Messrs. GUTERMAN & CAMPBELL, will be happy to answer any inquiries which may be addressed to them at the factory.

Patent Belt for Saving Life at S

A most important invention has just been patented purpose of saving life at sea. It was stated at the tim loss of the ill-fated s.s. "DRUMMOND CASTLE" that there reliable life-belt before the public, or one which met requirements expected of it in the event of sudden em at sea.

The newly patented life-belt under notice claims to b supply that deficiency. It has been thoroughly tested or at Hastings and Dover, and has proved satisfactory occasion. The Belt consists of an ordinary circular b this is attached four egg-shaped floats, covered with nett to be in front of the body and two behind. But this strictly perpendicular position can be maintained by a or non-swimmer, so much so that his head and shou above water, leaving the arms free to move about as The floats (specially prepared and matured) can be infl air by the wearer in a few seconds.

GUTERMAN & CAMPBELL, 55, Chalton Street, St. Pancras, London

TO COLLECTORS.

THE NAVAL SYNDICATE will obtain for subscribers photographs of any war-ships that are obtainable. No extra c profit is made on this, so all applications must be accompanied by stamped addressed envelope or wrapper.

ETTE INDEX OF EVERY MERCHANT LINER IN THE WORLD WITH A SPEED OF FIFTEEN KNOTS OR OVER.

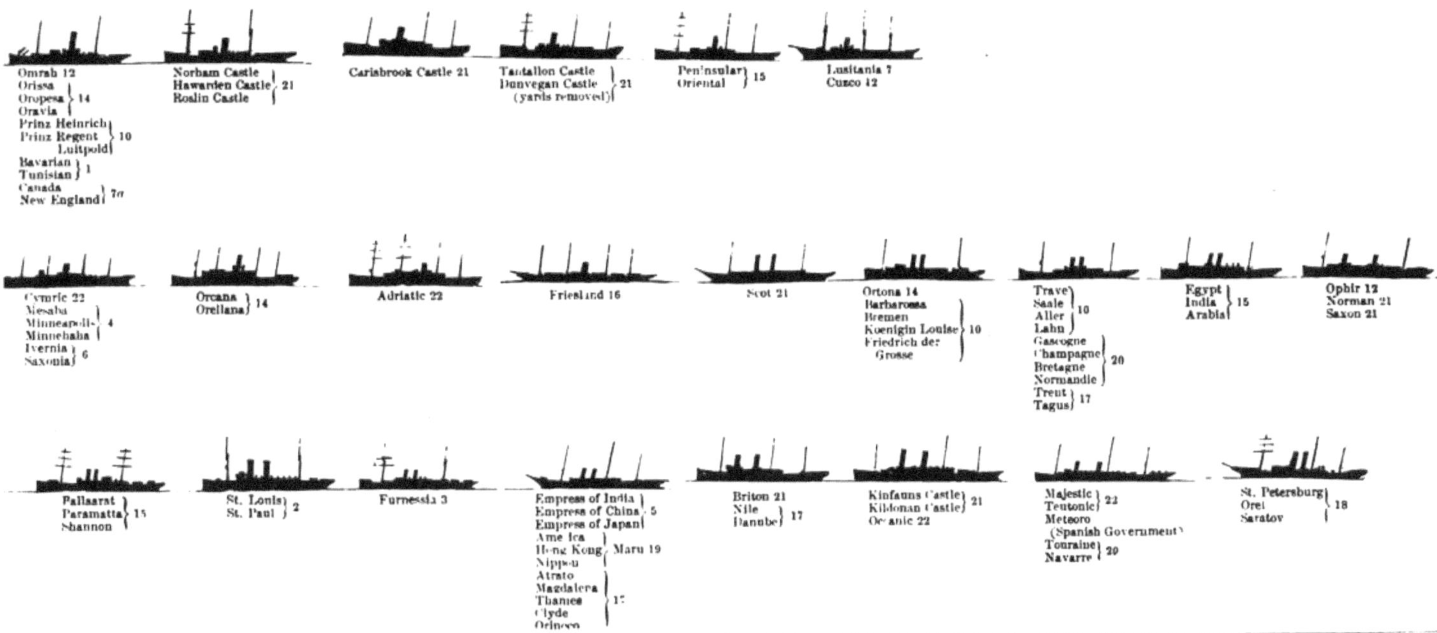

s: Allan = 1; American = 2; Anchor = 3; Atlantic Transport = 4; Canadian Pacific R. = 5; Cunard = 6; Elder, Dempster & Co. (Beaver Line) = 7; Dominion Line = 7a; Hamburg American = 8; Holland America = 9; Lloyd = 10; Oceanic Co. = 11; Orient = 12; Pacific Mail = 13; Pacific Steam Navigation Co. = 14; P. & O. = 15; Red Star 16; Royal Mail = 17; Russian Volunteer Fleet = 18; Toyo Co. = 19; Transatlantique Cie. Union Castle = 21; White Star = 22.

SILHOUETTE INDEX OF EVERY MERCHANT LINER IN THE WORLD WITH A SPEED OF FIFTEEN KNOTS OR OV

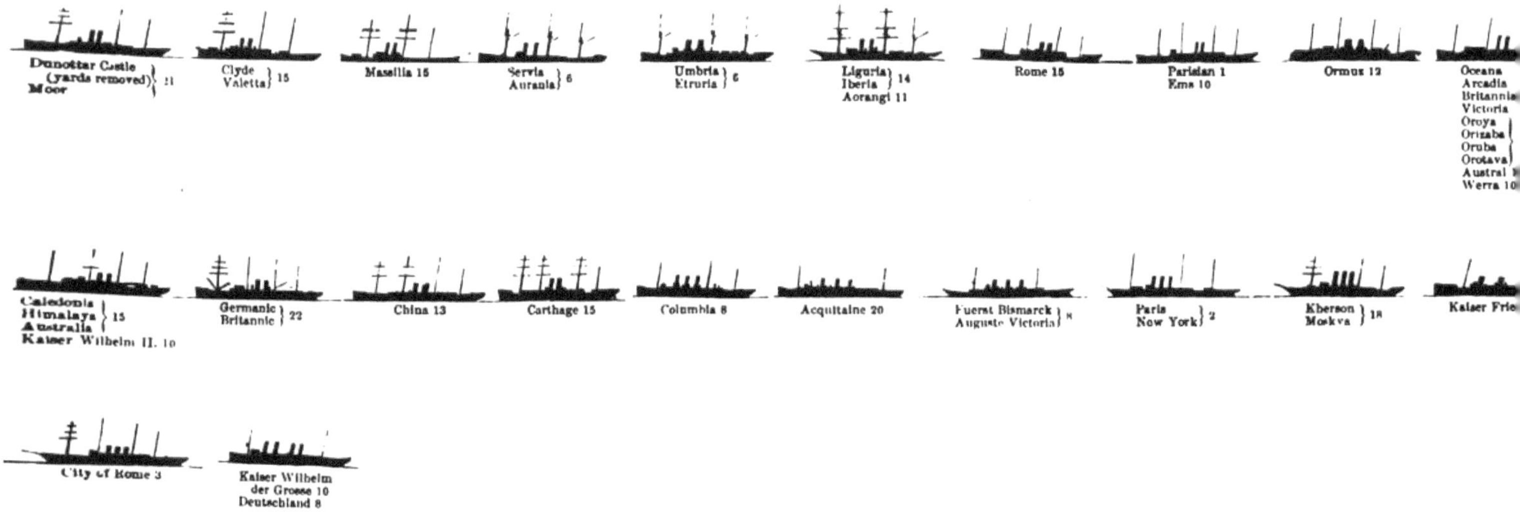

Reference to lines: Allan = 1: American = 2; Anchor = 3; Atlantic Transport = 4; Canadian Pacific R. = 5; Cunard = 6; Elder, Dempster & Co. (Beaver Line) = 7; Dominion Line = 7a; Hamburg American = 8; Holland Americ Nord Deutscher Lloyd = 10; Oceanic Co. = 11; Orient = 12; Pacific Mail = 13; Pacific Steam Navigation Co. = 14; P. & O. = 15; Red Star = 16; Royal Mail = 17; Russian Volunteer Fleet = 18; Toyo Co. = 19; Transatlantiq (French) = 20; Union Castle = 21; White Star = 22.

INDEX I.

…ers all the main divisions of the …he order in which they are placed, … upon which they may be found. …rranged primarily by funnels, then …asts and military tops. As a rule …eboard come before those of high …g-masts are placed before straight …s with low fighting-tops come …t carry the tops high up. Ships …ear together come before those …ide apart.

… is any uncertainty owing to a …eculiar appearance and difficult to … Greek *Psara*, or with a mast …and at other times down, as the …, an illustration of that ship will …h place to which it might seem to

…ne pair of funnels are placed after …funnel only: those with two pairs … two funnels, and not among those …e Italian *Sardegna* type will be …me group.

EXPLICATION À L'INDEX I.

Cet index donne toutes les divisions principales des illustrations des navires, l'ordre dans lequel ils sont placés, et indique les pages où ils se trouvent.

Les navires sont classés en premier lieu selon les cheminées, puis selon les mâts et les hunes militaires. Comme règle, tous les navires de franc-bord bas sont placés avant ceux de franc-bord haut; les navires aux mâts inclinés sont placés avant ceux aux mâts droits, et les navires ayant deux hunes militaires basses sont placés avant ceux qui les ont plus élevées. Les navires ayant les cheminées l'une près de l'autre sont placés avant ceux qui out les cheminées à une grande distance l'une de l'autre.

Là où il y a quelque doute quant à la classification d'un navire à cause de quelque singularité qu'il présente, comme par exemple le *Psara* grec, ou avec un mât quelquefois élevé et quelquefois non, comme le *Kaiser* allemand, on trouvera une illustration de tel navire à chaque catégorie à laquelle il parait appartenir.

Les navires ayant une paire de cheminées sont placés après ceux qui ont une seule cheminée: ceux avec deux paires de cheminées sont placés après les navires ayant deux cheminées, et non pas entre ceux qui en ont quatre. Le type du navire italien *Sardegna* se trouve donc dans ce cas.

ERLÄUTERUNG DES INDEX I.

Dieser Index umfasst alle Haupt-Theile der Abbildungen, die Ordnung in welcher sie erscheinen, und die Seiten auf welchen sie zu finden sind. Die Schiffe sind erstens nach den Schornsteinen, und zweitens nach Masten und Marsen geordnet. Im Allgemeinen kommen zuerst die Schiffe mit wenigem, nachher die mit höherem Freiborde, erst mit schrägen dann mit senkrechten Masten, erst die mit niedrig und dann die mit höher getragenen Marsen. Schiffe deren Schornsteine nahe zusammen sind kommen eher wie die mit weiter auseinander stehenden Schornsteinen.

Wenn man nicht recht weiss, in Folge des eigenthümlichen Aussehens eines Schiffes, wie es zu klassifizieren, zum Beispiel, die griechische *Psara*—oder mit einem Maste bald aufgestellt, und bald nicht, zum Beispiel der deutsche *Kaiser*—wird sich eine Abbildung eines solchen Schiffes in beiden Plätzen befinden.

Schiffe mit einem Paar Schornsteinen kommen nach denjenigen mit nur einem; die mit zwei Paar Schornsteinen nach denjenigen mit zwei Schornsteinen, und nicht unter diejenigen mit vier. Die italienische *Sardegna* Klasse wird sich in dieser letzteren Gruppe befinden.

SPIEGA DELL' INDICE I.

Quest' indice mostra le divisioni principali delle navi rappresentate, l' ordine in cui sono piazzate e la pagina ove ciascuna si trova.

Le navi sono ordinate primieramente secondo i fumaiuoli—poi secondo gli alberi e le coffe militari.

Come regola, i bastimenti di bordo basso vengono piazzati prima di quelli di bordo alto—gli alberi inclinati prima di quelli dritti, e le navi che hanno le coffe militari basse prima di quelle che le portano alte. Quelle navi che hanno i fumaiuoli l' uno accanto all' altro, vengono piazzate prima di quelle ove i fumaiuoli sono discosti l' uno dall' altro.

Ove vi sia incertezza a cagione dell' apparenza singolare di una nave, e difficile a classificare, come il *Psara* greco—ovvero con un albero talvolta elevato ed altra volta giù, come il *Kaiser* tedesco—un illustrazione di tal nave si troverà in ciascuna classe a cui sembra appartenere.

I bastimenti con un paio di fumaiuoli sono posti dopo di quelli che hanno solamente un fumaiuolo—quelli che ne hanno due paia, dopo le navi con due fumaiuoli, e non fra quelle che ne hanno quattro. Le navi del tipo della *Sardegna*, nave italiana, si trovano fra quest' ultima categoria.

(All the Ships in Index I. are on scale.)

SILHOUETTE INDEX.—"INDEX I."

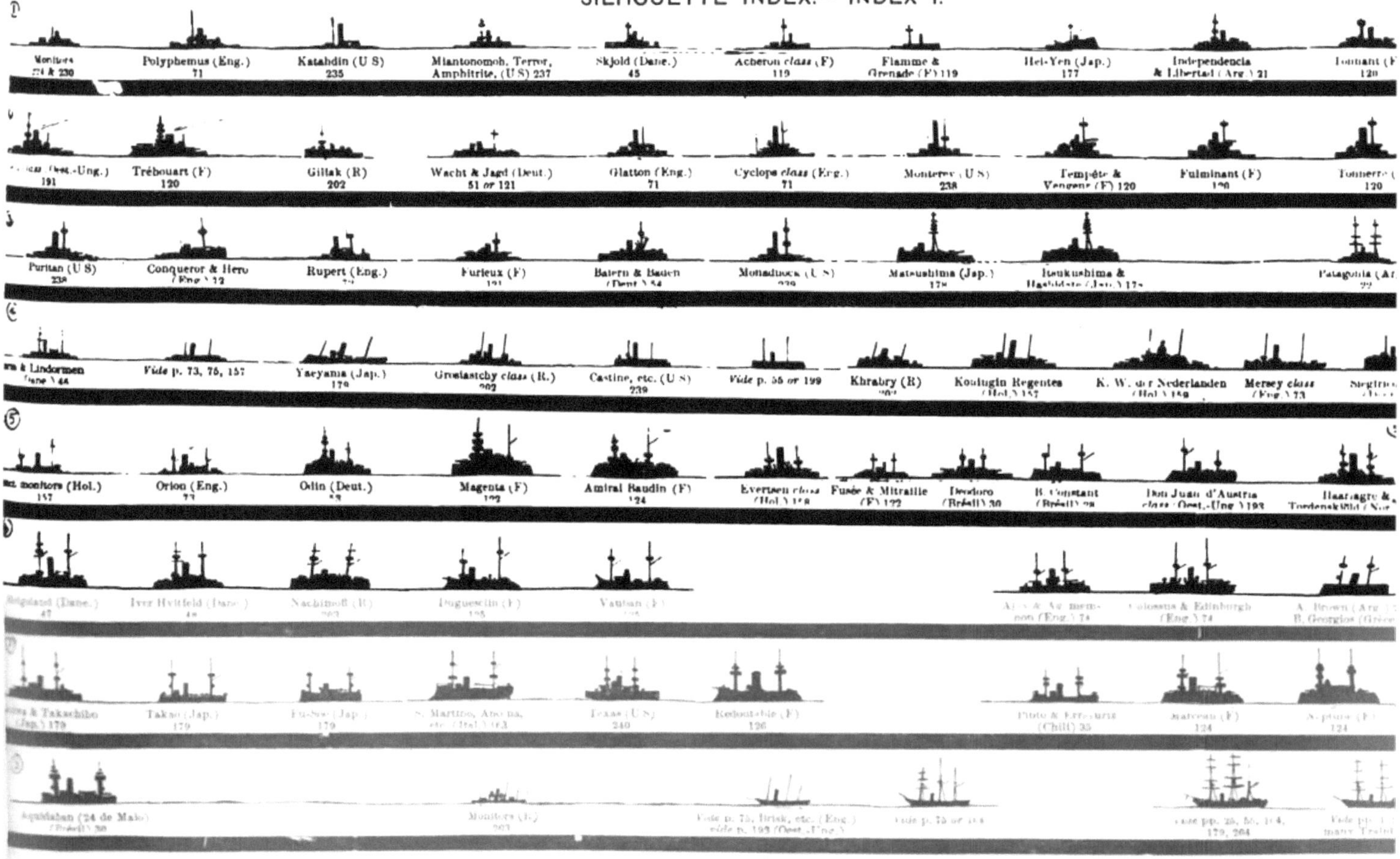

SILHOUETTE INDEX.—"INDEX I."

Vide p. 29 *or* 201 — *Vide* p. 42, 179, 199, 240 — *Vide* p. 55, 175, 204 — *Vide* p. 28 — *Vide* p. 49, 75, 159 — p. 116, 159, 190 — *Vide* p. 126, 204 — *Vide* p. 204 — *Vide* p. 116, 153, 204, 233 — *Vide* p. 224 (Suède)

Calabria (Ital.) 164 — Peter Veliki (R) 203 — Elizabetha (Roumania) 200 — Azizieh, *etc.* (Turquie) 232 — Audacious (Eng.) 76 — Monarch (Eng.) 76 — Chiyoda (Jap.) 180 — E. Albrecht (Oest.-Ung.) 194

funnels

...ble 27 — Trafalgar (Eng.) 77 — Nile (Eng.) 77 — Sanspareil (Eng.) 78 — Indomptable (F) 127 — Caïman *class* (F) 127 — Alabama *class* (U S) 241 — Hood (Eng.) 80 — Barfleur & Centurion (Eng.) 79 — Royal Sovereign *class* (Eng.) 80

Majestic *class* (Eng.) 82 — Hannibal *class* (Eng.) 82 — Courbet (F) 128 — Duperré (F) 128

Göta & Thule (Suède) 225 — Ekaterina II. & Tchesma (R) 205 — G. Pobedonosets & Sinope (R) 205 — Iowa (U S) 242 — Indiana *class* (U S) 243 — Jémappes & Valmy (F) 129 — Bouvines (F) 129

Oldenburg (Deut.) 57 — Imperieuse & Warspite (Eng.) 83 — Duilio (Ital.) 166 — Lauria (Ital.) 166 — Morosini & Doria (Ital.) 166 — Garibaldi (Arg.) 23 — S. Martin (Arg.) 23 — Belgrano & Pueyrredon (Arg.) 23 — E. Filiberto & A. di S. Bon (Ital.) 165 — Sardegna *class* (Ital.) 174

Benbow & Collingwood (Eng.) 85 — Navarin (R) 215 — Devastation (Eng.) 84 — Geiser (Dane.) 49 — Blitz & Pfeil (Deut.) 58 — Apollo *class* Pallas *class* Marathon *class* (Eng.) 87, 89, 90 — Talbot (Eng.) 87 — Crescent & Royal Arthur (Eng.) 91 — Blenheim & Blake (Eng.) 91 — Astræa *or* Edgar *cl.* (Eng.) *or* Holland *cl.* (Hol.) 90–92, 16...

Dunois (F) 131 — Vulcan (Eng.) 86 — Tiger, *etc.* Oest.-U. 195 — Chattanooga *or* Nashville U S 244 — Philadelphia (U S) 246 — Rio de la Plata (Esp.) 113 *or* Cincinnati U S 245 — Detroit *class* (U S) 245 — Minneapolis (U S) — (?) Téméraire — Surcouf (F) 131

SILHOUETTE INDEX.—"INDEX I."

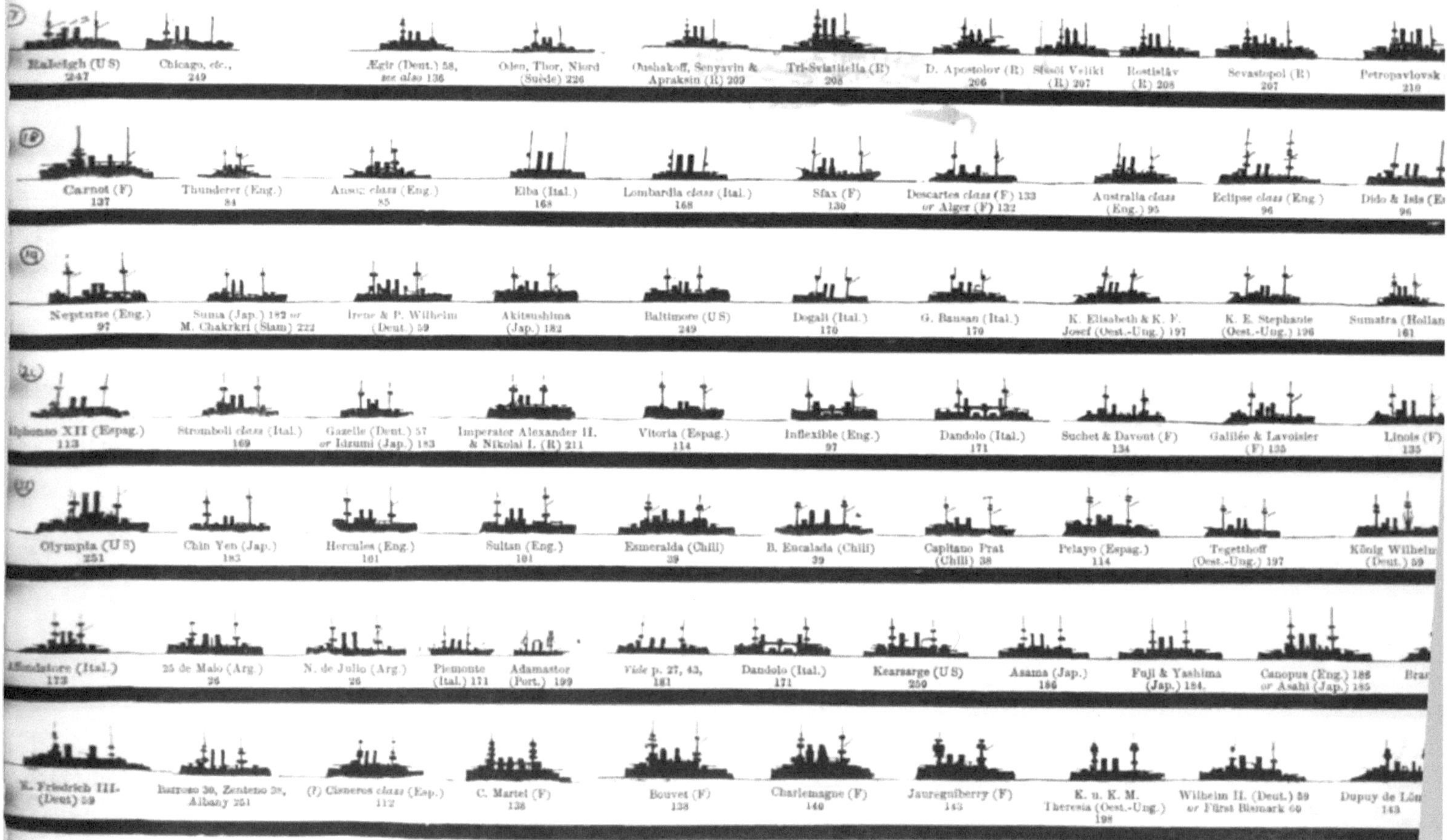

SILHOUETTE INDEX.—"INDEX I."

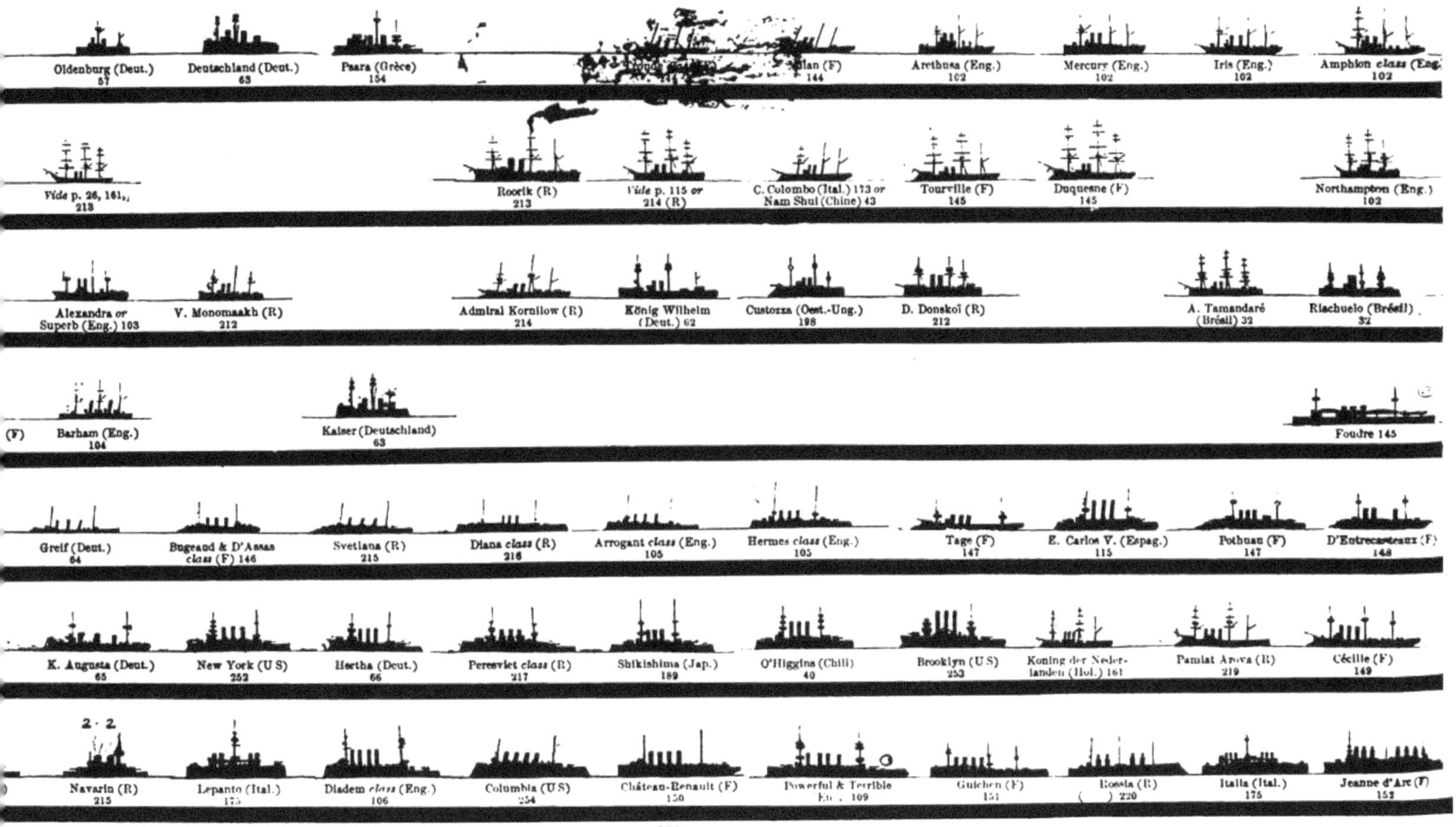

www.ingramcontent.com/pod-product-compliance
Ingram Content Group UK Ltd.
Pitfield, Milton Keynes, MK11 3LW, UK
UKHW062210071026
15591UKWH00020B/91
9781608881192